江苏大学英文教材基金资助出版

ENGINEERING MECHANICS

STATICS AND DYNAMICS

工程力学 静力学和动力学

主编
吴卫国
（Wu Weiguo）

江苏大学出版社
JIANGSU UNIVERSITY PRESS
镇 江

图书在版编目(CIP)数据

工程力学：静力学和动力学 = Engineering Mechanics: Statics and Dynamics：英文 / 吴卫国主编. — 镇江：江苏大学出版社，2020.5
ISBN 978-7-5684-1250-6

Ⅰ. ①工… Ⅱ. ①吴… Ⅲ. ①工程力学—高等学校—教材—英文②静力学—高等学校—教材—英文③材料力学—高等学校—教材—英文 Ⅳ. ①TB12②O312③TB301

中国版本图书馆 CIP 数据核字(2019)第 289463 号

工程力学:静力学和动力学

Engineering Mechanics: Statics and Dynamics

主　　编/吴卫国
责任编辑/郑晨晖
出版发行/江苏大学出版社
地　　址/江苏省镇江市梦溪园巷 30 号(邮编：212003)
电　　话/0511-84446464(传真)
网　　址/http://press.ujs.edu.cn
排　　版/镇江市江东印刷有限责任公司
印　　刷/虎彩印艺股份有限公司
开　　本/787 mm×1 092 mm　1/16
印　　张/19.75
字　　数/675 千字
版　　次/2020 年 5 月第 1 版　2020 年 5 月第 1 次印刷
书　　号/ISBN 978-7-5684-1250-6
定　　价/58.00 元

PREFACE

Engineering mechanics is an important basic course for engineering majors such as machinery, civil engineering, dynamics, aerospace, etc. It is one of the theoretical bases of engineering technology science. The main purpose of this book is to provide the students the fundamental concepts and principles of mechanics in the clearest and simplest form. And to help the students to develop problem solving skills in a systematic manner.

The book refers to the similar textbooks at home and abroad, develops out of many years of teaching experience gained by the author while giving courses on engineering mechanics to Chinese and foreign students of mechanical and civil engineer. In the course of writing, the author also referred to the basic requirements of engineering mechanics formulated by the Teaching Guidance Committee of basic mechanics under the Ministry of Education. The contents of the book correspond to the topics normally covered in courses on basic engineering mechanics at universities and colleges. The theory is presented in the simple and systematic form, and gives full consideration to the needs of the relevant major students. This approach makes the book accessible to students from different disciplines and allows for their different educational backgrounds, more convenient for the student self-study. Another aim of the book is to provide students with a solid foundation to help them bridge the gaps between undergraduate studies, advanced courses on mechanics and practical engineering problems.

The book contains still on the content of the layouts of three parts of statics, kinematics and kinetics, a total of 14 chapters, in which the principles are applied first to simple, then to more complicated situations. It is recommended as a textbook for undergraduate and graduate students in mechanical and civil engineering and aerospace as well as for researchers and engineers dealing with mechanics. It could also be used as a main reference for bilingual courses of engineering mechanics.

The book is edited by Wu Weiguo of Jiangsu University. And I wish to express my thanks to Prof. Sun Baocang who diligently checked all of the textbook.

I gratefully acknowledge many helpful comments and suggestions given by my colleagues, and in particular the contributions of Prof. Chen Daiheng, Prof. Wang Chenyuan and PhD. Jiang Shiping.

Special thanks go to the authors for the reference to their books in preparing.

I would greatly appreciate hearing from you if you have any comments, suggestions, or problems related to any matters regarding this edition.

Wu Weiguo

March, 2020

CONTENTS

Part 1 Statics

Part 3 Kinetics

Part 1 Statics

Mechanics is the oldest and the most developed branch of science that is concerned with the response of a body to external load. And it is the foundation of most engineering sciences and is an indispensable prerequisite to their study. In general, Mechanics can be divided into three branches: *Rigid bodies' mechanics*, *Deformable bodies' mechanics*, and *Fluids mechanics*. In this book, we will focus our attention on rigid bodies' mechanics since it is a basic requirement for the study of the deformable bodies' mechanics and the fluids mechanics. Furthermore, rigid body mechanics is essential for the design and analysis of many types of structural members and mechanical components in engineering.

Rigid-body mechanics is divided into two areas: *statics* and *dynamics*. Statics deals with the equilibrium of a rigid-body that is at rest or move with constant velocity under the action of forces. Dynamics is subdivided into *kinematics* and *kinetics*. Kinematics investigates the motion of bodies without referring to forces as a cause of the motion, namely it deals with the geometry of the motion in time and space. Kinetics relates the forces and the motion. We can consider statics as a special case of dynamics, in which the acceleration is zero.

In this book, we will study the simplest form of the rigid bodies' motion, namely the *mechanical motion*. The mechanical motion is defined as a body changes its position with respect to another body considered as reference system.

A basic concept of statics is the notion of equilibrium. If a body system acted on by a system of forces does not change its position during an arbitrarily long time, we say that it is in *equilibrium* under the action of that system of forces, and that system of forces must satisfy the so-called *equilibrium conditions*. The equilibrium can be considered as a special case of mechanical motion. During analysis of static equilibrium states, vector algebra and graphical methods are applied. In statics, we will study two fundamental problems as follows:

① The reduction of a force system, namely, which reduces a system of forces to a simpler, equivalent system of forces.

② Equilibrium condition of body acted by a system of forces and the equations of equilibrium.

Through experiments it was proven that time is not an absolute quantity as assumed by Newton; so, the equation of motion fails to predict the exact behavior of a particle when the particle's speed approaches the speed of light. Albert Einstein developed the *theory of relativity* and placed limitations on the use of Newton's second law for describing general motion of particle. The theory of *quantum mechanics* developed by Schrodinger and others indicates that conclusions drawn from using equation of motion are also invalid when particles are the size of an atom and move close to one another. However, for the most part, these requirements regarding particle speed and size are not encountered in engineering problems, so their effects will not be considered in this book.

Chapter 1

General Principles of Statics

Objectives

In this chapter, we introduce some fundamental concepts, formulas, and principles in statics. Of particular importance are the foundation axioms, support and reaction, and the free-body diagram, as they are used to solve most problems in statics.

1.1 Fundamental Concepts

1.1.1 Force

Force is the fundamental notion of mechanics and the concept of force can be taken from our daily life experience. For example, our muscle conveys a feeling of the force when pushing a block. Similarly, a stone can be accelerated by gravitational force during free fall. So the force is the interaction between bodies, this interaction can occur when there is direct contact between the bodies, or it can occur through a distance when the bodies are physically separated. Sometimes, the applied force may not be sufficient to move a body, e. g. , if we try to lift a stone weighing 500 kg, we fail to do so. In this case we exert a force, but no motion is produced. This shows that a force may not produce a motion in a body; but it may tend to do so. Therefore, the force can be defined:

Force is the mechanical interaction between bodies. This interaction can change or tend to change the state of motion or change the shape of a body.

The unit of force is a newton (N) or kilonewton (kN) in the International System of Units (SI).

(1) Characteristics of a Force

A single force is characterized by three properties: *magnitude*, *direction*, and *point of application*, i. e. , a force is a vector quantity and can be represented graphically in Fig. 1-1. A single force acts at a certain point of application. The quantitative effect of a force is given by its magnitude. The direction of the force can be described by its line of action and its sense of direction.

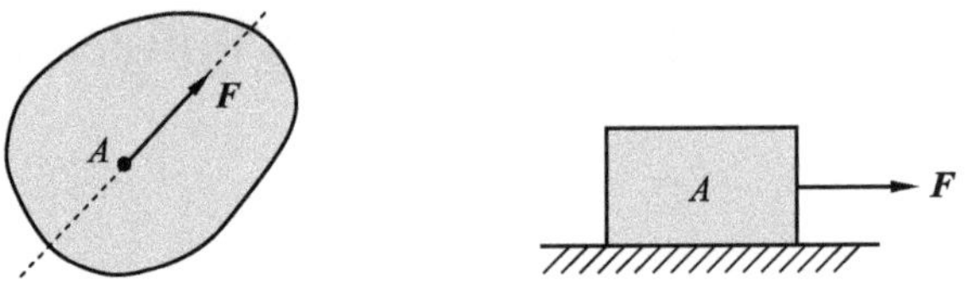

Fig. 1-1

(2) Classification of Forces

A *concentrated force* represents the effect of a loading which is assumed to act at a point on a body. This assumption holds that the area over which the load applies is very small compared to the overall size of the body. It is an idealization and an example would be the contact force between a wheel and the rail, Fig. 1-2a.

In reality, only *distributed forces* exist, which can be classified into three categories, i. e. , the volume force, area force and line force. A *volume force* is a force that is continuously distributed over the volume of a body. Weight is an example of a volume force, Fig. 1-2b. *Area force* occurs in the contact surface between two bodies. Examples of area forces include the water pressure $\boldsymbol{p}$ at a dam Fig. 1-2c. *Line force* comprises forces that are continuously distributed along a line. For example, if a blade is pressed against an object, the line force will act along the contact line, Fig. 1-2d. Distributed loads are characterized by their intensity $\boldsymbol{q}$, denoting the force per unit volume, surface area or per unit length.

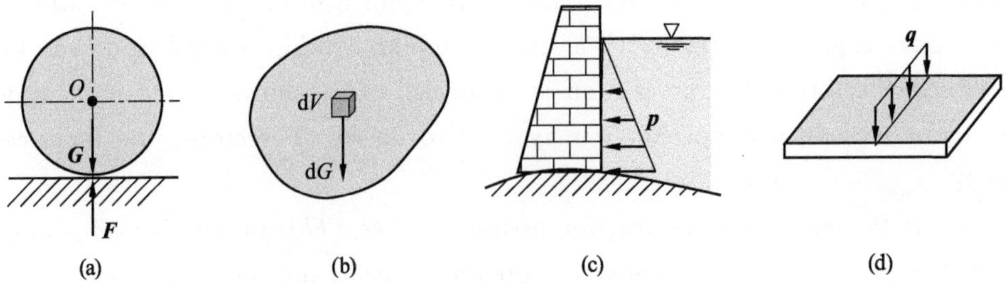

Fig. 1-2

The two most common distributions of loads are the *linear variation* and *uniform intensity* of the intensity, illustrated in Fig. 1-2c,d. The linear load intensity variation can be represented via triangular distributions. The uniform intensity load distributes same force over the even interval of the force.

Forces can also be classified according to other criteria. *Active forces* refer to the physically prescribed forces in a mechanical system, for example the weight, the pressure of the wind or the snow load on a roof. *Reaction forces* are generated if the freedom of movement of a body is constrained. For example, a falling stone is subjected only to an active force due to gravity if neglecting the air resistance. However, when the stone is held in the hand, its freedom of movement is constrained; the hand exerts a reaction force on the stone. Reaction forces can be visualized only if the body is separated from its geometrical constraints. This procedure is called *freeing* or isolating the body.

Another classification is introduced by distinguishing between *external forces* and *internal forces*. An external force acts from the outside on a mechanical system. Active forces as well as reaction forces are external forces. Internal forces act between two adjacent parts of a system. They also can be visualized only by imaginary cutting or *sectioning* of the body.

1.1.2 The System of Forces

Two or more forces acting on the same body are defined as a *system of forces*. Following systems of forces are important in the subject:

① *Coplanar forces*. The lines of action of all forces lie on the same plane.

② *Collinear forces*. The lines of action of all forces lie on the same line.

③ *Concurrent forces*. The lines of action of all forces intersect at one point.

④ *General system of forces*. The lines of action of all forces do not intersect at one point.

⑤ *Non-coplanar forces* (*spatial forces*). The lines of action of all forces do not lie on the same plane.

In what follows, the above systems of forces will be studied in detail.

1.1.3 The Rigid Body

A *rigid body*: the distances between any two points of the body remain constant when forces are acting on the body, i. e. , a rigid body does not deform under the action of the forces. This is an idealization of a real body. In reality, structures, machines, and mechanisms will deform slightly under the action of the forces. However, in many cases, such deformations are usually very small, and their effect on the studied bodies can therefore be neglected, which makes the rigid-body assumption suitable for analysis.

In the case of a deformable body, the effect of the force $\boldsymbol{F}$ depends on the point of application, the deformation of the body may be different as the action point changes, as shown in Fig. 1-3a. For a rigid body, the effect of the force $\boldsymbol{F}$ on the entire body is the same, regardless of the body being pulled or pushed, as shown in Fig. 1-3b.

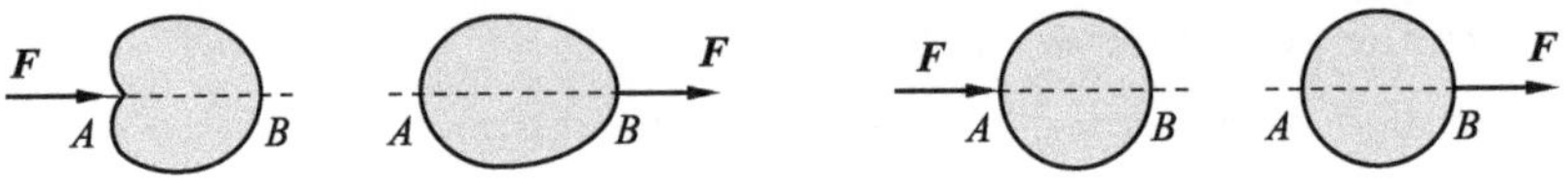

(a) deformable body (b) rigid body

Fig. 1-3

1.2 Fundamental Principles of Statics

There are some statements based on numerous observations and regarded as being known from experience. The conclusions drawn from these *axioms* are also confirmed by experience. The study of statics is based on these fundamental principles.

1.2.1 Axiom 1: Parallelogram Law of Forces

The effect of two nonparallel forces $\boldsymbol{F}_1$ and $\boldsymbol{F}_2$ acting at a point A of a body is equivalent to the effect of the single force $\boldsymbol{F}_R$ acting at the same point. The direction and magnitude of this force is determined by the *diagonal* of the parallelogram formed by $\boldsymbol{F}_1$ and $\boldsymbol{F}_2$, Fig. 1-4a. The force $\boldsymbol{F}_R$ is called the *resultant* of $\boldsymbol{F}_1$ and $\boldsymbol{F}_2$, and $\boldsymbol{F}_1$ and $\boldsymbol{F}_2$ are the two components of $\boldsymbol{F}_R$. The resultant vector:

$$\boldsymbol{F}_R = \boldsymbol{F}_1 + \boldsymbol{F}_2 \tag{1.1}$$

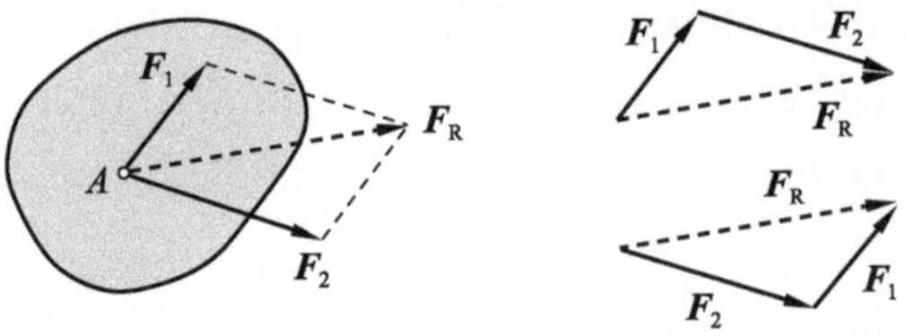

(a) (b)

Fig. 1-4

We can also add $\boldsymbol{F}_1$ and $\boldsymbol{F}_2$ using the *triangle rule*, which is a special case of the parallelogram law. The force $\boldsymbol{F}_2$ is added to force $\boldsymbol{F}_1$ in a "*head-to-tail*" fashion, i. e. , by connecting the head of $\boldsymbol{F}_1$ to the tail of $\boldsymbol{F}_2$, as shown in Fig. 1-4b. The resultant $\boldsymbol{F}_R$ extends from the tail of $\boldsymbol{F}_1$ to the head of $\boldsymbol{F}_2$. The force $\boldsymbol{F}_1$ and $\boldsymbol{F}_2$ are called sides of a force triangle and the force $\boldsymbol{F}_R$ is called a *closing vector* of the force triangle. We can use the rule of cosines or sines to obtain the magnitude of the resultant force and its direction.

The parallelogram rule can also be used to resolve a force $\boldsymbol{F}_R$ into two *components* $\boldsymbol{F}_1$ and $\boldsymbol{F}_2$, and these forces have intersecting lines of action. The rule of sines can be used to determine the magnitudes of the components, if $\boldsymbol{F}_R$ is given and the directions of $\boldsymbol{F}_1$ and $\boldsymbol{F}_2$ are specified.

1.2.2 Axiom 2: Two-Force Equilibrium

Two forces acting on a rigid body are in equilibrium if they are oppositely directed on the same line of action and have the same magnitude, Fig. 1-5. This means that the sum of the two forces, i. e., their resultant $\boldsymbol{F}_R$, is zero:

$$\boldsymbol{F}_R = \boldsymbol{F}_1 + \boldsymbol{F}_2 = 0 \tag{1.2}$$

◆ Two-Force Member

As shown in Fig. 1-6, a member is defined as a *two-force member* when it is subject to only two forces acting on two different points. Two-force member is irrelevant to the shape of bodies. For any two-force members, the equilibrium condition for such a system is that two forces acting on the member must have the same magnitude, act on the same straight line with opposite directions, directed along the *line joining the two points* where these forces act.

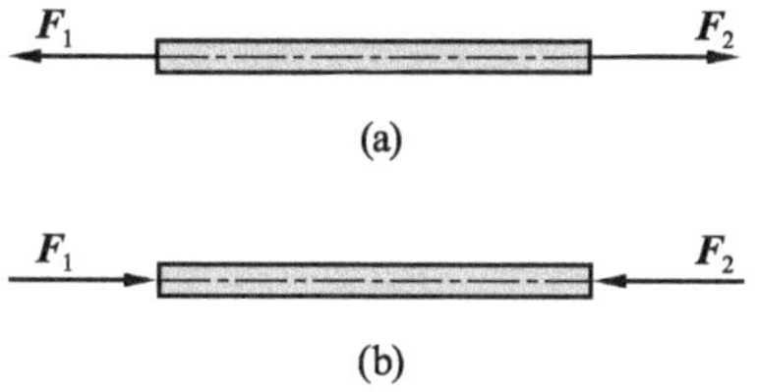

Fig. 1-5

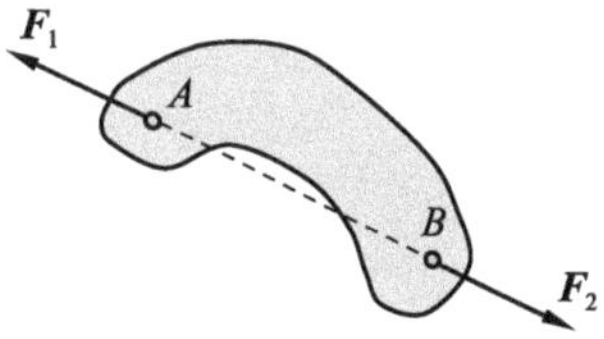

Fig. 1-6

1.2.3 Axiom 3: Plus or Minus an Equilibrium System of Forces

The action of a given system of forces on a rigid body remains unchanged if another equilibrium system of forces is added to or subtracted from the original system.

For example, in Fig. 1-7, the rigid body is subject to two different systems of forces, if $\boldsymbol{F}_1 = -\boldsymbol{F}_2$, two given systems of forces are equivalent, i. e., their resultant forces are identical.

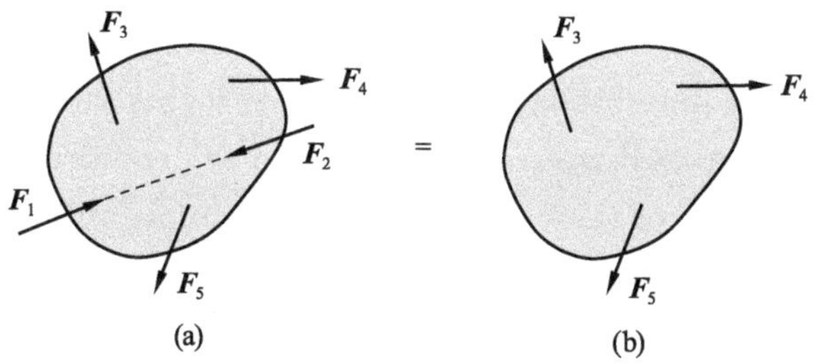

Fig. 1-7

(1) Principle of Transmissibility

A single force may be applied at any point on the line of action without changing its effect on the body.

Proof: We consider a rigid body that is subjected to a force acting on point A, Fig. 1-8a. According to axiom 3, we can add a pair of equilibrium system of forces $\boldsymbol{F}_2=-\boldsymbol{F}_1=\boldsymbol{F}$ at any point B on the line of action without changing the effect of the original system of forces, Fig. 1-8b. And it can be found that force $\boldsymbol{F}_1$ and $\boldsymbol{F}$ are a pair of equilibrium system of forces, so they can be subtracted from the original system according to axiom 3, Fig. 1-8c. The mentioned forces $\boldsymbol{F}$ and $\boldsymbol{F}_2$ are equivalent.

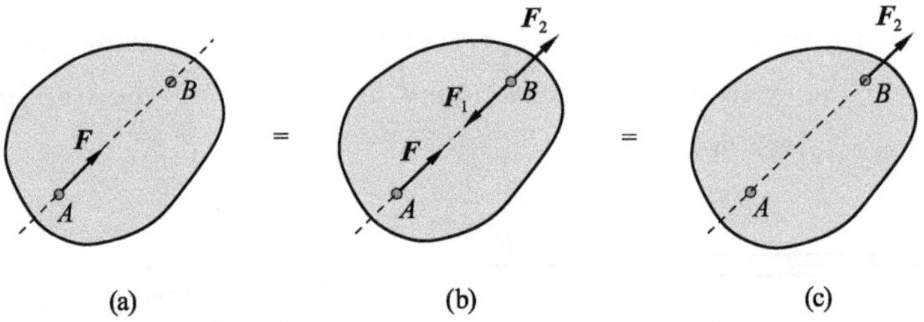

Fig. 1-8

In other words, the effect of a force on a rigid body is independent of its location on the line of action. The forces acting on rigid bodies are "*sliding vectors*", they can arbitrarily be moved along their action lines. When a rigid body is concerned, a force can be characterized by three elements: *magnitude*, *direction* and *line of action*.

(2) Three-Forces Theorem

If a rigid body remains in equilibrium under the action of only three non-parallel coplanar forces, then their lines of action must intersect at a single point, that is the system of concurrent forces.

To prove the above theorem, we consider three non-parallel forces $\boldsymbol{F}_1$, $\boldsymbol{F}_2$ and $\boldsymbol{F}_3$ acting on a rigid body. Using *Principle of Transmissibility*, any two of the forces, e. g. ,$\boldsymbol{F}_1$ and $\boldsymbol{F}_2$, can be moved to intersect at point O. Then the action of $\boldsymbol{F}_1$ and $\boldsymbol{F}_2$ can be replaced by their resultant $\boldsymbol{F}_{12}$ using *Parallelogram Law of Forces*. According to *Two-Force Equilibrium*, the body is in equilibrium only if $\boldsymbol{F}_{12}=-\boldsymbol{F}_3$ and they are collinear. Therefore, the lines of action of the three forces must intersect at a single point O, Fig. 1-9. Notes that if we are dealing with a system of three concurrent forces and a body under the action of these forces is in equilibrium, all of these forces must be *coplanar* (i. e. , lie in one plane).

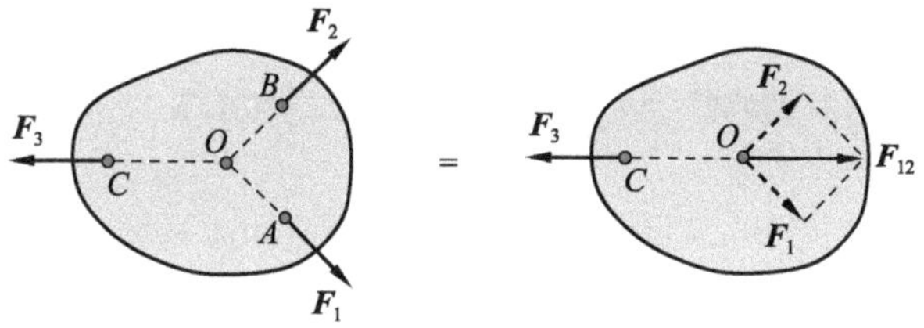

Fig. 1-9

1.2.4 Axiom 4: Law of Action and Reaction

The forces that two bodies exert upon each other are of the same magnitude but opposite directions and they lie on the same line of action. There exists an interaction between two bodies in physical contact.

This principle is a universally accepted law and is the *third Newton's Law*. It states that a force always has a counteracting force of the same magnitude but opposite direction, i. e. , a force can never exist alone. For example, if one pushes a block, the hand exerts a force $\boldsymbol{F}$ on the block, Fig. 1-10. And reaction force of the same magnitude will be applied on the hand due to the block. These forces can be made visible if the two bodies are separated at the area of contact. Note that the forces act upon two *different bodies*. It is independent of whether the bodies are at rest or in motion.

Fig. 1-10

1.2.5 Axiom 5: Principle of Solidification

If a deformable body subjected to the action of a force system is in equilibrium, the state of equilibrium will not be disturbed if the body *solidifies* (becomes rigid), Fig. 1-11a. The axiom indicates that the equilibrium of a balanced deformable body can be described by the theory for a rigid body. But it doesn't work in turn, Fig. 1-11b.

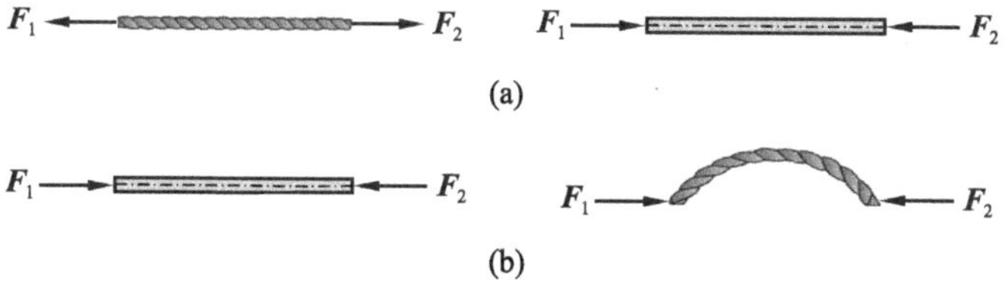

Fig. 1-11

1.3 Constraints and Support Reactions

A rigid body in space has six possibilities of motion, such a body we call *free*. Its contact with another body occurs by means of constraints are called *supports*. An arbitrary support produces a mechanical (supporting) interaction, that is, *support forces* and *moments* of support forces. Below we will characterize some of the supports and their reactions often used in mechanics systems.

1.3.1 Flexible Cable

In Fig. 1-12 shows a few examples of flexible cable such as a string, a chain or a strap. Flexible cable can only be *stretched*; Therefore, the reaction force applies at the point of contact and the sense of direction is *away from* the body in the direction of the flexible cable.

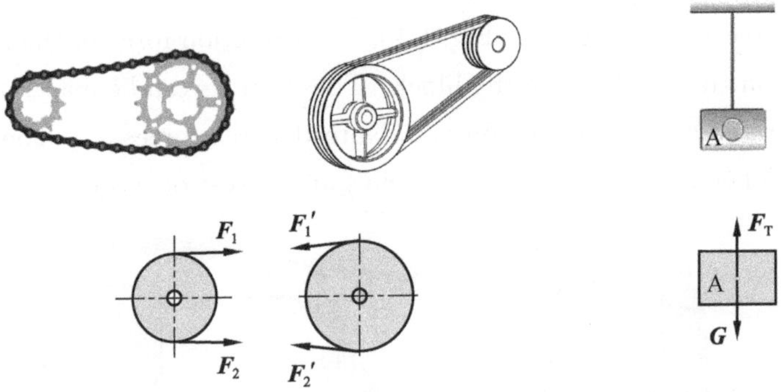

Fig. 1-12

In other words, the flexible cable can support only one tension or "pulling" force. And this force always acts in the direction of the cable.

1.3.2 Smooth Contact Surface

In Fig. 1-13 shows a few examples of smooth contact surface (friction is negligible). In the case the reaction is a force which acts *perpendicular* to the surface or along the common normal toward the body at the point of contact. Then we are dealing with only *one unknown*, that is, the magnitude of reaction $\boldsymbol{F}_N$, since its direction is defined.

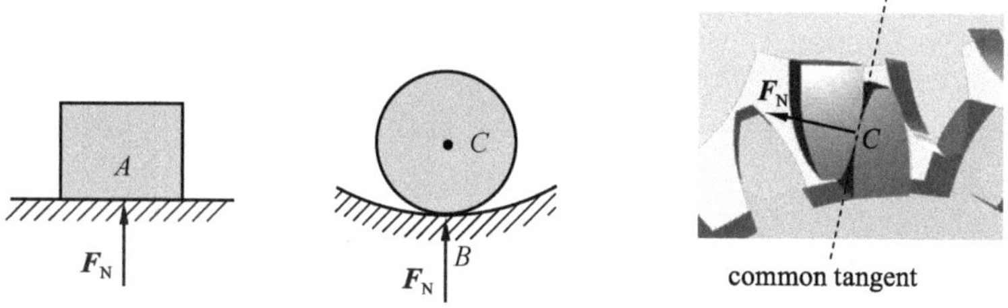

Fig. 1-13

1.3.3 Pin Support

The pin support is also present, where the pin passes through a hole in the body and two leaves, as shown in Fig.1-14a, b, c. Neglecting friction, the pin can prevent translation of the body in any direction, and so the pin must exert a force $\boldsymbol{F}$ (magnitude and direction of reaction unknowns) normal to the contact surface on the body in this direction (pass through the axes), Fig. 1-14d. For the purpose of analysis, the force $\boldsymbol{F}$ is represented generally by its *two rectangular components* $\boldsymbol{F}_x$ and $\boldsymbol{F}_y$, see Fig. 1-14e and f. The scissor is an examples of pin support, see Fig. 1-15.

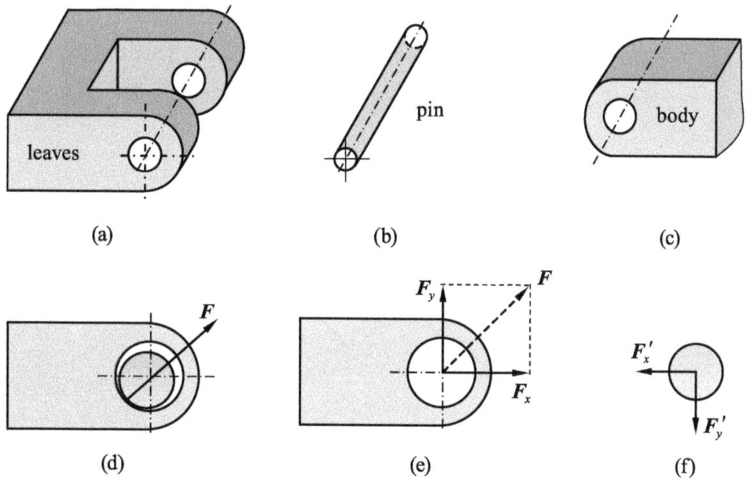

Fig. 1-14

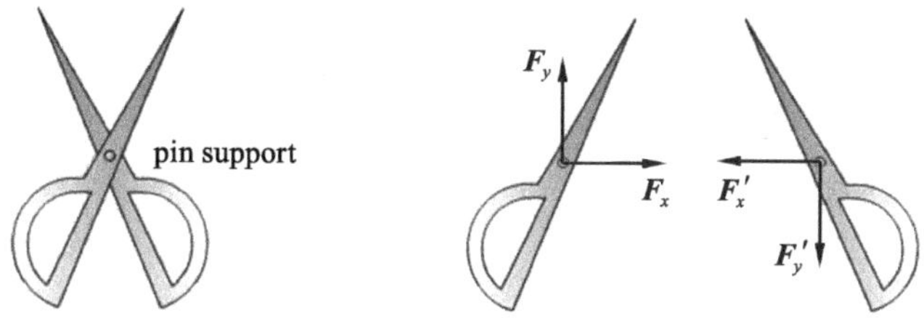

Fig. 1-15

(1) Fixed Pin Support

Example of this type of pin support is the *fixed pin support* or *hinged support*, as shown in Fig. 1-16a, which is depicted symbolically in Fig. 1-16b. The fixed pin support allows a rotation but not a displacement in any direction. Accordingly, it can transmit a reaction force $\boldsymbol{F}$ of arbitrary magnitude and arbitrary direction that can be resolved into its *horizontal and vertical components* $\boldsymbol{F}_x$ and $\boldsymbol{F}_y$.

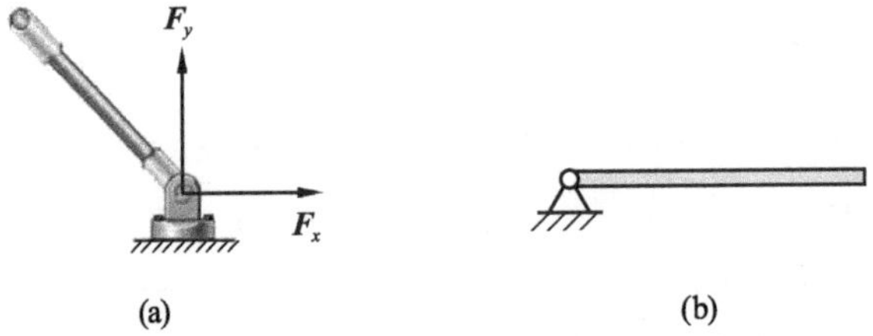

Fig. 1-16

(2) Radial Bearing

Another example of pin support is the *radial bearing*, as shown in Fig. 1-17a, which is depicted along with the corresponding reaction forces in Fig. 1-17b. The reaction force is resolved into two *orthogonal component* forces $\boldsymbol{F}_x$ and $\boldsymbol{F}_y$.

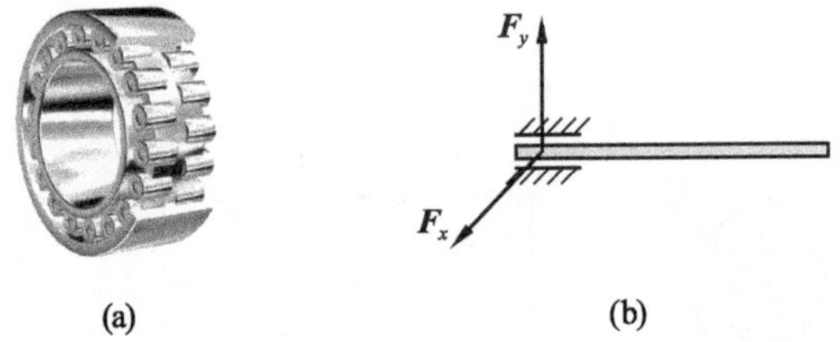

Fig. 1-17

1.3.4 Roller Support

The roller support is shown in Fig. 1-18a, which is depicted by the symbol shown in Fig. 1-18b. The horizontal translation and a rotation are not constrained by the support. If the contact areas are assumed as frictionless, the reaction force can be considered to act *perpendicular* to the contact surfaces. In this case, the direction of the reaction force is known (here vertical), and its magnitude is *unknown*. That is, roller supports can transmit only *one single reaction force*.

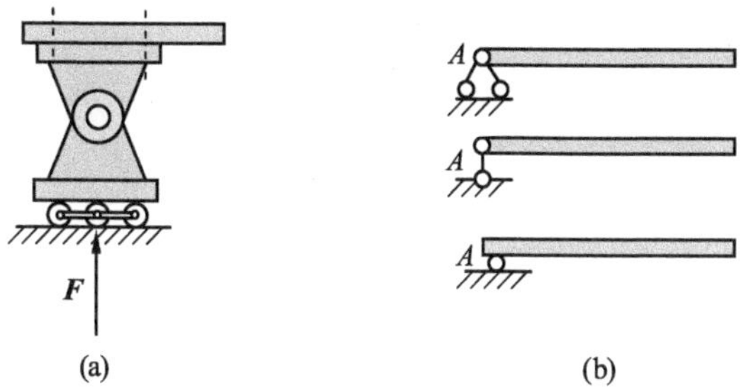

Fig. 1-18

All the steel trusses, or the bridges, have one of their ends as roller support. The main advantage of such support is that the beam can move easily towards left or right, on account of expansion or contraction due to change in temperature.

1.3.5 Fixed Support

In the case of a planar clamped support (fixed support) which is depicted symbolically in Fig. 1-19a. The free-body diagram in Fig. 1-19b shows that the fixed support can transmit a *reaction force* $\boldsymbol{F}$ of arbitrary magnitude and direction and a *couple* moment. That is, for the fixed support, we have three unknowns: two components $\boldsymbol{F}_x$ and $\boldsymbol{F}_y$ of reaction force and the reaction moment $\boldsymbol{M}_A$.

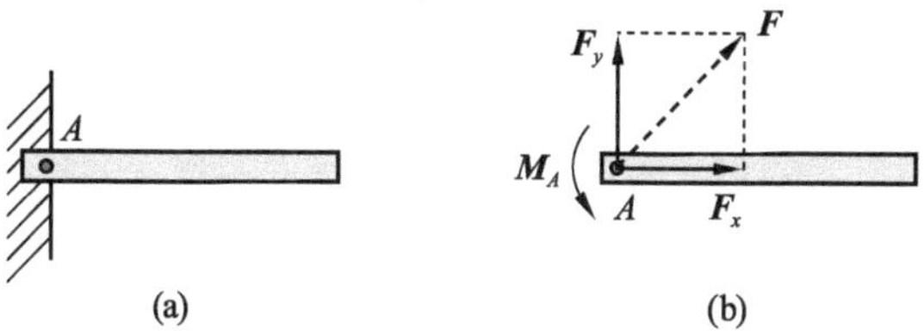

Fig. 1-19

1.3.6 Ball-and-Socket

In Fig. 1-20a, a ball-and-socket support is shown, it is an example of supports carrying three-dimensional systems of forces. If the connection between the rigid bodies is realized by ball-and-socket support, it prevents the translation of a body in every direction but allows for rotation about any axis passing through the center of the support (ball), so the reactions are three unknown rectangular force components $\boldsymbol{F}_x$, $\boldsymbol{F}_y$, and $\boldsymbol{F}_z$, as shown in Fig. 1-20b.

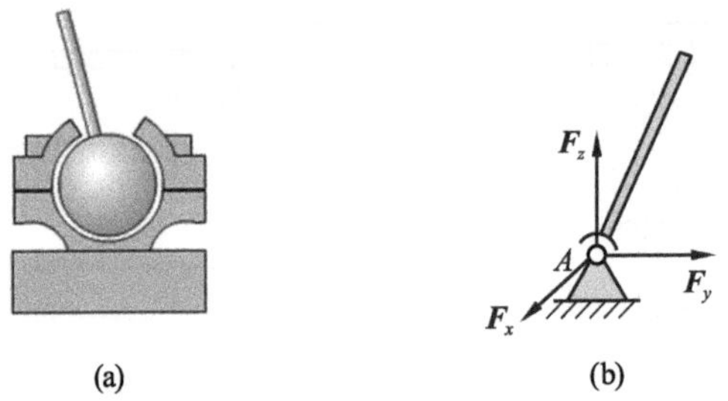

Fig. 1-20

1.4 Force Analysis of Rigid Body

We must account for all the known and unknown forces which act on the rigid body when using the equations of equilibrium. The best way to do this is to think of the body as isolated and "*free*" from its surroundings. And reaction forces can be visualized only if the body is separated from its geometrical constraints. This procedure is called *freeing* or *cutting free* or *isolating* the body. These reaction forces are made visible by this *freeing*, the relevant forces become accessible to analysis. *A drawing that shows the rigid body with all the forces that the surroundings exert on it is called a free-body diagram*(FBD). This diagram is a sketch of the outlined shape of the body.

A thorough understanding of how to draw a free-body diagram is of primary importance for solving problems in mechanics. To construct a free-body diagram, the following three steps are necessary.

(1) Draw Outlined Shape

Assume the rigid body to be isolated or cut "free" from its surroundings by drawing its outlined shape.

(2) Draw All Active Forces

On the sketch, draw and label all the active forces that act on the rigid body. These forces tend to set the rigid body in motion. Noting the direction and the line of action.

(3) Draw All Reactive Forces

Draw and label all the reactive forces that act on the rigid body. These forces are the result of the constraints or supports that tend to prevent motion. To avoid missing these forces, it is helpful to trace around the rigid body's boundary, carefully noting each force acting on it.

If we are dealing with *complex system of bodies*, we can divide the system into subsystems detaching one by one the bodies interacting until we isolate the body and have the forces and moments of forces coming from the interactions with other bodies.

According to Newton's third law, the internal forces are the effect of action and reaction and they are pairs of opposite forces, which means that they *cancel out* one another. If the free-body diagram for the body is drawn, the forces that are internal to the body *are never shown on the free-body diagram since they occur in equal but opposite collinear pairs and therefore cancel out.*

Example 1-1

The running pulley in Fig. 1-21a with weight **G** is subjected to a towing force **F** and is supported as shown. Draw free-body diagram of the running pulley.

Solution:

The free-body diagram of the running pulley is shown in Fig. 1-21b. The first step is to draw the sketch of the running pully. The second step is to draw the active forces, there are two active forces acting on the pulley, that is, the weight **G**, and the towing force **F**. Finally drawing the reactive forces, assuming all contacting surfaces are *smooth*, there are two reactive forces $\boldsymbol{F}_{NA}$ *and* $\boldsymbol{F}_{NB}$ act in a direction *normal* to the tangent at their surfaces of contact point *A* and *B*.

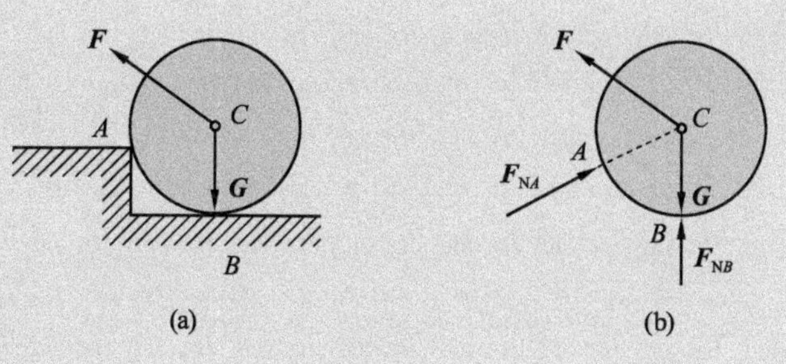

Fig. 1-21

Example 1-2

The simply supported beam *AB* shown in Fig. 1-22a is loaded by uniformly distributed force **q** and the concentrated force **F** which acts under an angle α. Draw the free-body diagram of the beam *AB*.

Solution:

The free-body diagram of the beam *AB* is shown in Fig. 1-22b. In the sketch of the beam, draw two active forces of concentrated force **F** and uniformly distributed force **q**. Since the support at *A* is a *fixed pin*, there are two reactions acting at *A*, denoted as $\boldsymbol{F}_{Ax}$ and $\boldsymbol{F}_{Ay}$, drawn in an orthogonal direction with assumed sense. Support at B is a *roller*, there is one reaction $\boldsymbol{F}_B$ acts *perpendicular* to the contact surfaces under an angle α.

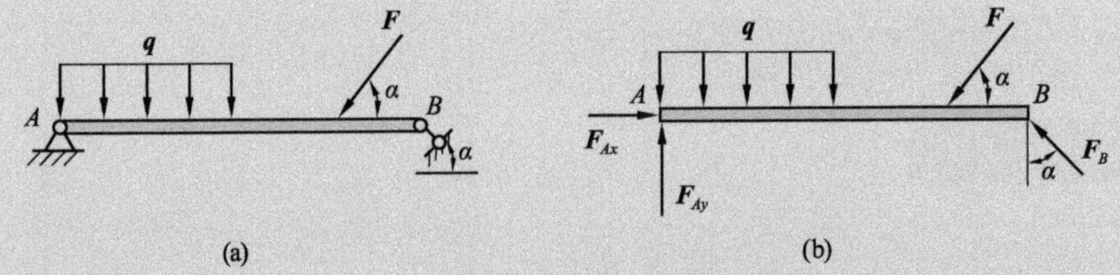

Fig. 1-22

Example 1-3

The *three-hinged* arch shown in Fig. 1-23a, consists of two arches, which are connected by the hinge C, irrespective of the self-weight and friction, draw the free-body diagrams of the arch AC, BC and both arches together.

Solution:

The free-body diagram of arch BC is shown in Fig. 1-23c. If irrespective of the self-weight and friction, the arch BC is a *two-force member*. There are two reactions $\boldsymbol{F}_B$ and $\boldsymbol{F}_C$ acting respectively on point B and C, these forces are oppositely directed with assumed sense along the line joining point B and C.

The free-body diagram of arch AC is shown in Fig. 1-23d. Force $\boldsymbol{F}$ is an active force. In particular, note that $\boldsymbol{F}_C'$, representing the force of arch BC on AC, is equal and opposite to $\boldsymbol{F}_C$, Fig. 1-23c. This is a consequence of *Law of Action and Reaction*. Since the support at A is a *fixed pin*, there are two reactions acting at A, denoted as $\boldsymbol{F}_{Ax}$ and $\boldsymbol{F}_{Ay}$, drawn in an orthogonal direction with assumed sense.

The free-body diagram of both arches combined ("system") is shown in Fig. 1-23b. Here the contact force $\boldsymbol{F}_C$, which acts between arch AC and BC, is considered as an *internal force* and hence is not shown on the free-body diagram of system. That is, it represents a pair of equal but opposite collinear forces which cancel out each other.

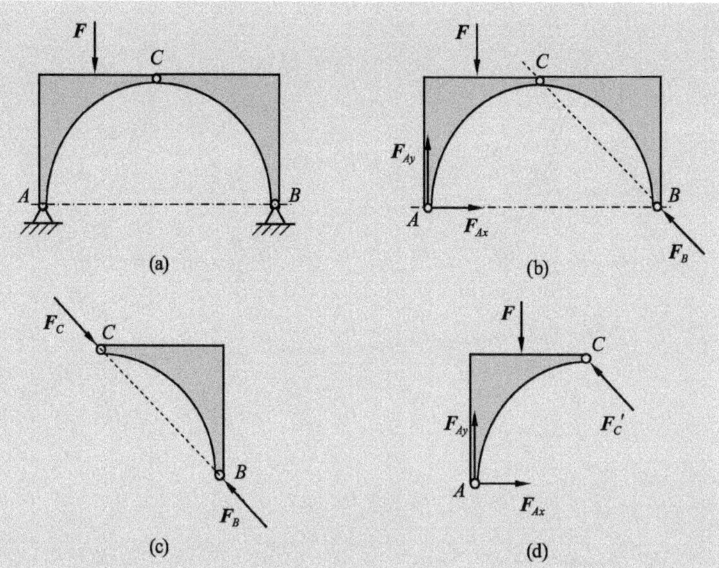

Fig. 1-23

Example 1-4

The stepladder shown in Fig. 1-24a, consists of two ladders AB and AC connected at the hinge A and fixed by the rope DE, there is load $\boldsymbol{F}$ acting on point H. Irrespective of the friction, draw the force diagrams of the ladder AB, AC and the stepladders.

Solution:

The free-body diagram of ladder AB is shown in Fig. 1-24c. Force $\boldsymbol{F}$ on point H is an active force. If contacting surfaces are *smooth*, the reactive force $\boldsymbol{F}_B$ acts in a direction *normal* to the tangent at contacting surfaces. Since point A is a hinge, there are two components $\boldsymbol{F}_{Ax}$ and $\boldsymbol{F}_{Ay}$ acting on A. On point D, there is a tension $\boldsymbol{F}_D$ from the rope.

The free-body diagram of ladder AC is shown in Fig. 1-24e. Note that $\boldsymbol{F}'_{Ax}$ and $\boldsymbol{F}'_{Ay}$, representing the reaction force of AB on BC, is equal and opposite to $\boldsymbol{F}_{Ax}$ and $\boldsymbol{F}_{Ay}$. There is a tension $\boldsymbol{F}_E$ from the rope on point E. Since the contacting surfaces are *smooth*, the reactive forces $\boldsymbol{F}_C$ act *perpendicular* to the contacting surfaces.

The free-body diagram of rope DE is shown in Fig. 1-24d. A pair of balanced forces is opposite to $\boldsymbol{F}_D$ and $\boldsymbol{F}_E$, respectively.

The free-body diagram of stepladders ("system") is shown in Fig. 1-24b. Here the forces $\boldsymbol{F}_{Ax}$ and $\boldsymbol{F}_{Ay}$, which act between ladder AB and AC, are considered as an *internal force* and hence are not shown on the free-body diagram.

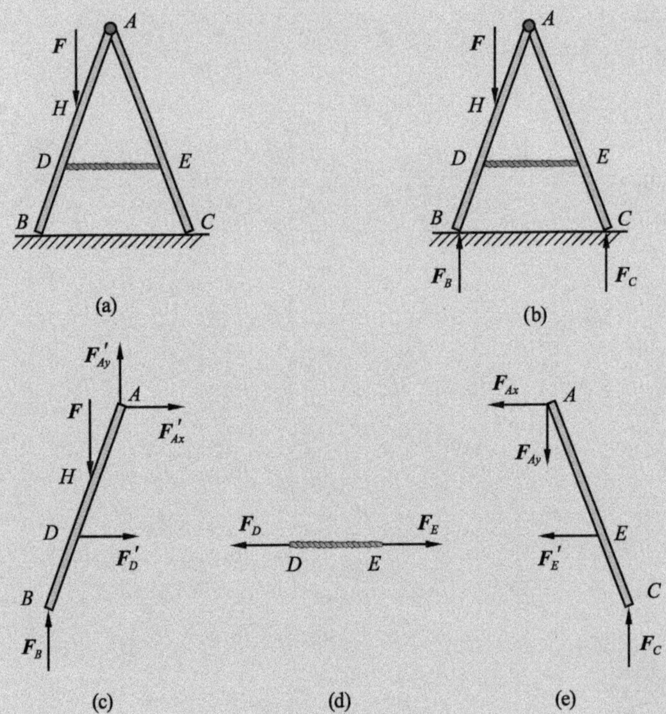

Fig. 1-24

Exercises

1-1 As shown in Fig. 1-25, the ball is hung by a rope, draw the force diagram of the ball, assume that all contact surfaces are smooth.

1-2 Draw the force diagram of the uniform bar in Fig. 1-26, assume that all contact surfaces are smooth.

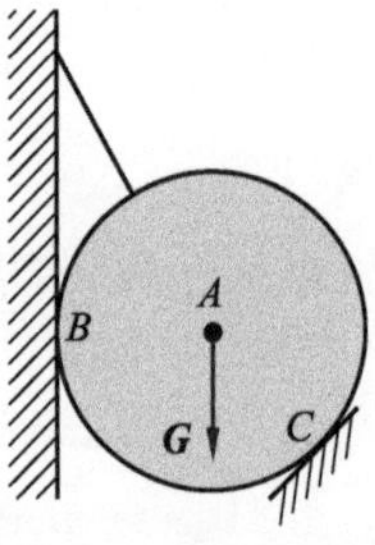

Fig. 1-25

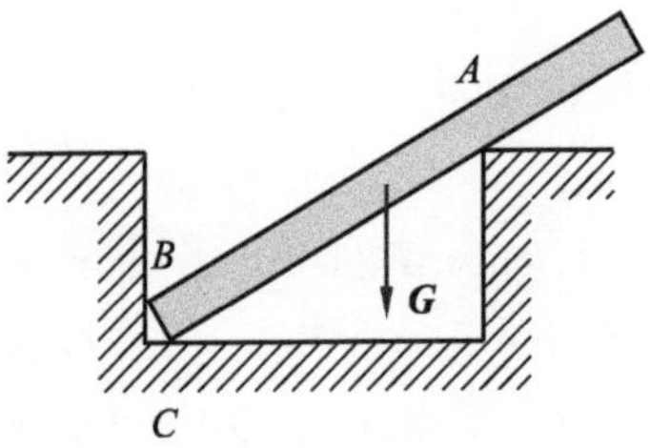

Fig. 1-26

1-3 Draw the force diagram of the uniform arch in Fig. 1-27. The arch is supported by a fixed pin at A and a roller at B.

1-4 Draw the free-body diagram of the beam in Fig. 1-28, which is fixed pin-supported at A and rests on the smooth incline at B.

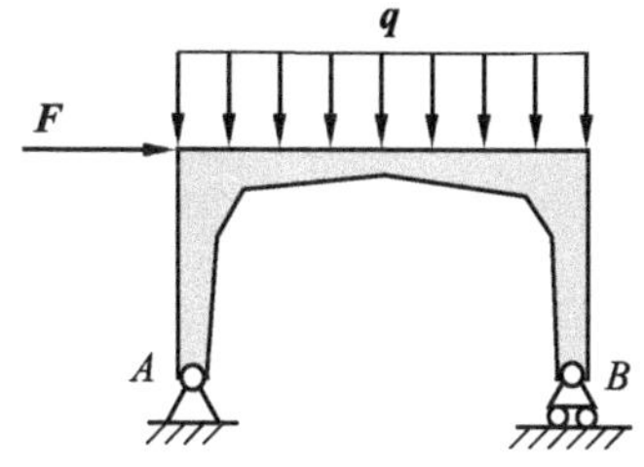

Fig. 1-27

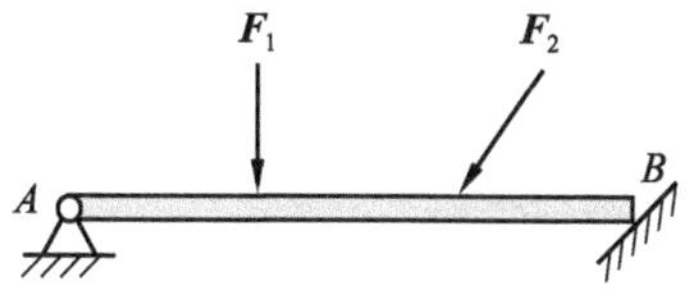

Fig. 1-28

1-5 Draw the free-body diagram of member ABC in Fig. 1-29, which is fixed pin-supported at A and roller support at C.

1-6 Draw the free-body diagram of the beam AB in Fig. 1-30, which is fixed pin-supported at B and rests on the smooth surface at A.

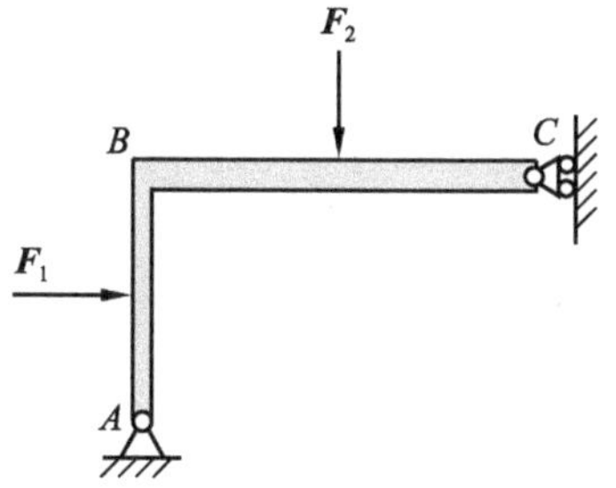

Fig. 1-29

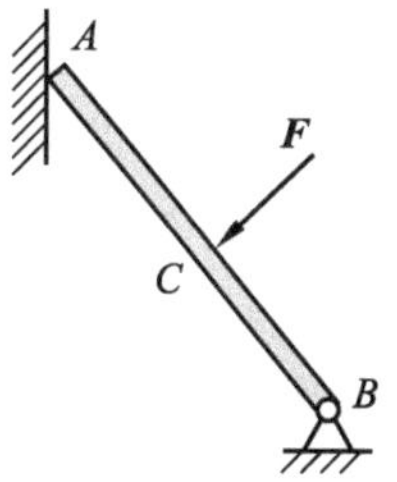

Fig. 1-30

1-7 As shown in Fig. 1-31, the structure consists of two bars, which are connected by the hinge E, irrespective of the self-weight and friction, draw the force diagrams of the members AB,CDE and whole system of bodies.

1-8 As shown in Fig. 1-32, the structure consists of two bars, which are connected by the hinge C, irrespective of the self-weight and friction (not drawing), draw the force diagrams for each component (except the pin) and whole system of bodies.

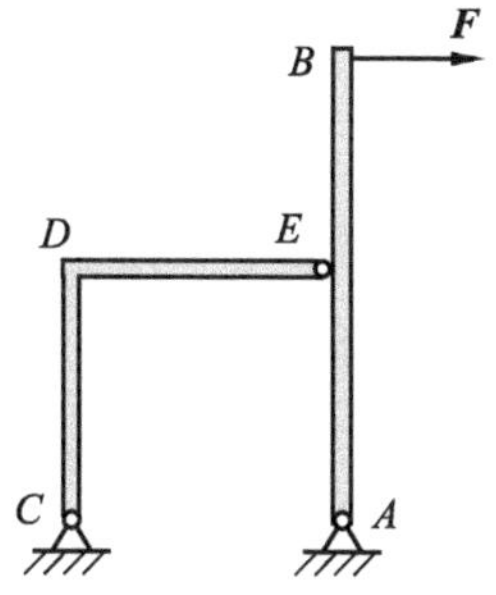

Fig. 1-31

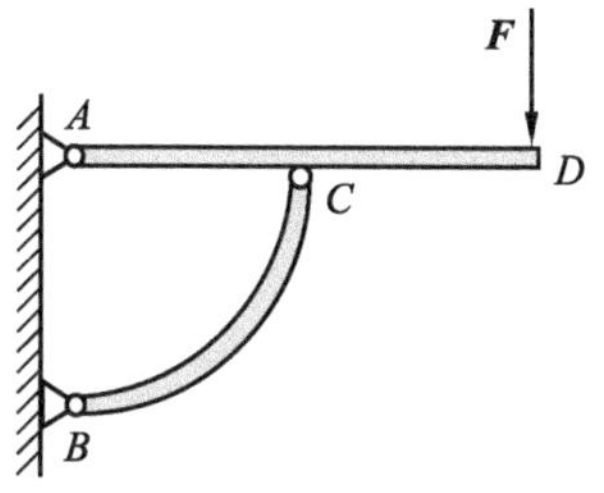

Fig. 1-32

1-9 The structure shown in Fig.1-33, points E and D are pin supports, irrespective of the self-weight and friction, draw the free-body diagrams of the members AB, CD and the pulley.

1-10 The roller shown in Fig. 1-34 is put on smooth surface, irrespective of the self-weight of bar AB and friction, draw the free-body diagrams of the bar AB and the roller O.

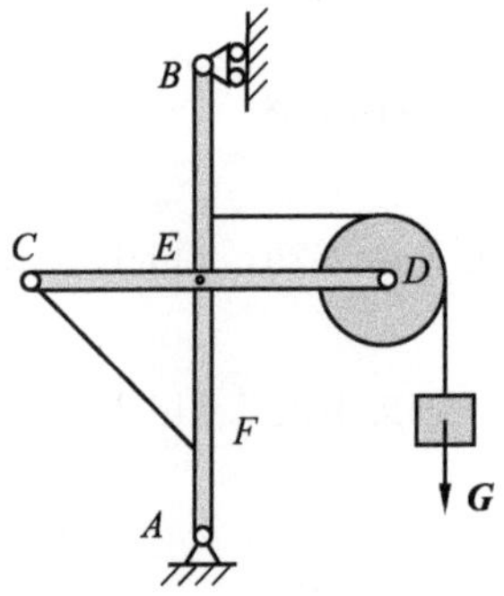

Fig. 1-33

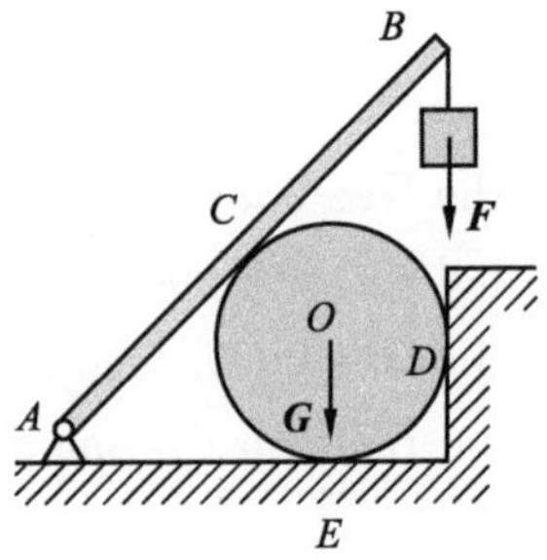

Fig. 1-34

1-11 The structure shown in Fig. 1-35 consists of two bars, joined by the hinge D and supported at A and C by fixed pin supports. Draw the force diagrams of the bars AB, CD and both bars together.

1-12 The multi-part structure shown in Fig. 1-36 consists of beams ABC and CD, connected by the hinge C, and it is supported by the fixed pin support at A and roller at B and D, irrespective of the self-weight and friction, draw the force diagrams of the beam ABC, CD and both beams together.

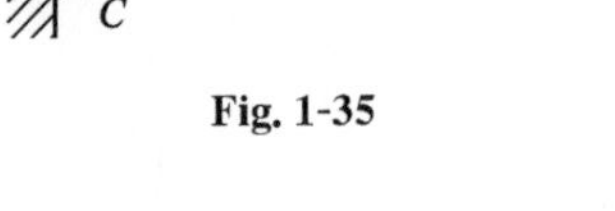

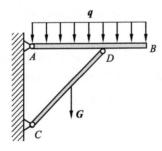

Fig. 1-35

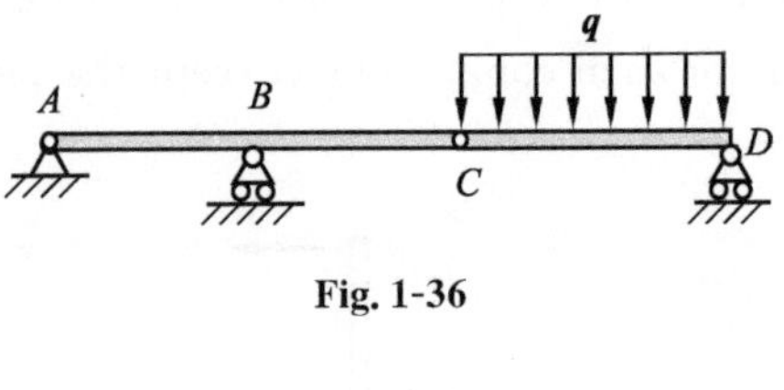

Fig. 1-36

Chapter 2

System of Forces in a Plane

Objectives

In this chapter, coplanar forces are considered that include *concurrent system of forces*, *couple and moment of a force*, and *general system of forces*. Students will learn how coplanar system of forces can be reduced and under which conditions they are in equilibrium. They should also learn how to apply the method of sections to formulate the conditions of equilibrium. In addition, they will be familiar with methods to determine the internal forces in a statically determinate *truss*.

2.1 Coplanar System of Concurrent Forces

2.1.1 Addition of Concurrent Forces in a Plane

If all the forces acting on a body lie in one plane, they are called *coplanar forces*. Now consider a system of n forces that all lie in a plane and whose lines of action intersect at point A, as shown in Fig. 2-1a. Such a system is called a *coplanar* system of *concurrent forces*. The resultant can be obtained through successive application of the *parallelogram law* of forces. Mathematically, the vector summation may be written in the form of the following vector equation:

$$\boldsymbol{F}_{\mathrm{R}} = \boldsymbol{F}_1 + \boldsymbol{F}_2 + \cdots + \boldsymbol{F}_n = \sum \boldsymbol{F}_i \tag{2.1}$$

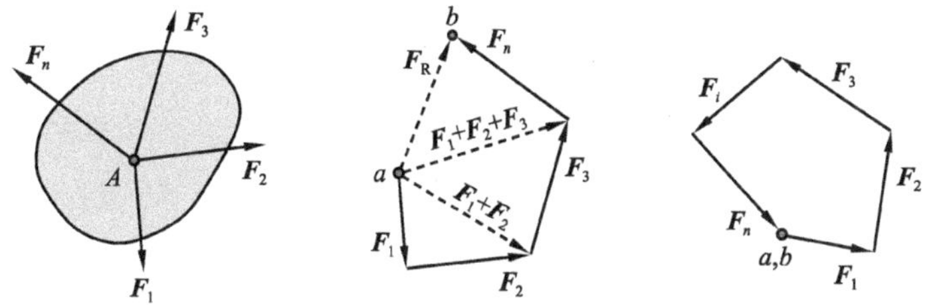

Fig. 2-1

The system of forces is reduced to a single force, this process is called *reduction*. Since the forces acting on a rigid body are sliding vectors, they do not have to act at point A; only their lines of action intersect at this point. We can draw a *force triangle* to graphically determine the resultant instead of drawing parallelogram, as shown in Fig. 2-1b, n forces are added head-to-tail. The sequence of the addition is arbitrary. The resultant $\boldsymbol{F}_R$ is the vector that points from the initial point a to the end point b of the *force polygon*.

In the case of a graphical solution, the force polygon must be drawn using a scale. Sometimes problems can be solved using the graphic-analytical method, e.g., the solution can be obtained by trigonometry.

Example 2-1

The end of a bar in Fig. 2-2a is subjected to four concurrent and coplanar forces ($\boldsymbol{F}_1=12$ kN, $\boldsymbol{F}_2=8$ kN, $\boldsymbol{F}_3=18$ kN, $\boldsymbol{F}_4=4$ kN) under given angles ($\alpha_1=45°$, $\alpha_2=100°$, $\alpha_3=205°$, $\alpha_4=270°$) with respect to the horizontal. Determine the magnitude and direction of the resultant.

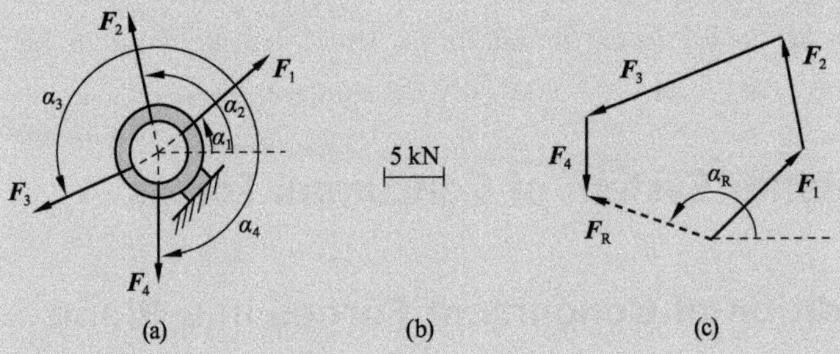

Fig. 2-2

Solution:

The problem can be solved graphically. First, a scale is chosen, Fig. 2-2b. Then the force polygon is drawn by adding the given vectors head-to-tail, Fig.2-2c. By measuring the length of resultant $\boldsymbol{F}_R$, within the limit of the accuracy of the drawing, the result is obtained.

$$\boldsymbol{F}_R = 10.5 \text{ kN}, \ \alpha_R = 155° \qquad \textit{Ans.}$$

There are various possible methods to draw the force polygon. Depending on the choice of the first vector and the sequence of the others. But they all yield the same resultant $\boldsymbol{F}_R$.

2.1.2 Resolution of a Force in a Plane

In coplanar problems, a force can be resolved into *two components* in different directions by an infinite number of ways. For analytical work, we usually resolve forces $\boldsymbol{F}$ into its *rectangular components* $\boldsymbol{F}_x$ and $\boldsymbol{F}_y$, which lie along the x and y axes, respectively, Fig. 2-3. Although the axes are horizontal and vertical, they may in general be directed at any inclination, as long as they remain perpendicular to each other. The directions of the components may be given by the axes x and y of a Cartesian coordinate system, Fig. 2-3. With the unit vectors $\boldsymbol{i}$ and $\boldsymbol{j}$, the components of $\boldsymbol{F}$ are written as

$$\boldsymbol{F}_x = F_x\boldsymbol{i}, \ \boldsymbol{F}_y = F_y\boldsymbol{j} \tag{2.2}$$

and the force $\boldsymbol{F}$ is expressed as

$$\boldsymbol{F} = F_x\boldsymbol{i} + F_y\boldsymbol{j} \tag{2.3}$$

The quantities $\boldsymbol{F}_x$ and $\boldsymbol{F}_y$ are called the *coordinates* of the force vector $\boldsymbol{F}$. From Fig. 2-3, they can be determined from

$$F_x = F\cos\alpha, \ F_y = F\sin\alpha \tag{2.4}$$

For a system of concurrent forces, its resultant can be obtained by simply adding the respective coordinates of the forces. This procedure is demonstrated in Fig. 2-4a, with the aid of the example of two forces, each force is first resolved into its x and y components and represented as vector,

$$\boldsymbol{F}_1 = F_{1x}\boldsymbol{i} + F_{1y}\boldsymbol{j}, \boldsymbol{F}_2 = F_{2x}\boldsymbol{i} + F_{2y}\boldsymbol{j} \tag{2.5}$$

The vector resultant is therefore:

$$\begin{aligned}\boldsymbol{F}_{\mathrm{R}} &= \boldsymbol{F}_1 + \boldsymbol{F}_2 = F_{1x}\boldsymbol{i} + F_{1y}\boldsymbol{j} + F_{2x}\boldsymbol{i} + F_{2y}\boldsymbol{j} \\ &= (F_{1x} + F_{2x})\boldsymbol{i} + (F_{1y} + F_{2y})\boldsymbol{j}\end{aligned} \tag{2.6}$$

Hence, the coordinates of the resultant are obtained as

$$F_{\mathrm{R}x} = F_{1x} + F_{2x}, \ F_{\mathrm{R}y} = F_{1y} + F_{2y} \tag{2.7}$$

In a general case, the coordinates of the resultant $\boldsymbol{F}_{\mathrm{R}}$ can be represented by the algebraic summation of the x and y components of all the forces, i. e.,

$$\begin{aligned}F_{\mathrm{R}x} &= F_{1x} + F_{2x} + \cdots + F_{nx} = \sum F_x \\ F_{\mathrm{R}y} &= F_{1y} + F_{2y} + \cdots + F_{ny} = \sum F_y\end{aligned} \tag{2.8}$$

This is known as the *theorem of projections*: *the projection, on an any axis, of the resultant force of a system of concurrent forces is equal to the sum of all projections of the forces from the system on the same axis.*

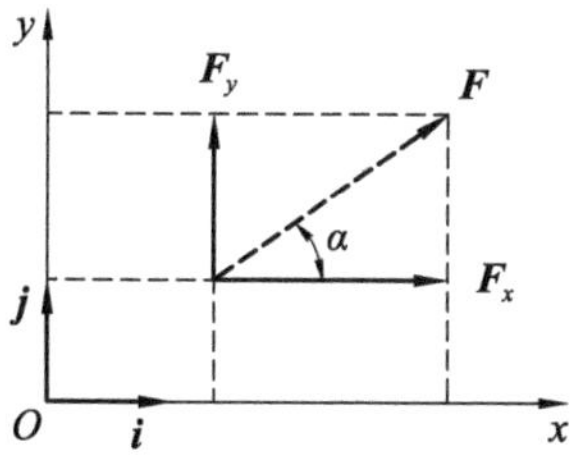

Fig. 2-3

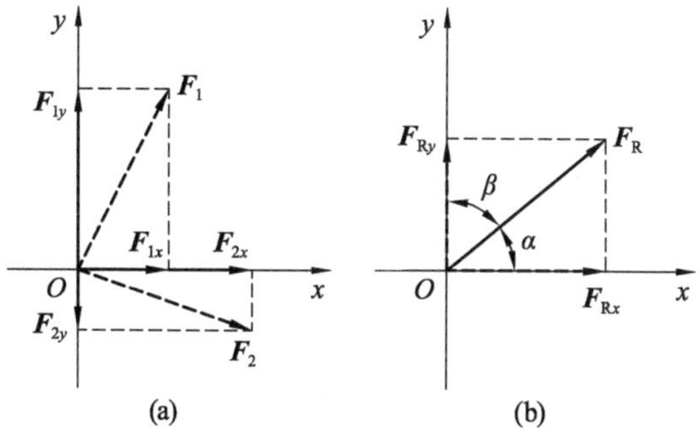

Fig. 2-4

From the Fig. 2-4b, the magnitude and direction of the resultant are given by

$$F_R = \sqrt{F_{Rx}^2 + F_{Ry}^2} = \sqrt{\left(\sum F_x\right)^2 + \left(\sum F_y\right)^2} \tag{2.9}$$

$$\cos\alpha = \frac{F_{Rx}}{F_R} = \frac{\sum F_x}{F_R}, \quad \cos\beta = \frac{F_{Ry}}{F_R} = \frac{\sum F_y}{F_R} \tag{2.10}$$

Example 2-2

An eyebolt is subjected to four forces: $F_1=200$ N, $F_2=300$ N, $F_3=500$ N, $F_4=400$ N, that act under given angles ($\alpha_1=30°$, $\alpha_2=45°$, $\alpha_3=0°$, $\alpha_4=60°$) with respect to the horizontal axis, Fig. 2-5a. Determine the magnitude and direction of the resultant force.

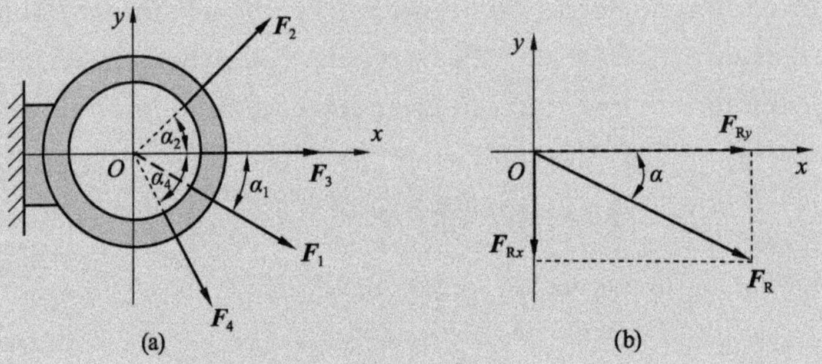

Fig. 2-5

Solution:

We choose the coordinate system shown in Fig. 2-5a, the x-axis coincides with the horizontal. The angles are measured from this axis. Each force is resolved into its x and y components. Summing the x components, we have

$$\begin{aligned} F_{Rx} &= F_{1x}+F_{2x}+F_{3x}+F_{4x} = F_1\cos\alpha_1 + F_2\cos\alpha_2 + F_3\cos\alpha_3 + F_4\cos\alpha_4 \\ &= 200\cos 30° + 300\cos 45° + 500\cos 0° + 400\cos 60° \\ &= 1\,085 \text{ N} \end{aligned}$$

Summing the y components yields

$$\begin{aligned} F_{Ry} &= F_{1y}+F_{2y}+F_{3y}+F_{4y} = -F_1\sin\alpha_1 + F_2\sin\alpha_2 + F_3\sin\alpha_3 - F_4\sin\alpha_4 \\ &= -200\sin 30° + 300\sin 45° + 500\sin 0° - 400\sin 60° \\ &= -234 \text{ N} \end{aligned}$$

The magnitude of the resultant follows from Eq. (2.9),

$$F_R = \sqrt{F_{Rx}^2 + F_{Ry}^2} = \sqrt{1\,085^2 + (-234)^2} = 1\,110 \text{ N} \qquad \textit{Ans.}$$

Direction of the resultant follows from Eq. (2.10),

$$\cos\alpha = \frac{F_{Rx}}{F_R} = \frac{1\,085}{1\,110} = 0.977\,5, \quad \alpha = 12.2° \qquad \textit{Ans.}$$

Since F_{Rx} is positive and F_{Ry} is negative, therefore resultant lies between 270° and 360°, the actual angle of the resultant $=0°-12.2°=-12.2°$. Shown in Fig. 2-5b.

2.1.3 Equilibrium in a Plane

We now investigate the conditions under which a body is in equilibrium when subjected to the action of a system of concurrent forces, Fig. 2-6a. According to Newton's first law, a system of concurrent forces is in equilibrium if the *resultant force* is zero. This condition can be stated mathematically as

$$\boldsymbol{F}_{\mathrm{R}} = \sum \boldsymbol{F} = 0 \tag{2.11}$$

The Eq.(2.11) is not only a necessary condition for equilibrium, but also a sufficient condition.

(1) Geometrical Equilibrium Condition

In a general case for an arbitrary system of planar forces, the geometrical interpretation of Eq.(2.11) is that of a closed force polygon, i.e., the initial point a and the terminal point b have to coincide, Fig. 2-6b. The final geometrical equilibrium condition for a two-dimensional forces system is: *A system of planar forces is equivalent to zero if the force polygon is closed.*

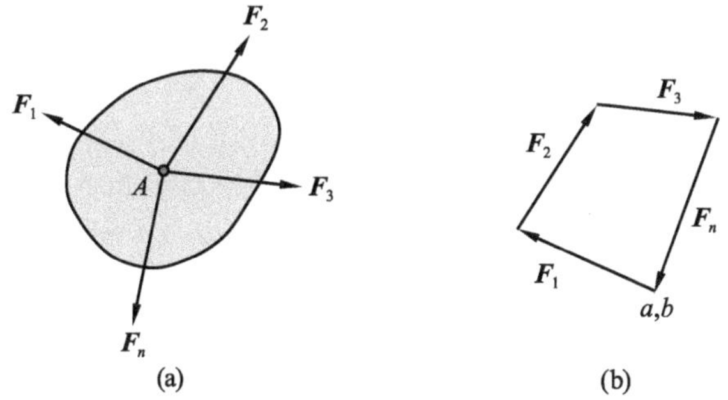

Fig. 2-6

Example 2-3

A wheel with weight $G=600$ N is held between a smooth inclined plane and a smooth wall, Fig. 2-7a. Determine the contact forces between the plane, wall and the wheel.

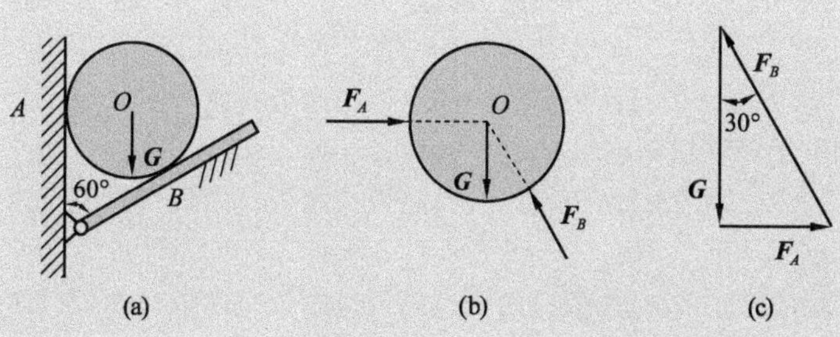

Fig. 2-7

Solution:

The wheel is separated from the system of bodies and its free-body diagram is shown in Fig. 2-7b. Three forces, which include active force weight $\boldsymbol{G}$ and reaction forces: the contact force $\boldsymbol{F}_A$ and $\boldsymbol{F}_B$, are the concurrent forces; we solve the problem using graphic-analytical solution by sketching the geometrical equilibrium condition, namely, a closed force triangle, Fig. 2-7c. The law of sines yields:

$$F_A = G\tan 30° = 600 \times \tan 30° = 416\ \text{N} \qquad \textit{Ans.}$$

$$F_B = \frac{G}{\cos 30°} = \frac{600}{\cos 30°} = 693\ \text{N} \qquad \textit{Ans.}$$

(2) Analytical Equilibrium Condition

It is also known from Section 2.1 that a system of n concurrent forces is in equilibrium if the resultant is zero. The resultant force is zero if its components are zero.

$$F_R = \sqrt{F_{Rx}^2 + F_{Ry}^2} = \sqrt{\left(\sum F_x\right)^2 + \left(\sum F_y\right)^2} = 0$$

For this equation to be satisfied, in the case of a coplanar system of forces, both the forces components $\boldsymbol{F}_{Rx}$ and $\boldsymbol{F}_{Ry}$ must be equal to zero. Hence,

$$\begin{aligned} \sum F_x &= 0 \\ \sum F_y &= 0 \end{aligned} \qquad (2.12)$$

These two equations can be solved for at most two unknowns. Thus, the analytical equilibrium conditions of the concurrent force: *A coplanar system of concurrent forces is in equilibrium if the sums of the respective coordinates of the force vectors (here the x and y coordinates) is equal to zero.*

It is important to note that if a force has an *unknown magnitude*, then the arrowhead sense of the force on the free-body diagram can be *assumed*. If the *solution* yields a *negative scalar*, this indicates that the sense of the force is opposite to that which was assumed.

Example 2-4

The 80 kN cylinder in Fig. 2-8a is suspended using the cables AB and BC. Determine the tension in cables AB and BC necessary to support the cylinder.

Solution:

The forces in cables AB and BC can be determined by investigating the equilibrium of ring B, whose free-body diagram is shown in Fig. 2-8b. There are three forces acting on it. The magnitude of $\boldsymbol{F}_{BD}$ is equal to the weight of the cylinder. The magnitudes of $\boldsymbol{F}_{AB}$ and $\boldsymbol{F}_{BC}$ are unknown, but their directions are known. Applying the equations of equilibrium along the x and y axes, we have

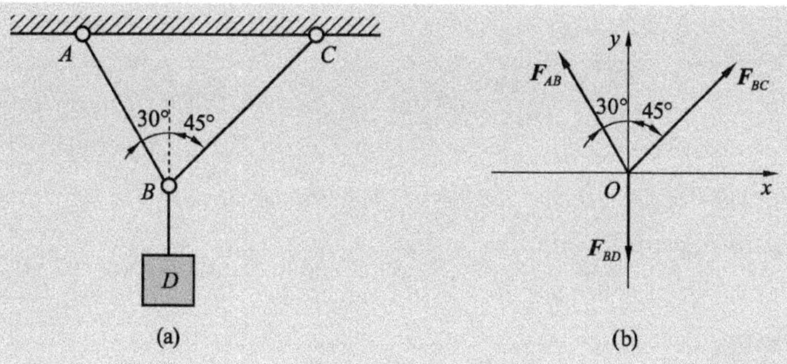

Fig. 2-8

$$\sum F_x = 0,\ -F_{AB}\sin 30^\circ + F_{BC}\sin 45^\circ = 0$$

$$\sum F_y = 0,\ F_{AB}\cos 30^\circ + F_{BC}\cos 45^\circ - F_{BD} = 0$$

These are two equations for the two unknowns, substituting $F_{BD} = 80$ kN into above equations, solving, we obtain

$$F_{AB} = 58.6 \text{ kN},\ F_{BC} = 41.4 \text{ kN} \qquad \textit{Ans.}$$

Example 2-5

Two bars *AC* and *BC* are attached at *A* and *B* to a wall by fixed pins supports. They are pin-connected at *C* and subjected to a weight $G = 60$ kN, as shown in Fig. 2-9a. Determine the forces in two bars.

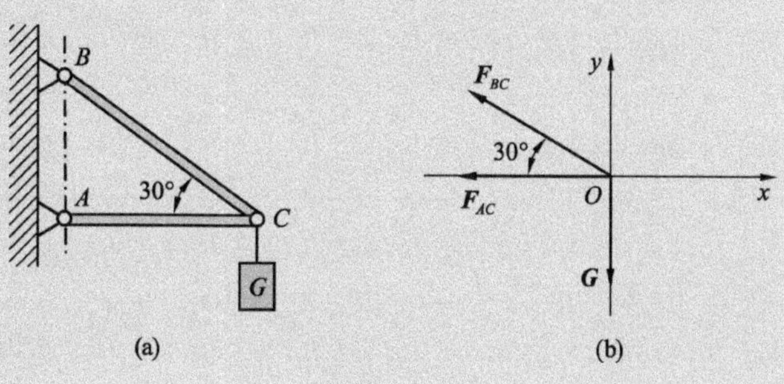

Fig. 2-9

Solution:

Pin *C* is isolated from the bars by an imaginary cut. The free-body diagram is shown in Fig. 2-9b. Bars *AC* and *BC* are *two-force member*, so the *resultant forces* on bars *AC* and *BC* must be equal, opposite, and collinear, respectively. The orientations of the forces along their action lines may be chosen arbitrarily in the free-body diagram. If we choose a coordinate system shown in Fig. 2-9b, the equilibrium conditions in the horizontal and vertical directions

$$\sum F_x = 0,\ -F_{AC} - F_{BC}\cos 30^\circ = 0$$

$$\sum F_y = 0,\ F_{BC}\sin 30^\circ - G = 0$$

Solving, we get

$$F_{BC}=120\ \text{kN},\ F_{AC}=-104\ \text{kN} \qquad \textit{Ans.}$$

Since F_{AC} is negative, the orientation of the vector F_{AC} along its action line is opposite to the orientation chosen in the free-body diagram. Therefore, in reality bar AC is subjected to compression.

Example 2-6

The crane is shown in Fig. 2-10a, two bars AB and BC are attached at A and B to a wall by fixed pins. They are pin-connected at B with a pulley and subjected to a weight $G=20$ kN. Assuming pulley B and bar AB and BC is frictionless and weightless, neglect the size of the pulley B. Determine the forces in the bars AB and BC.

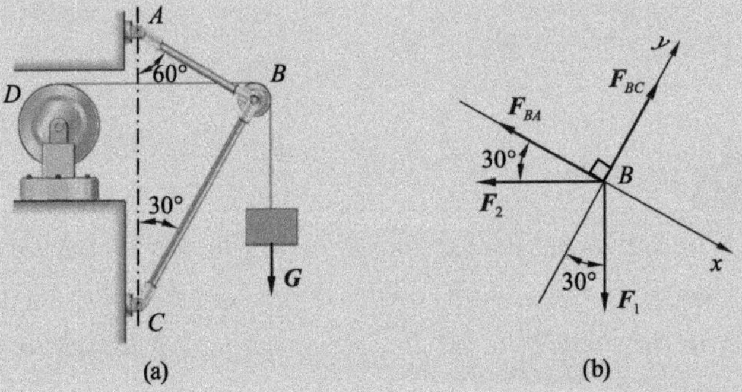

Fig. 2-10

Solution:

Pin B is isolated from the surrounding and its free-body diagram is shown in Fig. 2-10b. The bars AB and BC are *two-force member*. Although the magnitudes of the *resultant forces* on bars AB and BC are unknown, the line of action is along the line joining two points. Here $F_1=F_2=G=20$ kN. Then we choose a coordinate system (see Fig. 2-10b, two axes are not in the horizontal and vertical directions), since it involves the least number of unknown in the equation. Applying the equations of equilibrium along the x and y axes, we have

$$\sum F_x=0,\ -F_{BA}+F_1\sin 30^\circ-F_2\sin 60^\circ=0$$

$$\sum F_y=0,\ F_{BC}-F_1\cos 30^\circ-F_2\cos 60^\circ=0$$

These are two equations for the two unknowns, substituting $F_1=F_2=20$ kN into above equations, solving, we obtain

$$F_{BC}=27.32\ \text{kN},\ F_{BA}=-7.32\ \text{kN} \qquad \textit{Ans.}$$

Since F_{BA} is negative, the orientation of the vector F_{BA} along its action line is opposite to the orientation chosen in the free-body diagram. Therefore, bar BA is subjected to compression.

2.2 Moment of a Force in a Plane

2.2.1 Moment of a Force about a Point

The *moment* of a force about a point or axis provides a measure of the tendency of the force to cause a body to rotate about the point or axis. For example, consider the horizontal force $\boldsymbol{F}$, which acts on the handle of the wrench and is located a perpendicular distance h from point O (or the z axis) to the line of action of force $\boldsymbol{F}$, Fig. 2-11a. It is seen that this force tends to turn the bolt about the z axis. The larger the force $\boldsymbol{F}$ or the distance h, the greater the turning effect. This tendency for rotation caused by $\boldsymbol{F}$, is called the *moment of a force* or simply the *moment* ($\boldsymbol{M}_O$), and is defined by its magnitude

$$\boldsymbol{M}_O(\boldsymbol{F}) = \pm Fh \tag{2.13}$$

where the superscript (O) indicates the reference point. The perpendicular distance h from point O to the action line is called the *moment arm* of force $\boldsymbol{F}$ with respect to O. The direction of M_O is defined by its moment axis, which is perpendicular to the plane that contains the force $\boldsymbol{F}$ and its moment arm h, Fig. 2-11b. In two dimensions, the rotational sense of the moment is given by the rotational sense of force $\boldsymbol{F}$ about O, that is, the moment is positive if it tends to rotate the body *counterclockwise* when viewed from above. It should be noted that the magnitude and rotational sense of the moment of a force depend on the point O.

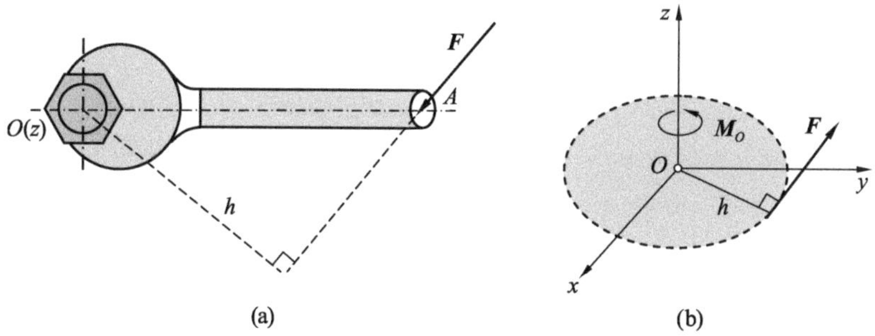

Fig. 2-11

2.2.2 Resultant Moment

For a system of coplanar forces, where all the forces lie within the x-y plane, Fig. 2-12. The resultant moment $\boldsymbol{M}_{RO}$ about point O (the z axis) can be determined by finding the *algebraic sum* of the moments caused by all the forces in the system. If we consider positive moments as counterclockwise and clockwise moments as negative, the directional sense of each moment can be represented by a plus or minus sign. Then the resultant moment in Fig. 2-12 is therefore

$$M_{RO} = \sum M_{iO} = F_1 h_1 + F_2 h_2 + \cdots + F_n h_n \tag{2.14}$$

If the numerical result of this sum is a positive scalar. $\boldsymbol{M}_{RO}$ will be a counterclockwise moment.

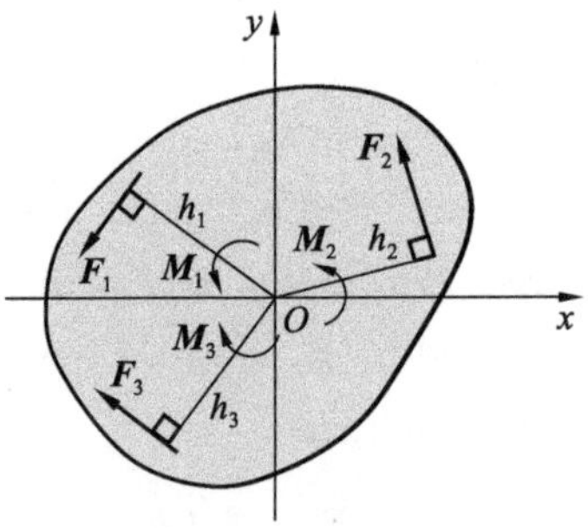

Fig. 2-12

2.2.3 Principle of Moments

Consider the moment of a force $\boldsymbol{F}$ about a point O as a planar problem, shown in Fig. 2-13, the force $\boldsymbol{F}$ located in the x-y plane of a rectangular coordinate system with origin O. The components of the force $\boldsymbol{F}$ in this coordinate system are F_x and F_y. The moment of the force $\boldsymbol{F}$ about the point O can be calculated as the sum of the moments of the force's components F_x and F_y about the same point.

$$M_O = xF_y - yF_x \tag{2.15}$$

This concept often used in mechanics is the *Principle of Moments*: *The moment of a resultant force of a system of concurrent forces about an arbitrary point O is equal to the sum of individual moments of each force of the system about that point.* This procedure, where the moment is calculated from suitable components instead of using lengths and angles of the corresponding vectors, is often easier than finding the same moment using Eq.(2.13). And the property holds for an arbitrary number of forces.

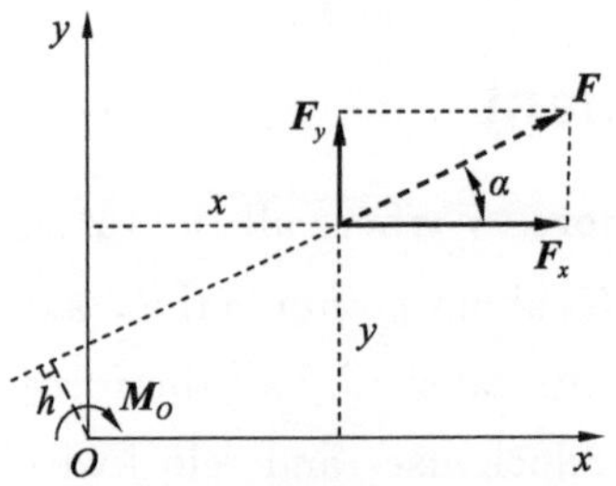

Fig. 2-13

Example 2-7

Determine the moments of the 200 N force acting on the end of wrench in Fig. 2-14a about point O.

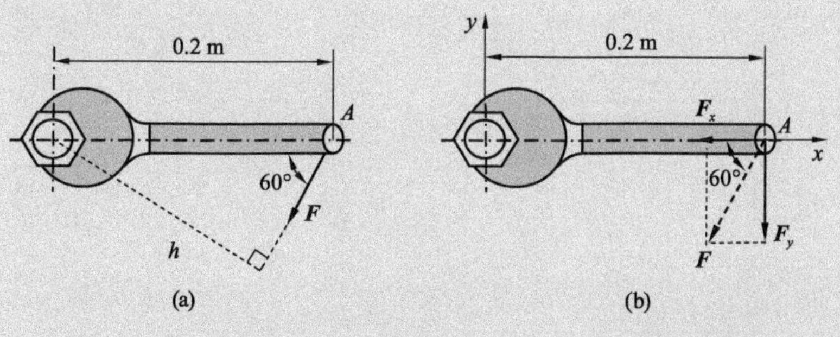

Fig. 2-14

Solution Ⅰ:

The moment arm h in Fig. 2-14a can be found from trigonometry.

$$h = 0.2 \times \sin 60° = 0.173 \text{ m}$$

Thus

$$M_O(F) = F \cdot h = 200 \times 0.173 = 34.6 \text{ N} \cdot \text{m} \qquad \textit{Ans.}$$

Since the force tends to rotate the bolt clockwise about point O, the moment is directed into the page.

Solution Ⅱ:

The x and y axes can be set parallel and perpendicular to the rod's axis as shown in Fig.2-14b. The x and y components of the force F can be found from trigonometry.

$$F_x = F \cdot \cos 60° = 100 \text{ N}$$

$$F_y = F \cdot \sin 60° = 173.2 \text{ N}$$

Considering counterclockwise moments as positive, and applying the principle of moments, here F_x produces no moment about point O since its line of action passes through this point. Therefore,

$$M_O(F) = xF_y - yF_x = 173.2 \times 0.2$$

$$= 34.6 \text{ N} \cdot \text{m} \qquad \textit{Ans.}$$

This example illustrates two alternative methods of computing the moment of a force in a plane: ① straightforward calculation using given components, and ② applying the principle of moments.

Example 2-8

Force $F=400$ N acts at the end A of the bracket shown in Fig. 2-15a. Determine the moment of the force about point O.

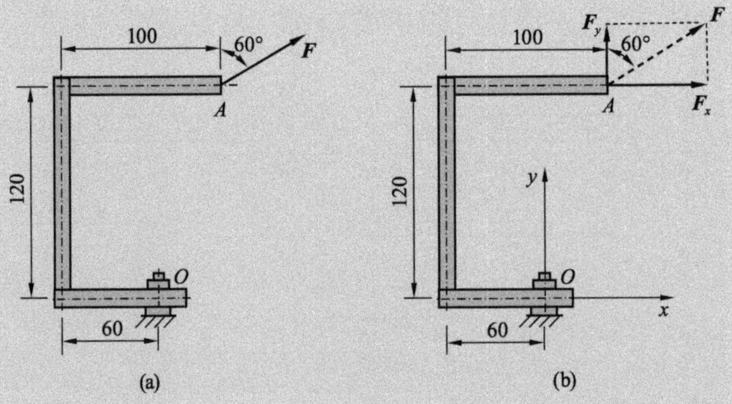

Fig. 2-15

Solution:

We choose the rectangle coordinates as shown in Fig. 2-15b, the force $\boldsymbol{F}$ is resolved into its x and y components that can be found from trigonometry.

$$F_x = F\sin 60° = 400\sin 60° = 346.4 \text{ N}$$

$$F_y = F\cos 60° = 400\cos 60° = 200 \text{ N}$$

Location of point A:

$$x = 100 - 60 = 40 \text{ mm} = 0.04 \text{ m}, \quad y = 120 \text{ mm} = 0.12 \text{ m}$$

Applying the principle of moments, we have

$$M_O(F) = xF_y - yF_x = 0.04 \times 200 - 0.12 \times 346.4 = -33.57 \text{ N} \cdot \text{m} \qquad \textit{Ans.}$$

Since the moment about the point O is negative, the sense of M_O is clockwise.

2.3 Couple and Moment of a Couple

2.3.1 Couple and Its Moment

A couple is defined as two parallel forces that have the same magnitude, opposite directions, and are separated by a perpendicular distance h, as shown in Fig. 2-16a. Since the resultant force is zero, a couple can *not* be reduced to a resultant single force, the only effect of a couple is to produce a rotation or tendency of rotation in a specified direction. Fig. 2-16b,c show two examples of couples: ① a wheel which is to be turned and ② a screw tap tapping on the workpiece.

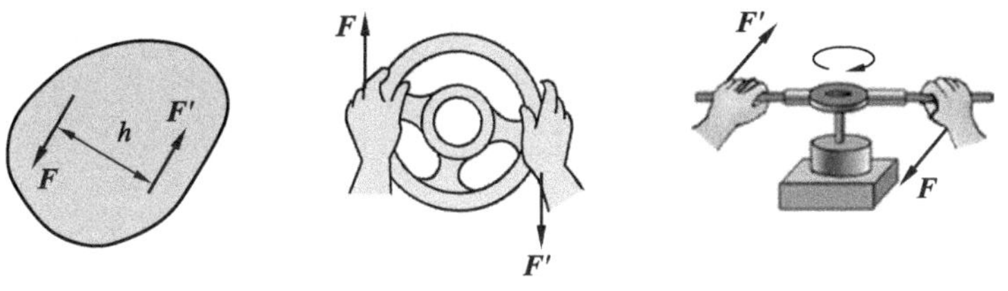

(a) (b) (c)

Fig. 2-16

The couple in Fig. 2-16 tends to *rotate* the rigid body on which it acts. This effect can be determined by its *moment* which is defined as having a magnitude of

$$M(F, F') = \pm Fh \tag{2.16}$$

where F is the magnitude of the force and h is the perpendicular distance between the lines of action of the forces and is called the "*arm of a couple*". The *direction* and sense of the couple moment are determined by the right-hand rule, as shown in Fig. 2-17, where the thumb indicates the direction when the fingers are curled with the sense of rotation caused by the couple force. The sense of rotation is represented by a curved arrow. The moment of a couple is always perpendicular to the plane containing the couple forces.

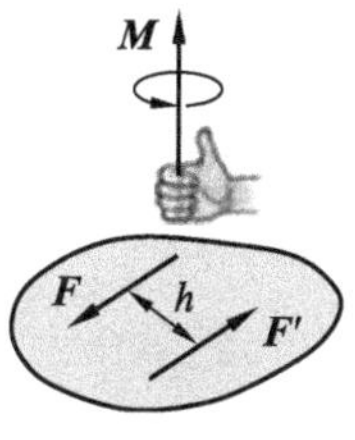

Fig. 2-17

2.3.2 The Characteristics of Coplanar Couple

What follows are some theorems regarding a couple (proofs are left to the reader):

① A couple cannot be reduced to a resultant single force and can only be equilibrium to a couple of opposite sense.

② If two couples produce the moment with the *same magnitude and direction*, then these two couples are *equivalent*.

③ Two forces of the couple can be replaced by the couple moment, that is, in the figures we can replace the two forces by a curved arrow, as shown in Fig. 2-18.

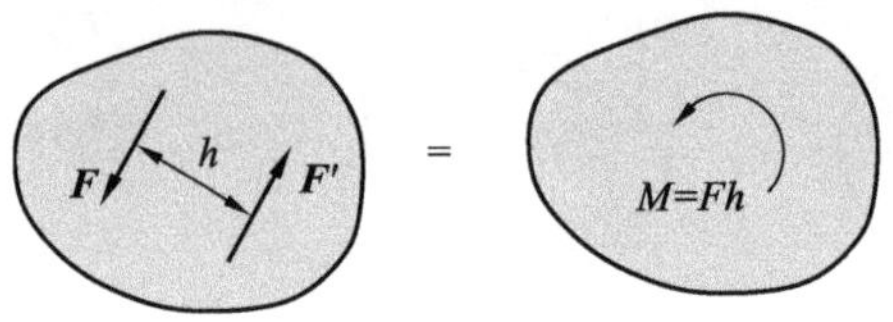

Fig. 2-18

④ The action of an arbitrary couple on a rigid body is invariant when the rotation on the couple plane through an arbitrary angle, as shown in Fig. 2-19, that is, the couple is free vector.

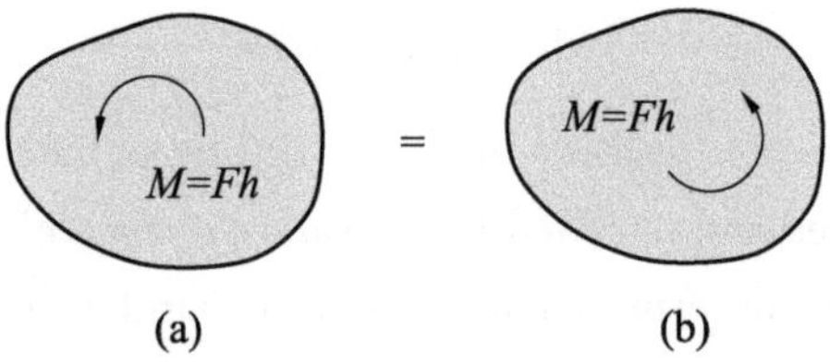

Fig. 2-19

⑤ The action of an arbitrary couple on a rigid body is invariant if the couple moment remains unchanged, that is, it is possible to vary the magnitude of force and its arm as long as couple moment remains constant, as shown in Fig. 2-20.

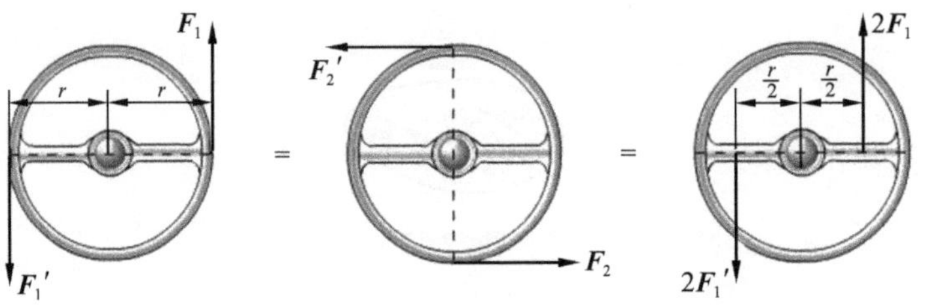

Fig. 2-20

2.3.3 Resultant Couples Moment

If several coplanar couples act on a rigid body, they may be appropriately moved and rotated and then added to form a resultant couple. The resultant couple moment is the algebraical sum of the moments of each couple considering their algebraic signs (given by their respective senses of rotation). For example, two couples acting on same plane of the body in Fig. 2-21a,

$$M_1 = F_1 h_1, M_2 = -F_2 h_2$$

may be replaced by their corresponding couple moments M_1 and M_2, as shown in Fig. 2-21b,

$$M_1 = F_1 h_1 = F_3 h, M_2 = -F_2 h_2 = -F_4 h$$

and then these couples may be moved to the *arbitrary point* and added to obtain the *resultant couple* moment shown in Fig. 2-21c,

$$M = M_1 + M_2$$

If more than two couple moments act on the plane, we may generalize this concept,

$$M = M_1 + M_2 + \cdots + M_n = \sum M_i \tag{2.17}$$

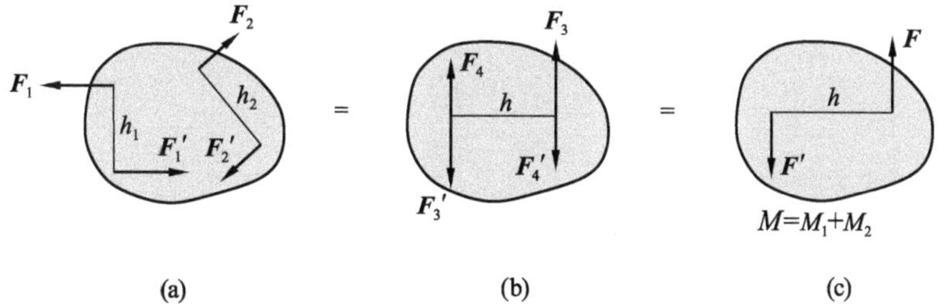

Fig. 2-21

2.3.4 Equilibrium of a Coplanar Couples System

Considering a rigid body, there are only couples acting on it with no resultant force. If the body is to be at rest or in equilibrium, the moment of the resultant couple causing rotation must be zero. In other words, the resultant moment of all the forces must be zero. Thus, the *equilibrium condition* for a system of couple moments is

$$\sum M_i = 0 \tag{2.18}$$

For a coplanar couples system, there is only one equation being solved for one unknown.

Example 2-9

Three couples with forces of $F_1 = 200$ N, $F_2 = 300$ N, $F_3 = 500$ N act on the plate in Fig. 2-22. Determine the resultant couple moment of the three couples.

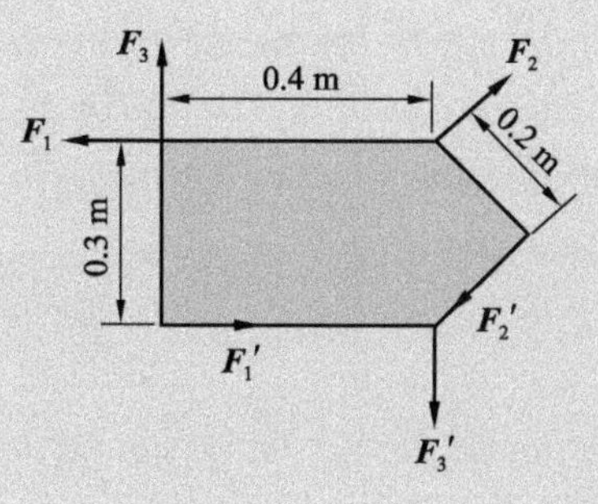

Fig. 2-22

Solution:

As shown the perpendicular distances between each pair of couple forces are $h_1 = 0.3$ m, $h_2 = 0.2$ m, and $h_3 = 0.4$ m, the couple moment M_i, developed by each pair of forces, can easily be determined from a scalar formulation.

$$M_1 = F_1 h_1 = 200 \times 0.3 = 60 \text{ N} \cdot \text{m}$$

$$M_2 = F_2 h_2 = 300 \times 0.2 = 60 \text{ N} \cdot \text{m}$$

$$M_3 = F_3 h_3 = 500 \times 0.4 = 200 \text{ N} \cdot \text{m}$$

Considering counterclockwise couple moments as positive, using the Eq.(2.17), we have

$$M_R = M_1 + M_2 + M_3 = 60 - 60 - 200$$

$$= -200 \text{ N} \cdot \text{m} \qquad \textit{Ans.}$$

The negative sign indicates that resultant couple moment M_R has a clockwise rotational sense.

Example 2-10

A couple with moment M acts on the beam AB, the length of beam is l, as shown in Fig. 2-23a. Neglect the weight of the beam, determine the reactions of the fixed pin support A and roller B.

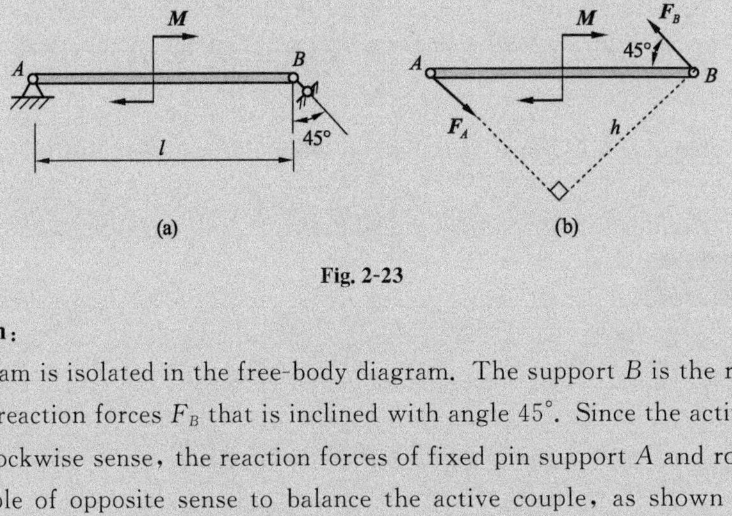

Fig. 2-23

Solution:

The beam is isolated in the free-body diagram. The support B is the roller, there is only one reaction forces F_B that is inclined with angle 45°. Since the active force is a couple of clockwise sense, the reaction forces of fixed pin support A and roller B must form a couple of opposite sense to balance the active couple, as shown Fig. 2-23b. Then, using Eq.(2.18), the equilibrium equation from the couple moment equilibrium condition is given by

$$\sum M_i = 0, \ F_A h - M = 0$$

The couple arm is

$$h = l \sin 45° = \frac{\sqrt{2}}{2} l$$

Instituting this result into the equation, and $F_A = F_B$ lead to the result

$$F_A = F_B = \frac{M}{l \sin 45^\circ} = \frac{\sqrt{2}M}{l} \qquad Ans.$$

2.4 General System of Forces in a Plane

Forces whose lines of action do not intersect at a point are called *general system of forces*. In order to reduce a general system of forces acting on a rigid body to a simpler form with an equivalent system, we firstly need to move the action lines of the forces without changing their effect, that is, *translation theorem of a force*.

2.4.1 Translation Theorem of a Force

With the aid of the couple moment, we will investigate how a force may be moved to a parallel line of action.

Consider a force $\boldsymbol{F}$ acting at the point A on a rigid body, in Fig. 2-24a. The line of action f of the force is assumed to be moved to the line f', which is parallel to f and passes through point B. The perpendicular distance of the two parallel lines is given by d. Then we can attach a pair of equal but opposite forces $\boldsymbol{F}$ and $-\boldsymbol{F}$ to point B along the line f', as shown in Fig. 2-24b. One of the forces $-\boldsymbol{F}$ and the originally given force $\boldsymbol{F}$ (action line f) form a couple. The couple moment is given by its magnitude $M = Fd$ and the sense of rotation, in Fig. 2-24c. Therefore, the force $\boldsymbol{F}$ is moved from A to B with action line f', but a couple moment M is added to maintain an equivalent to force F with action line f.

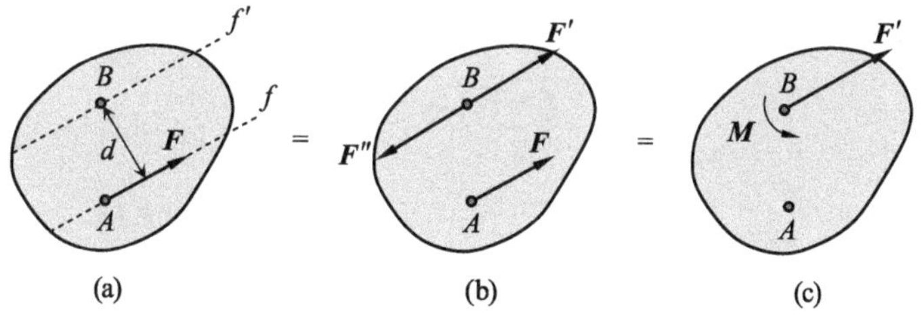

Fig. 2-24

Translation theorem of a force: *A force acting on a rigid body may be moved to a parallel lines of action to any point of the body, if we add a couple with a moment equal to the moment of the force about the point to which it is translated.*

2.4.2 Resultant of General System of Coplanar Forces

Consider a rigid body that is subjected to a general system of coplanar forces, as shown in Fig. 2-25a. To investigate how this system can be reduced to a simpler system, a reference point O is chosen firstly. Using the above *translation theorem*, the forces can be moved to point O without changing their directions. To avoid changing the effect of the forces on the body, the respective moments of the forces about O must be added to the body. Hence, the given *general system of forces* is equivalent to a *system of concurrent forces* and a *system of moments*, as shown in Fig. 2-25b. If we sum the forces and couple moments, these two systems can be reduced to a resultant force $\boldsymbol{F}_R$ (main vector) with the components F_{Rx} and F_{Ry} and a resultant moment M_O (main moment), Fig. 2-25c.

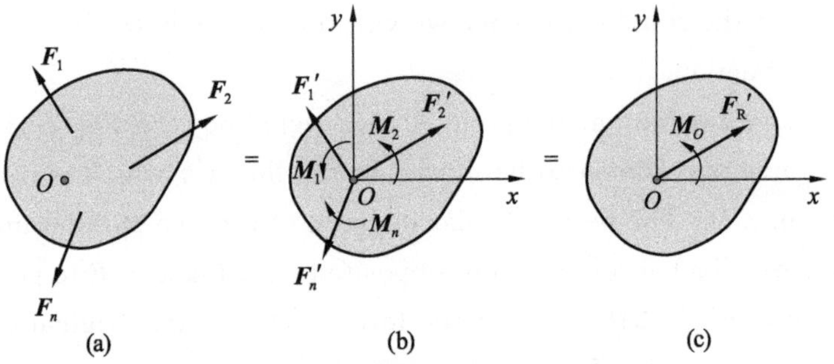

Fig. 2-25

If the forces system lays in the x-y plane, according to Eq. (2.9) and Eq. (2.14), the scalar equations of equivalent resultant force and resultant couple moment are given by

$$F'_{Rx} = F'_{1x} + F'_{2x} + \cdots + F'_{nx} = \sum F_x$$
$$F'_{Ry} = F'_{1y} + F'_{2y} + \cdots + F'_{ny} = \sum F_y \tag{2.19a}$$

$$\begin{aligned} M_O &= M_1 + M_2 + \cdots + M_n \\ &= M_O(F_1) + M_O(F_2) + \cdots + M_O(F_n) \\ &= \sum M_O(F) \end{aligned} \tag{2.19b}$$

The first Eq. (2.19a) states that the resultant force of the system is equivalent to the sum of all the forces; and the second Eq. (2.19b) states that the resultant couple moment of the system is equivalent to the sum of the moments about point O of all the forces. The magnitude and direction of the resultant force can be calculated from

$$F'_R = \sqrt{F'^2_{Rx} + F'^2_{Ry}} = \sqrt{(\sum F_x)^2 + (\sum F_y)^2} \tag{2.20a}$$

$$\cos\alpha = \frac{F'_{Rx}}{F'_R} = \frac{\sum F_x}{F'_R}, \quad \cos\beta = \frac{F'_{Ry}}{F'_R} = \frac{\sum F_y}{F'_R} \tag{2.20b}$$

Note that main vector $\boldsymbol{F}_R$ is independent of the location of point O; however, main moment M_O depends upon the location since the moments of each force are determined using the position distance d.

> **The criterion of equivalence can be stated as a theorem:** *The necessary and sufficient condition of equivalence of two force systems, acting on a rigid body, with respect to a certain point is that these systems have identical main force vectors and main moment of force vectors with respect to that point.*

◆ Fixed support

Fixed support that we have analyzed in Section 1. 3 can prevent both *translation* and *rotation*. With the aid of above method, we will investigate how to obtain the reaction force acting on the rigid body.

For a planar fixed end of rigid body, Fig. 2-26a, there is a general system of coplanar forces acting on it. Using the above reduction result, this general system of coplanar forces can be reduced to a main vector $\boldsymbol{F}_A$ and main moment M_A, Fig. 2-26b, the main vector $\boldsymbol{F}_A$ can be resolved into two rectangular components $\boldsymbol{F}_{Ax}$ and $\boldsymbol{F}_{Ay}$, so for the coplanar fixed support, we have three unknows of reactions, the couple moment and the two force components, in Fig. 2-26c.

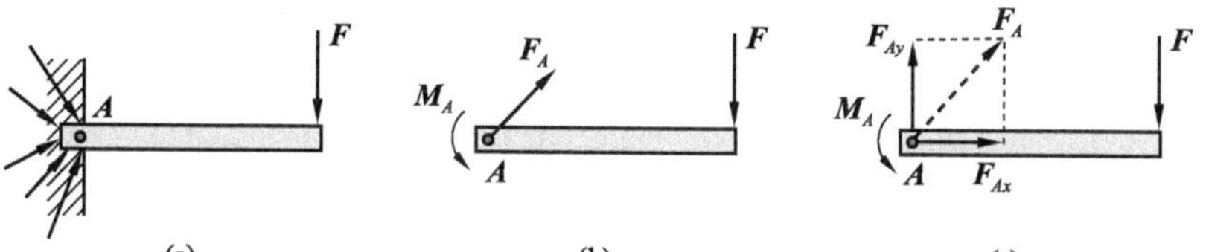

(a) (b) (c)

Fig. 2-26

2.4.3 Further Reduction of a Force and Couple System

The system of the resultant $\boldsymbol{F}_R$ (action line through O) and the moment M_O may be further simplified. One can always reduce the system to one of the following four cases:

(1) $\boldsymbol{F}'_R \neq 0, \boldsymbol{M}_O \neq 0$

The resultant moment M_R can be replaced by a pair of equal but opposite forces $\boldsymbol{F}_R$ and $-\boldsymbol{F}''_R$, Fig. 2-27. The force $\boldsymbol{F}'_R$ and $-\boldsymbol{F}''_R$ is a pair of equilibrium forces and can be subtracted from the body without changing the effect. Finally, we reduce the system to single resultant $\boldsymbol{F}_R$ with a perpendicular distance d away from point O, as shown in Fig. 2-27c. This perpendicular distance d can be determined from the scalar equation

$$d=\frac{M_O}{F'_R} \tag{2.21}$$

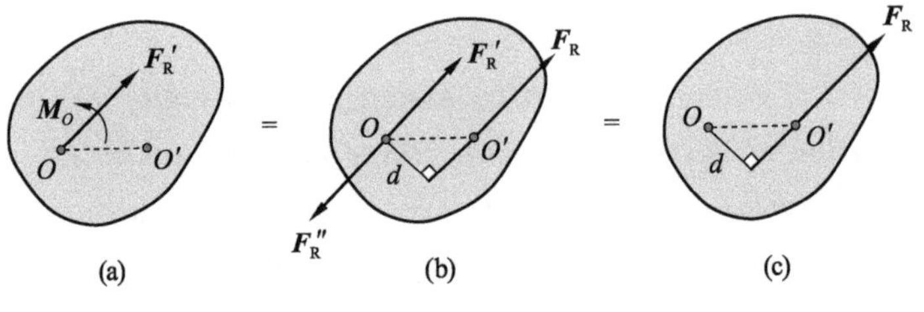

Fig. 2-27

(2) $F'_R \neq 0, M_O = 0$

In this case, Eq. (2.21) gives $d=0$, the system is reduced to single resultant $\boldsymbol{F}_R$ and its action line passes through reference point O, as shown in Fig. 2-28a.

(3) $F'_R = 0, M_O \neq 0$

In this case, further reduction is not possible: the general system of forces is reduced to only a moment (i. e., a couple), which is independent of the choice of the reference point, as shown in Fig. 2-28b.

(4) $F'_R = 0, M_O = 0$

In this case, the system is in *equilibrium*, Fig. 2-28c.

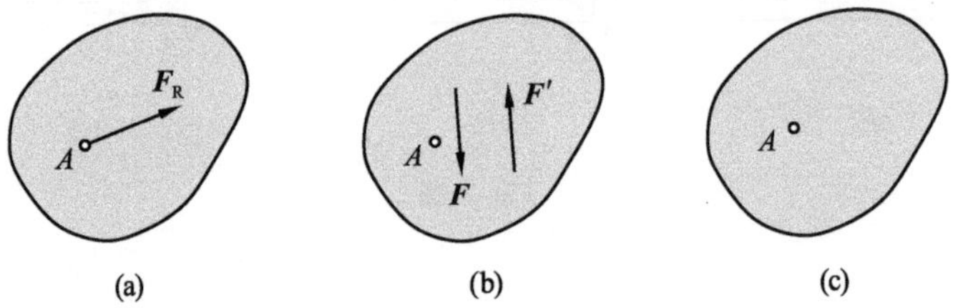

Fig. 2-28

Example 2-11

A plate is subjected to four forces as shown in Fig. 2-29a. The forces have the given magnitudes F or $2F$, respectively. Determine the magnitude and direction of the resultant and the location of its line of action.

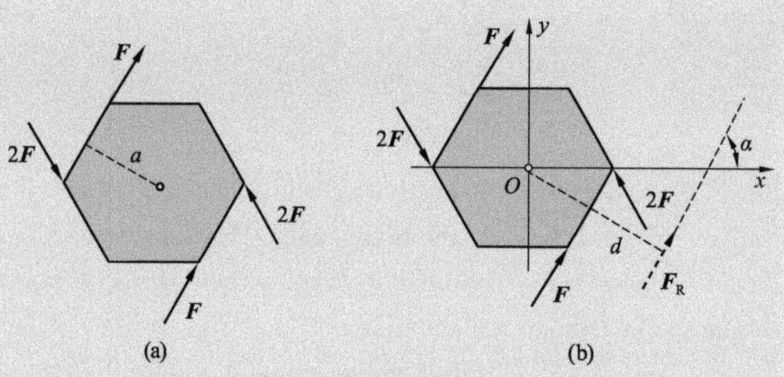

Fig. 2-29

Solution:

We choose a coordinate system x, y, Fig. 2-29b, and its origin O is taken as the reference point. According to the sign convention, positive moments tend to rotate the disk counterclockwise. Thus, from Eq. (2.19), summing the force components, we obtain

$$F'_{Rx}=\sum F_x=2F\cos 60°+F\cos 60°+F\cos 60°-2F\cos 60°=F$$

$$F'_{Ry}=\sum F_y=-2F\sin 60°+F\sin 60°+F\sin 60°+2F\sin 60°=\sqrt{3}F$$

The magnitude of main vector F'_R is

$$F'_R=\sqrt{F'^2_{Rx}+F'^2_{Ry}}=\sqrt{(\sum F_x)^2+(\sum F_y)^2}=2F \qquad Ans.$$

The direction angle α is

$$\tan\alpha=\frac{F'_{Ry}}{F'_{Rx}}=\sqrt{3},\ \alpha=60° \qquad Ans.$$

Summing the moments of the force, we obtain the main moment

$$M_O=\sum M_O=2aF+aF+2aF-aF=4aF$$

Using the Eq. (2.21), the perpendicular distance d of the resultant from reference point O

$$d=\frac{M_R}{R}=\frac{4aF}{2F}=2a \qquad Ans.$$

2.4.4 Equilibrium Conditions

As shown in former Section 2.4.2, a general system of coplanar forces can be reduced to an equivalent resultant force and resultant couple moment with respect to an arbitrary reference point. If this resultant force and resultant couple moment are both equal to zero, then the body is said to be in *equilibrium*. That is, the system will be subjected to the *conditions of* Eq. (2.12) and Eq. (2.18), respectively. Hence, in a coplanar problem, a rigid body under the action of a *general system of coplanar forces* is in equilibrium if the following equilibrium conditions are satisfied:

$$\sum F_x = 0$$
$$\sum F_y = 0 \qquad (2.22)$$
$$\sum M_O(\mathbf{F}) = 0$$

Here $\sum F_x$ and $\sum F_y$ represent, respectively, the *algebraic sums* of the x and y components of all the forces acting on the body, and $\sum M_O$ represents the *algebraic sum* of the moments of all the force components about the arbitrary point O. We can determine three unknowns from three equations.

Instead of using *two* force conditions and *one* moment condition, two alternative sets of three independent equilibrium equations may also be used. One such set is *one* force condition and *two* moment conditions.

$$\sum F_x = 0$$
$$\sum M_A = 0 \qquad (2.23)$$
$$\sum M_B = 0$$

When using these equations, it is required that a line passing through points A and B is *not perpendicular* to the x axis.

A second alternative set is that three points A, B and C may be chosen and only *three* moment equations of equilibrium used

$$\sum M_C = 0$$
$$\sum M_A = 0 \qquad (2.24)$$
$$\sum M_B = 0$$

Here it is necessary that the points A, B, and C are *not lying on a straight line*.

In principle, it is irrelevant whether one applies the equilibrium conditions (2.22), (2.23) or (2.24) to solve a given problem. However, it may be advantageous to use one form or the other. To use a moment equilibrium condition, it is necessary to specify a reference point and a positive sense of rotation (e.g., counterclockwise). If the solution of the equilibrium equations yields a negative scalar for a force or couple moment magnitude, this indicates that the sense is opposite to that which was assumed on the free-body diagram.

Example 2-12

The cantilever beam shown in Fig. 2-30a is loaded by the uniformly distributed force q, the couple $M = ql^2$ and the force $F = ql$ which acts under an angle 45°. Determine the reaction forces at the fixed support A. Neglect the weight of the beam.

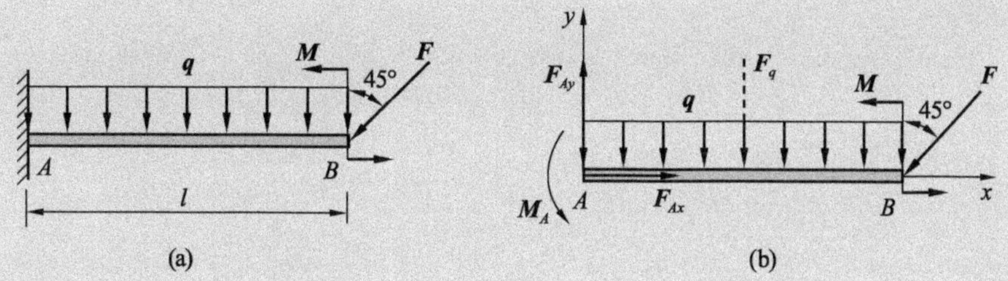

Fig. 2-30

Solution:

We free the beam from its supports and make the reaction forces visible in the free-body diagram, as shown in Fig. 2-30b. Since the support at A is fixed, the wall exerts three reactions $\boldsymbol{F}_{Ax}$, $\boldsymbol{F}_{Ay}$ and $\boldsymbol{M}_A$ on the beam. The magnitudes of these reactions are *unknown*, and their sense has been assumed. The forces $\boldsymbol{F}_{Ax}$ and $\boldsymbol{F}_{Ay}$ can be determined directly from the force equations of equilibrium. The force $\boldsymbol{F}$ can be represented by its x and y components, summing forces in the x and y direction yields

$$\sum F_x = 0,\ F_{Ax} - F\sin 45^\circ = 0$$

$$\sum F_y = 0,\ F_{Ay} - F\cos 45^\circ - ql = 0$$

The moment M_A can be determined by summing moments about point A

$$\sum M_A = 0,\ M_A + M - ql \times \frac{l}{2} - Fl\cos 45^\circ = 0$$

Note that M_A must be included in this moment summation. There are *three* independent equilibrium equations for *three unknowns*, solving, we get

$$F_{Ax}=0.707ql,\ F_{Ay}=1.707ql,\ M_A=0.207ql^2 \qquad \textit{Ans.}$$

Example 2-13

The simple support beam AB shown in Fig. 2-31a is loaded by the uniformly distributed force q, the couple $\boldsymbol{M}$ and the force $\boldsymbol{F}$ which acts on the middle point C. Determine the reaction forces at the fixed pin support A and roller B. Neglect the weight of the beam.

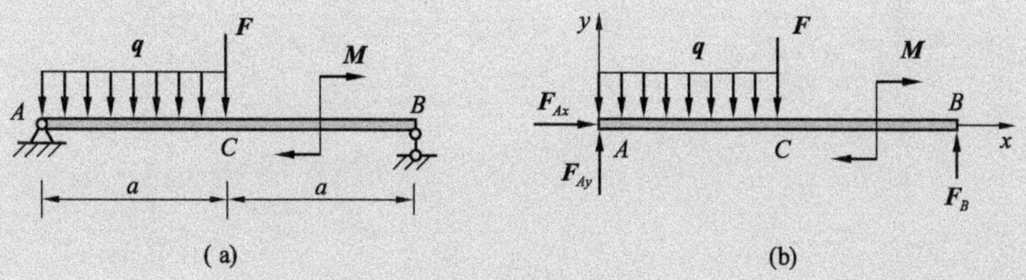

Fig. 2-31

Solution:

The free-body diagram of beam AB is shown in Fig. 2-31b. The fixed pin support A transmits two reactions and roller support B one reaction. In total, the three unknown reaction forces F_{Ax}, F_{Ay} and F_B exist. The reaction F_B can be obtained *directly* by summing moments about point A since F_{Ax} and F_{Ay} produce no moment about A.

$$\sum M_A = 0,\ F_B \cdot 2a - M - \frac{1}{2}qa^2 - Fa = 0$$

$$F_B = \frac{F}{2} + \frac{1}{4}qa + \frac{M}{2a} \qquad Ans.$$

Summing forces in the x and y direction yields

$$\sum F_x = 0,\ F_{Ax} = 0 \qquad Ans.$$

$$\sum F_y = 0,\ F_{Ay} + F_B - F - qa = 0$$

Substituting the result of F_B into above equation, solving, we get

$$F_{Ay} = \frac{F}{2} + \frac{3}{4}qa - \frac{M}{2a} \qquad Ans.$$

Although only *three* independent equilibrium equations can be written for a rigid body. We can check this result by summing moments about point B.

$$\sum M_B = 0,\ -F_{Ay} \cdot 2a - M + \frac{3}{2}qa^2 + Fa = 0$$

$$F_{Ay} = \frac{F}{2} + \frac{3}{4}qa - \frac{M}{2a} \qquad Ans.$$

Example 2-14

The structure shown in Fig. 2-32a consists of two beams AB and CD, connected by the hinge C and supported in A and D by fixed pin supports. The system is loaded by the force $\boldsymbol{F}$ on the point B. If the weight of the members is negligible, determine the reaction forces at the fixed pin support A and the hinge C.

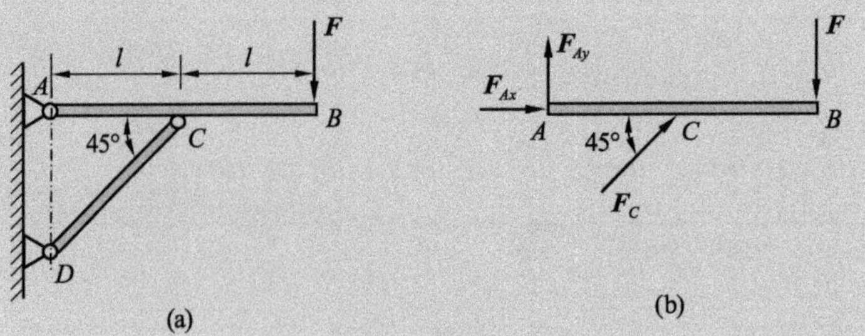

Fig. 2-32

Solution:

The beam *AB* is isolated from the surrounding and its free-body diagram is shown in Fig. 2-32b, if irrespective of the self-weight and friction, the bar *CD* is a *two-force member*, so the resultant forces at pins *D* and *C* must be equal, opposite, and collinear. Although the magnitude of the force is unknown, the line of action is known since it passes through *C* and *D*. So, there are three unknown reaction forces $\boldsymbol{F}_{Ax}$, $\boldsymbol{F}_{Ay}$ and $\boldsymbol{F}_C$. Summing moments about point *A* to eliminate F_{Ax}, and F_{Ay}, we have

$$\sum M_A = 0,\ F_C \cos 45° \cdot l - M - F \cdot 2l = 0$$

$$F_C = 28.28\ \text{kN} \qquad \textit{Ans.}$$

The force $\boldsymbol{F}_C$ can be represented by its x and y components, hence, the forces equilibrium conditions are given by

$$\sum F_x = 0,\ F_{Ax} + F_C \cos 45° = 0$$

$$\sum F_y = 0,\ F_{Ay} + F_C \sin 45° - F = 0$$

Substituting the result of F_C into above equation, solving, we get

$$F_{Ax} = -20\ \text{kN},\ F_{Ay} = -10\ \text{kN} \qquad \textit{Ans.}$$

The negative sign indicates that $\boldsymbol{F}_{Ax}$, and $\boldsymbol{F}_{Ay}$, have the opposite sense to that shown on the free-body diagram.

2.4.5 Coplanar System of Parallel Forces

Consider now the special case of a coplanar system of parallel forces. If all action lines of the forces are parallel and in one plane, for example, in the x-y plane, they form a *coplanar system of parallel forces*, as shown in Fig.2-33. Then the general equilibrium conditions Eq. (2.22) may be simplified.

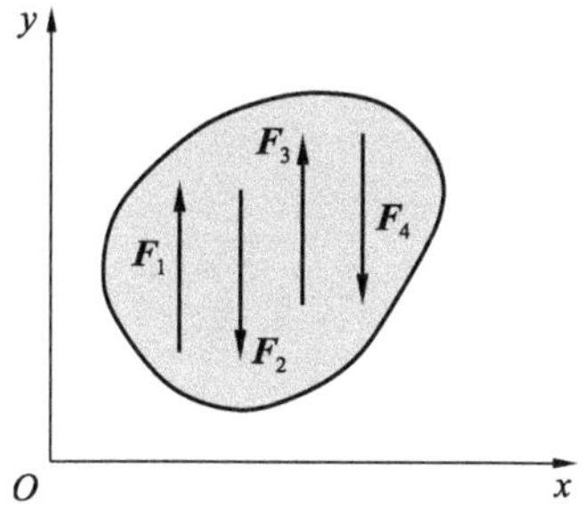

Fig. 2-33

For the case of parallel forces of different magnitudes and arbitrary senses, the projections of all the forces in *one* axis (such as x axis) are zero.

$$\sum F_x \equiv 0$$

Then the equilibrium conditions of the coplanar system of parallel forces is reduced to

$$\sum F_y = 0 \qquad \sum M_A(\boldsymbol{F}) = 0$$
$$\text{or} \tag{2.25}$$
$$\sum M_O(\boldsymbol{F}) = 0 \qquad \sum M_B(\boldsymbol{F}) = 0$$

In this case, the equilibrium conditions in the x- directions for the forces and the moment equation about the axis which is parallel to the y-axis are identically satisfied. So, we have two independent equilibrium equations for two unknowns.

Example 2-15

As shown in Fig. 2-34a, the weight of tower crane is $G_2 = 220$ kN, the maximum hoisting weight is $G_1 = 50$ kN, the distance between two rails A and B is 4 m, determine the weight G_3 of counterbalance that keep the crane equilibrium when it works.

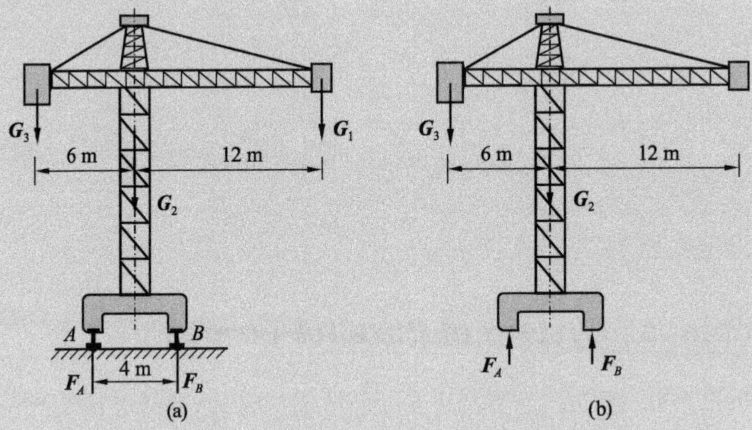

Fig. 2-34

Solution:

The free-body diagram for the crane is shown in Fig. 2-34a. When the crane lifts the maximum hoisting weight, the limit condition is $F_A = 0$ if the crane is in the critical state to turn right about the point B. The minimum of $\boldsymbol{G}_3$ is determined by the moment equilibrium conditions

$$\sum M_B = 0,\ G_{3\min} \cdot (6+2) + G_2 \cdot 2 - G_1 \cdot (12-2) = 0$$

$$G_{3\min} = 7.5 \text{ kN}$$

When the crane is empty $\boldsymbol{G}_1 = 0$ in Fig. 2-34b, the limit condition is $F_B = 0$ if the crane is in the critical state to turn left about the point A. The maximum of $\boldsymbol{G}_3$ is determined by the equilibrium conditions

$$\sum M_A = 0,\ G_{3\max} \cdot (6-2) - G_2 \cdot 2 = 0$$

$$G_{3\max} = 110 \text{ kN}$$

Consequently, the range of magnitude of G_3 to keep the crane equilibrium is

$$7.5 \text{ kN} \leqslant G_3 \leqslant 110 \text{ kN} \qquad \textit{Ans.}$$

2.5 Equilibrium of the System of Bodies

2.5.1 Statically Determinacy and Statically Indeterminate

A structure is called *statically determinate* if all the unknown forces, i. e., the forces in the members and the forces at the supports, can be determined from the equilibrium conditions. In this case, the number of independent equations is more than the number of unknown quantities (which can be solved), shown in Fig. 2-35a.

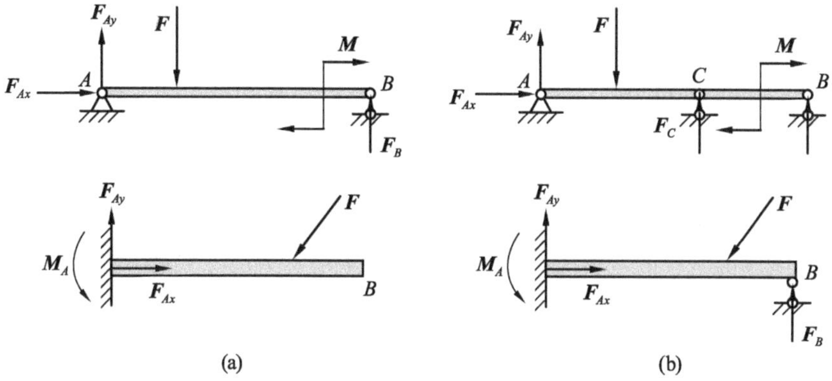

Fig. 2-35

If additional supports are attached to a statically determinate structure, more than three support reactions exist, which can no longer be determined solely from the three equilibrium conditions, such a structure is called *statically indeterminate*. Statically indeterminate means that there will be more unknown loads on the body than equations of equilibrium available. For example, the two-dimensional problem, Fig. 2-35b case, there are four unknowns, for which only three equilibrium equations can be written. The additional equations needed to solve indeterminate problems of the type shown in Fig. 2-35b are generally obtained from the deformation conditions at the point of support.

2.5.2 Equilibrium of the System of Bodies

Structures often consist not only of one single part but of *a number of rigid bodies* that are appropriately connected. The connecting members transfer forces and moments, respectively, which can be made *visible* by cutting through the connections.

Up to now, we have been discussing "*one-body*" problem, each problem required only one free-body diagram. In the equilibrium analysis of a *system of bodies*, we must be able to construct the appropriate free-body diagrams. The primary difference between one-body and *multi-part* problems is that the later often requires that you analyze *more than one* free-body diagram and solve more than one set of equilibrium equations.

In order to determine the support reactions and the forces and moments transferred by the connecting members, the *method of sections* is applied, we *free* the different bodies of the structure by removing all of the joints and supports and replace them by support reactions. Three equilibrium conditions can be formulated for each body of the structure. Therefore, there are in total $3n$ equations if the structure consists of n bodies.

Example 2-16

The structure shown in Fig. 2-36a consists of two beams, joined by the hinge B. The support in A is fixed support and B is roller support in an incline with angle θ. The system is loaded by the single couple M, determine the reaction forces at the support A, C and hinge B. The weight of the members is negligible.

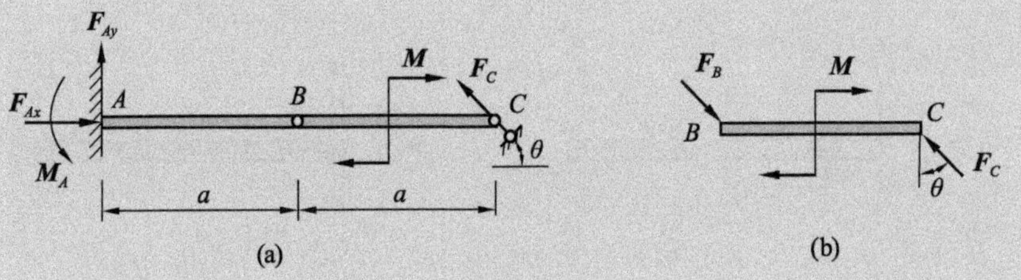

Fig. 2-36

Solution:

The system is statically determinate. Firstly, the beam BC is separate from the structure and the free-body diagram is shown in Fig. 2-36b. The active force is a clockwise couple, so the reaction forces $\boldsymbol{F}_B$ and $\boldsymbol{F}_C$ form an opposite couple to balance the active couple, and can be obtained by couple moment equilibrium equation

$$\sum M = 0, \quad F_C \cos\theta \cdot a - M = 0$$

$$F_C = F_B = \frac{M}{a \cdot \cos\theta} \qquad \textit{Ans.}$$

Secondly, the free-body diagram of the complete system is shown in Fig. 2-36a, there are *three unknown* $\boldsymbol{F}_{Ax}$, $\boldsymbol{F}_{Ay}$ and $\boldsymbol{M}_A$, the equilibrium conditions for the complete system are used

$$\sum F_x = 0, \quad F_{Ax} - F_C \sin\theta = 0$$

$$\sum F_y = 0, \quad F_{Ay} + F_C \cos\theta = 0$$

$$\sum M_A = 0, \quad M_A - M + F_C \cos\theta \cdot 2a = 0$$

These are *three* equations for the *three* unknown forces. Using the result of $\boldsymbol{F}_C$, solving, we obtain

$$M_A = M, \quad F_{Ax} = \frac{M}{a}\tan\theta, \quad F_{Ay} = -\frac{M}{a} \qquad \textit{Ans.}$$

Example 2-17

The structure shown in Fig. 2-37a consists of two beams, which are connected by the hinge C. The supported in A is fixed pin support and B, D are roller supports. The system is loaded by the couple of moment $M = 40$ kN · m and uniformly distributed force q=10 kN/m, determine the reaction forces at the supports A, B, C and D. The weight of the members is negligible.

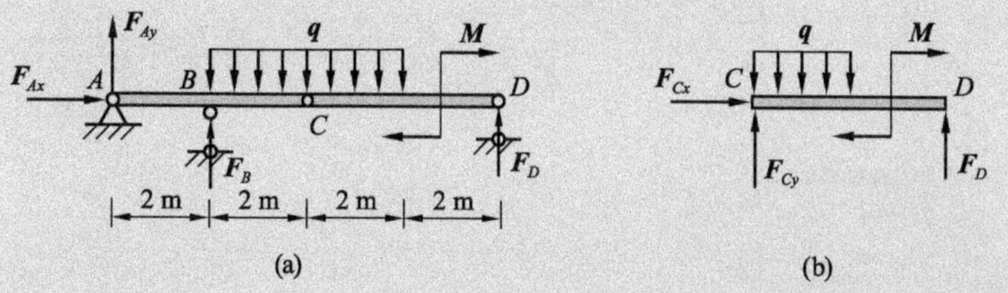

Fig. 2-37

Solution:

Firstly, we separate the beam CD from the structure and draw the free-body diagram, Fig. 2-37b, there are three unknown $\boldsymbol{F}_{Cx}$, $\boldsymbol{F}_{Cy}$ and $\boldsymbol{F}_D$. The uniformly distributed load can be replaced by its equivalent resultant force acting in the middle of distributed load. Summing moments about C, we obtain a direct solution for $\boldsymbol{F}_D$.

$$\sum M_C = 0,\ 4F_D - M - \frac{1}{2}q \times 2^2 = 0$$

$$F_D = 15 \text{ kN} \qquad Ans.$$

Using this result, summing the forces in the x and y direction yields

$$\sum F_x = 0,\ F_{Cx} = 0 \qquad Ans.$$

$$\sum F_y = 0,\ F_{Cy} + F_D - 2 \cdot q = 0$$

$$F_{Cx} = 0,\ F_{Cy} = 5 \text{ kN} \qquad Ans.$$

Now only three support reactions $\boldsymbol{F}_{Ax}$, $\boldsymbol{F}_{Ay}$ and $\boldsymbol{F}_B$ are unknown (see Fig. 2-37a), they can be determined by applying the equilibrium conditions to the complete system,

$$\sum F_x = 0,\ F_{Ax} = 0 \qquad Ans.$$

$$\sum M_A = 0,\ F_B \times 2 - q \times 4^2 - M + F_D \times 8 = 0$$

$$\sum F_y = 0,\ F_{Ay} + F_B - q \times 4 + F_D = 0$$

Using the result of $\boldsymbol{F}_D$, solving above equations, we obtain

$$F_B = 40 \text{ kN},\ F_{Ay} = -15 \text{ kN} \qquad Ans.$$

Example 2-18

Determine the reactions of the fixed pin support A and B for the structure shown in Fig. 2-38a. The members BC and ADE are connected by a hinge D. Point C is also a pin. The pulley is frictionless, neglect weight of numbers.

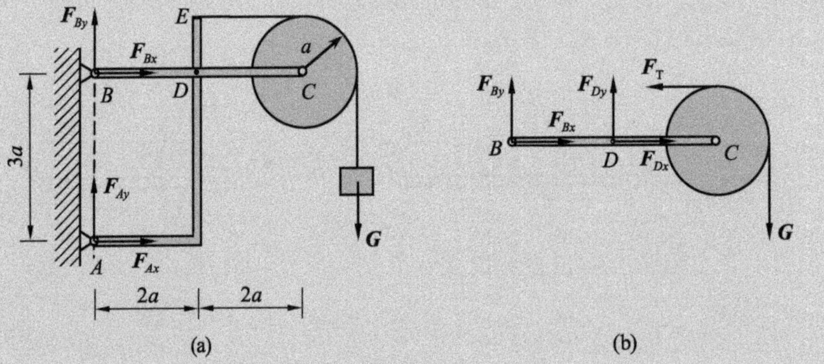

Fig. 2-38

Solution:

Firstly, we can determine some external reactions at A and B. The free-body diagram of the complete structure is shown in Fig. 2-38a, there are four unknowns $\boldsymbol{F}_{Ax}$, $\boldsymbol{F}_{Ay}$, $\boldsymbol{F}_{Bx}$ and $\boldsymbol{F}_{By}$. Three equilibrium conditions for the complete system are used and two constraints forces can be determined.

$$\sum M_A = 0, -F_{Bx} \cdot 3a - G \cdot 5a = 0 \tag{a}$$

$$\sum M_B = 0, F_{Ax} \cdot 3a - G \cdot 5a = 0 \tag{b}$$

$$\sum F_y = 0, F_{Ay} + F_{By} - G = 0 \tag{c}$$

Solving Eq. (a) and Eq. (b), we obtain

$$F_{Ax} = \frac{5}{3}G, F_{Bx} = -\frac{5}{3}G \qquad Ans.$$

Secondly, isolating the beam BC and pulley from the structure and the free-body diagram is shown in Fig. 2-38b. Summing moments about point D to eliminate F_{Dx}, and F_{Dy}, we have

$$\sum M_D = 0, -F_{By} \cdot 2a + F_T \cdot a - G \cdot 3a = 0$$

Here $F_T = G$, we get

$$F_{By} = -G \qquad Ans.$$

Substituting the result into Eq. (c) to obtain F_{Ay}, we have

$$F_{Ay} = 2G \qquad Ans.$$

Example 2-19

The *three-hinged arch* structure shown in Fig. 2-39a consists of two arches of weight $G=30$ kN, which are connected by the hinge C. The supported in A and B are fixed pin support. Given that: $l=32$ m, $h=10$ m, determine reaction forces at the support A, B and hinge C.

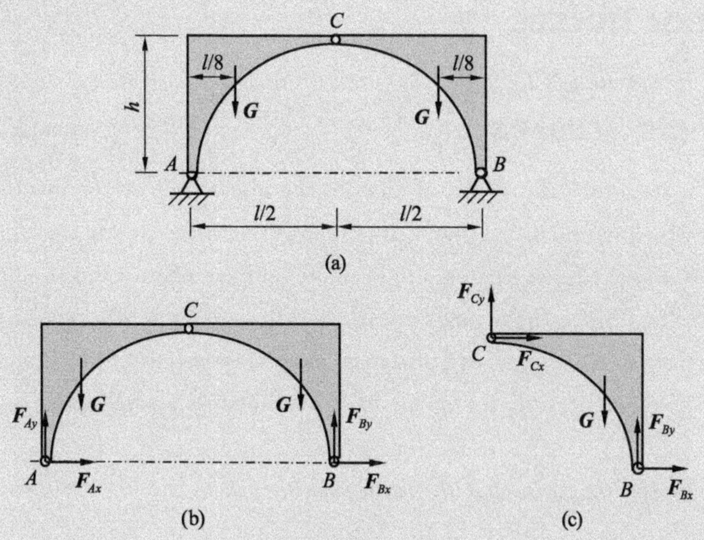

Fig. 2-39

Solution:

Firstly, we can determine some external reactions at A and B. The free-body diagram of the entire structure is shown in Fig. 2-39b, there are four unknowns $\boldsymbol{F}_{Ax}$, $\boldsymbol{F}_{Ay}$, $\boldsymbol{F}_{Bx}$ and $\boldsymbol{F}_{By}$. Three equilibrium conditions for the complete system are used and two constraints forces can be determined.

$$\sum M_A = 0,\ F_{By}l - G\frac{7}{8}l - G\frac{1}{8}l = 0 \tag{a}$$

$$\sum M_B = 0,\ -F_{Ay}l + G\frac{7}{8}l + G\frac{1}{8}l = 0 \tag{b}$$

$$\sum F_x = 0,\ F_{Ax} + F_{Bx} = 0 \tag{c}$$

Solving Eq. (a) and Eq. (b), we obtain direct solution for

$$F_{Ay} = F_{By} = 300 \text{ kN} \qquad \textit{Ans.}$$

Secondly, isolating the arch BC and the free-body diagram is shown in Fig. 2-39c, the equilibrium equations for the arch are used

$$\sum M_C = 0,\ F_{Bx}h + F_{By}\frac{l}{2} - G\frac{3}{8}l = 0$$

$$\sum F_x = 0,\ F_{Cx} + F_{Bx} = 0$$

$$\sum F_y = 0,\ F_{Cy} + F_{By} - G = 0$$

Using the result of $\boldsymbol{F}_{By}$, solving the system of equations yields

$$F_{Bx}=-120\ \text{kN},\ F_{Cx}=120\ \text{kN},\ F_{Cy}=0 \qquad \textit{Ans.}$$

Substituting the result of F_{Bx} into Eq. (c), we get

$$F_{Ax}=120\ \text{kN} \qquad \textit{Ans.}$$

2.6 Planar Trusses

2.6.1 Simple Trusses

A *truss* is a structure composed of slender members that are connected at their ends by *joints*. The truss is one of the most important structures in engineering applications. In particular, *planar* trusses lie in a single plane and are often used to support roofs, as shown in Fig. 2-40a. Since these loads act in the same plane on the truss, Fig. 2-40b, the analysis of the forces developed in the truss members will be two-dimensional. To be able to determine the internal forces in each member, the following assumptions are made:

① *The members are connected through smooth pins* (frictionless joints).

② *All loadings are applied at the joints*. Usually the weight of the members is neglected because the force supported by each member is much larger than its weight. If the weight is to be included in the analysis, it is general to apply it as a vertical force with half of its magnitude applied at each end of the member.

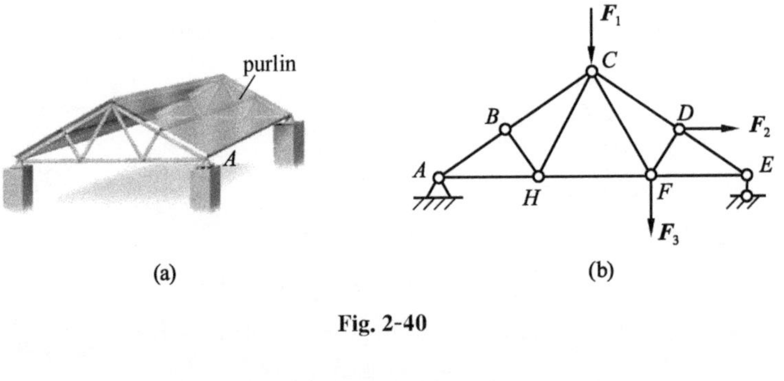

Fig. 2-40

A truss that satisfies these assumptions is called an *ideal truss*. Each truss member will act as *two-force member*, and therefore the force acting at each end of the member will be directed along the axis of the member. That is, its members are subjected to tension or to compression only.

In real trusses, these ideal conditions are not exactly satisfied. For example, the joints may not be frictionless, or the ends of the members may be welded. Even then, the assumption of frictionless *pin-jointed* connections yields satisfactory results if the axes of the members are concurrent at the joints.

In order to analyze or design a truss, it is necessary to determine the force in each of its member. In the following, two methods to determine the internal forces in each members of a statically determinate truss will be discussed. In both methods, the conditions of equilibrium are applied to suitable free-body diagrams.

2.6.2 Zero-Force Members

In practice, it is often convenient first to identify those members of the truss that support *no loading*. These members are called *zero-force members*. These *zero-force members* are used to increase the stability of the truss during construction. If the zero-force members are recognized in advance, the number of unknowns is reduced, which simplifies the analysis.

The zero-force members can generally be found by *inspection* of each of the joints. The following rules are useful in identifying zero-force members.

① If two members are not collinear at an unloaded joint, Fig. 2-41a, then *both* members are zero-force members.

② Let two members be connected at a loaded joint, Fig. 2-41b. If the action line of the external force F *coincides with* the direction of one of the members, then the other member is a zero-force member.

③ Let three members be connected at an unloaded joint, Fig. 2-41c. If two members have the *same direction*, the *third* member is a zero-force member.

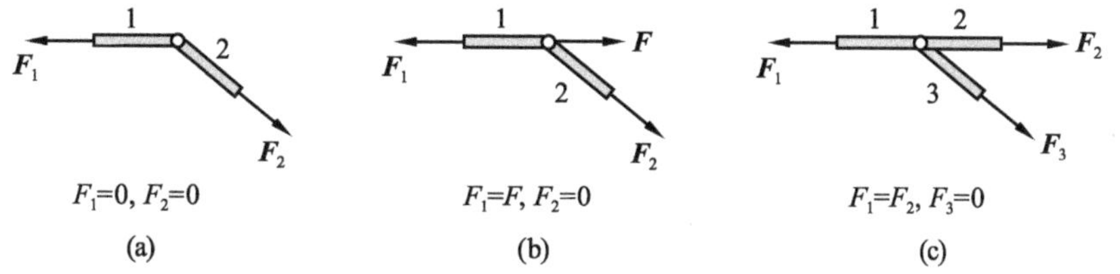

Fig. 2-41

2.6.3 The Method of Joints

The *method of joints* is to consider each joint as separated from the trusses and apply the equilibrium conditions to the free-body diagram of each joint. Since the members of a *plane truss* are straight two-force member lying in a single plane, each joint is subjected to a force system that is *coplanar and concurrent*, as shown in Fig. 2-42. Therefore, if the free-body diagram of each joint is drawn, two force equilibrium equations can be used to obtain the member forces acting on each joint.

It is not necessary to determine by inspection whether a member is subject to tension or compression. We shall always assume that all the unknown members' forces acting on the joint are under tension. If the answer gives a negative value for the force in a member, this member is in reality subject to compression.

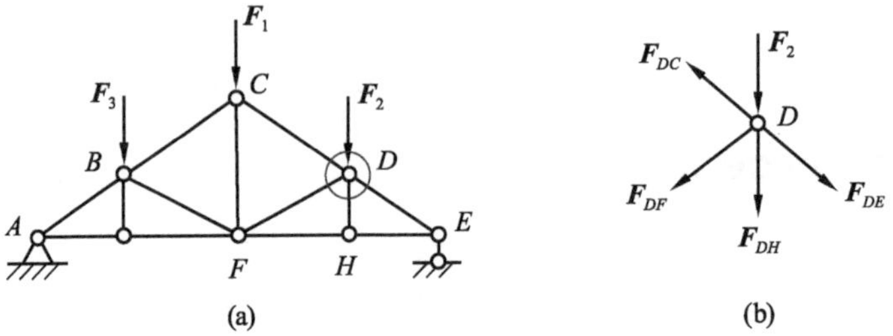

Fig. 2-42

Example 2-20

Using the method of joints, determine all the zero-force members of the planar trusses shown in Fig. 2-43a. Assume all joints are pin connected.

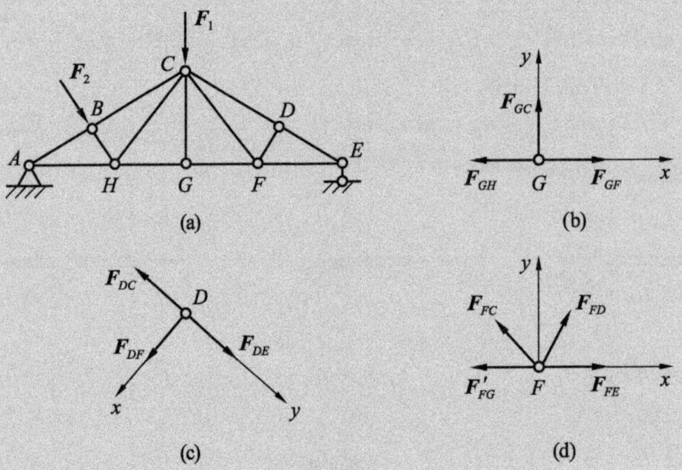

Fig. 2-43

Solution:

Inspecting the joint of the truss, there are three members for which two are collinear. We analyze one by one. The free-body diagram of joint G is shown in Fig. 2-43b, summing forces in y direction, we have

$$\sum \boldsymbol{F}_y = 0, \ \boldsymbol{F}_{GC} = 0$$

This means that the load at C must be supported by members CB, CH, CF, and CD.

The free-body diagram of joint D is shown in Fig. 2-43c, summing forces in x direction, we have

$$\sum \boldsymbol{F}_x = 0, \ \boldsymbol{F}_{DF} = 0$$

The free-body diagram of joint F is shown in Fig. 2-43d, using the result of F_{DF}, summing forces in y direction, we have

$$\sum \boldsymbol{F}_y = 0, \ \boldsymbol{F}_{FC} = 0$$

Note: Since there is an active force F_2, the member HB is *not* a zero-force member.

Example 2-21

The planar trusses shown in Fig. 2-44a is loaded by an external force F on point D. Using the method of joints, determine the forces in each member of the trusses.

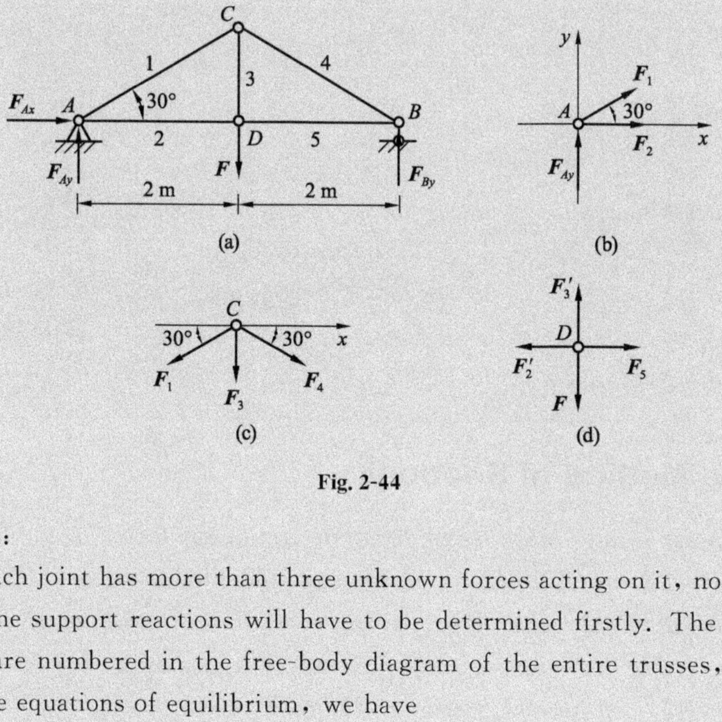

Fig. 2-44

Solution:

Since each joint has more than three unknown forces acting on it, no joint can be analyzed. The support reactions will have to be determined firstly. The members of the trusses are numbered in the free-body diagram of the entire trusses, Fig. 2-44a. Applying the equations of equilibrium, we have

$$\sum F_{xi} = 0, F_{Ax} = 0$$

$$\sum F_{yi} = 0, F_{Ay} + F_{By} - F = 0$$

$$\sum M_A(F) = 0, F_{By} \cdot 2a - F \cdot a = 0$$

Solving, we obtain

$$F_{Ay} = F_{By} = \frac{F}{2}$$

The analysis can start at either joint A or B. The choice is arbitrary since there are one known and two unknown member forces acting at each of these joints. The free-body diagram of joint A is shown in Fig. 2-44b. Applying the equations of joint equilibrium, we have

$$\sum F_{xi} = 0, \ F_1 \cos 30° + F_2 = 0$$

$$\sum F_{yi} = 0, \ F_{Ay} + F_1 \sin 30° = 0$$

Solving, we obtain

$$F_1 = -F, \ F_2 = \frac{\sqrt{3}F}{2} \qquad \textit{Ans.}$$

Using the result for F_1, from the free-body diagram of joint C, Fig. 2-44c, we have

$$\sum F_{xi} = 0, \ -F_1 \cos 30° + F_4 \cos 30° = 0$$

$$\sum F_{yi} = 0, \ -(F_1 + F_4)\sin 30° - F_3 = 0$$

Solving, we obtain

$$F_3 = F, \ F_4 = -F \qquad \textit{Ans.}$$

Fig. 2-44d shows the free-body diagram of the joint D, using the result for F_2, summing forces in horizontal direction, we have

$$\sum F_{xi} = 0, \ -F_2 + F_5 = 0$$

$$F_5 = \frac{\sqrt{3}F}{2} \qquad \textit{Ans.}$$

The negative values for the members indicate that these members act in the opposite sense to that shown.

2.6.4 The Method of Sections

In practice, it is not always necessary to determine the forces in all of the members of a trusses. If we need to determine the force in only a few members of a truss, it may be advantageous to use the *method of sections* instead of the *method of joints*. In this case, the truss is *divided into two parts by a cut*. For example, consider a planar truss shown in Fig. 2-45a, an imaginary section, indicated by the dotted line, can be used to cut the members of the entire truss into two parts and thereby expose each internal force as external, Fig. 2-45b. If the section passes through the truss and the free-body diagram of either of its two parts is drawn for the analysis, we can apply the equations of equilibrium to determine the member forces at the cut section. Since only *three* independent equilibrium equations can be applied to the free-body diagram of any segment, then we should select a section that passes through not more than *three* members in which the forces are unknown. In practice, the part that involves a smaller number of forces will usually lead to a simpler calculation.

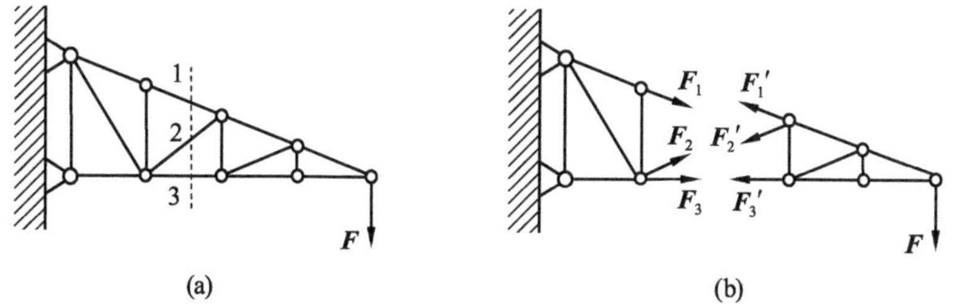

Fig. 2-45

In many cases, the *method of sections* can be applied without having to determine the forces at the supports. The example shown in Fig. 2-45, the forces in members 1～3 can be obtained immediately from the equilibrium conditions for the part of the truss on the right as shown in Fig. 2-45b. Similar to the method of joints, we always assume that all the unknown members' forces at the cut section are tensile forces.

Example 2-22

A truss is loaded by two forces, $F_1=100$ kN and $F_2=70$ kN, as shown in Fig. 2-46a. The length of all members is 1 m, determine the forces in the members 1, 2 and 3 of the trusses.

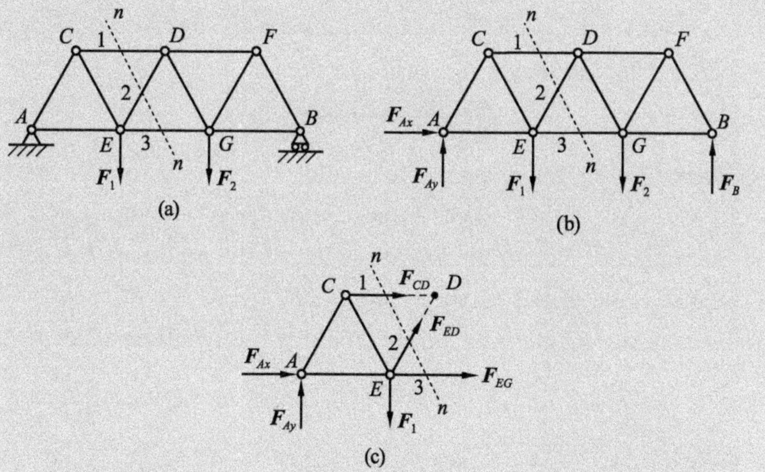

Fig. 2-46

Solution:

First, we need to determine the external reactions at the supports. A free-body diagram of the entire truss is shown in Fig. 2-46b. Applying the equilibrium conditions, we have

$$\sum F_x = 0,\ F_{Ax} = 0$$

$$\sum F_y = 0,\ F_{Ay} + F_{By} - F_1 - F_2 = 0$$

$$\sum M_B = 0,\ F_1 \times 2 + F_2 \times 1 - F_{Ay} \times 3 = 0$$

Solving, we get

$$F_{Ay} = 90\ \text{kN},\ F_{By} = 80\ \text{kN}$$

Then an imaginary section passes through the members 1~3 and divided the truss into two parts, Fig. 2-46b. The free-body diagram of the left portion of the sectioned truss is shown in Fig. 2-46c, since it involves the least number of forces. Summing moments about point D eliminates F_{CD} and F_{ED} and yields a direct solution for F_{EG}.

$$\sum M_D = 0,\ F_1 \times 0.5 + F_{EG} \times 1 \times \sin 60° - F_{Ay} \times 1.5 = 0$$

$$F_{EG} = 9.82\ \text{kN} \qquad \textit{Ans.}$$

In the same way, by summing moments about point E we obtain a direct solution for F_{CD}.

$$\sum M_E = 0,\ -F_{CD} \times 1 \times \sin 60° - F_{Ay} \times 1 = 0$$

$$F_{CD} = -10.4\ \text{kN} \qquad \textit{Ans.}$$

Summing forces in the y direction yields F_{ED}.

$$\sum F_y = 0,\ F_{Ay} + F_{ED} \sin 60° - F_1 = 0$$

$$F_{ED} = 1.16\ \text{kN} \qquad \textit{Ans.}$$

Here each of the corresponding equations contains one unknown force only and can be solved easily.

Exercises

2-1 As shown in Fig. 2-47, a particle is subjected to four forces: $F_1 = 20$ kN, $F_2 = 25$ kN, $F_3 = 10$ kN, $F_4 = 30$ kN, these forces act under given angles with respect to the horizontal. Determine the magnitude and direction of the resultant force.

2-2 The hook is subjected to three forces ($F_1 = 180$ N, $F_2 = 50$ N, $F_3 = 30$ N), Fig. 2-48. Determine the magnitude and direction of the resultant force.

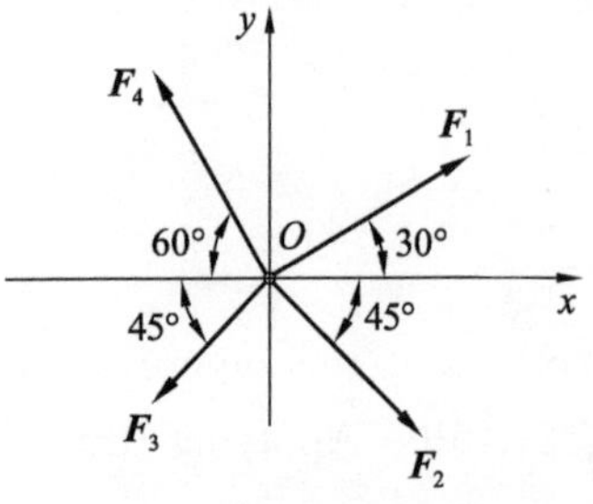

Fig. 2-47

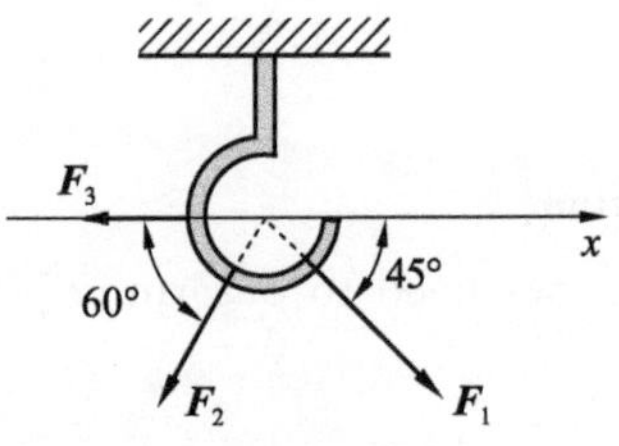

Fig. 2-48

2-3 A smooth sphere of weight $G=20$ N, radius $r=0.2$ m, is suspended by a rope of length $l=0.6$ m, Fig. 2-49, determine the magnitude of tensile force $\boldsymbol{F}_T$ in the rope and reaction force $\boldsymbol{F}_N$ from the wall.

2-4 Determine the force in cables AB and BC necessary to support the $G=200$ kN block, Fig. 2-50. Cord BC remains horizontal, and AB has a length of 1.5 m.

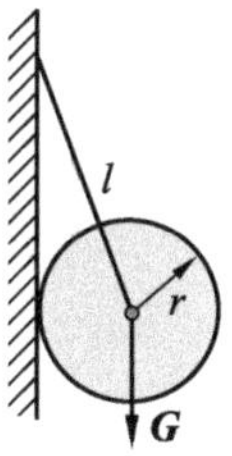

Fig. 2-49

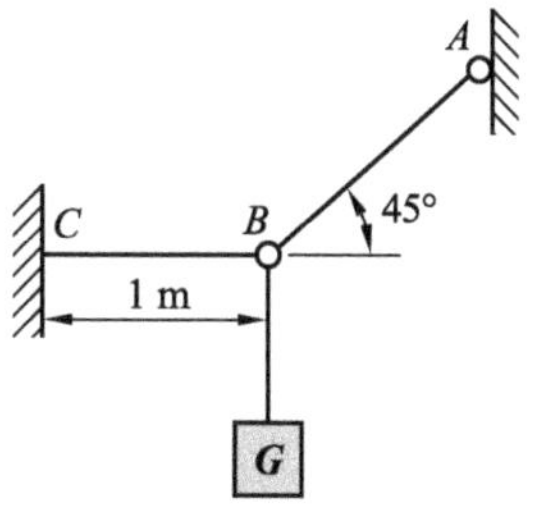

Fig. 2-50

2-5 As shown in Fig. 2-51, two bars AB and BC of the crane are attached at A and B to a wall by fixed pins. They are pin-connected at B with a pulley and subjected to a weight $G=20$ kN. Assuming pulley B and bar AB and BC is frictionless and weightless, neglect the size of the pulley B. Determine the forces in the bars AB and BC.

2-6 The pile pulling structure in Fig. 2-52, one end of rope AC is attached to stake A, another end is attached to fixed point C. Another rope BE is tied on the point B of rope AC, the end of it is attached on point E, the force $F=800$ N acts on point D of rope BE, $\theta=0.1$ rad (when θ is tiny, $\tan\theta\approx\theta$). Determine the tension on the stake A.

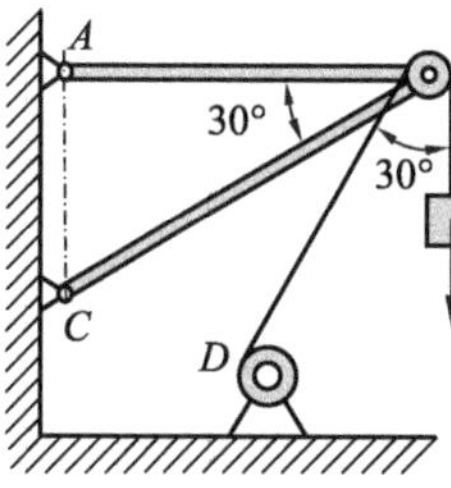

Fig. 2-51

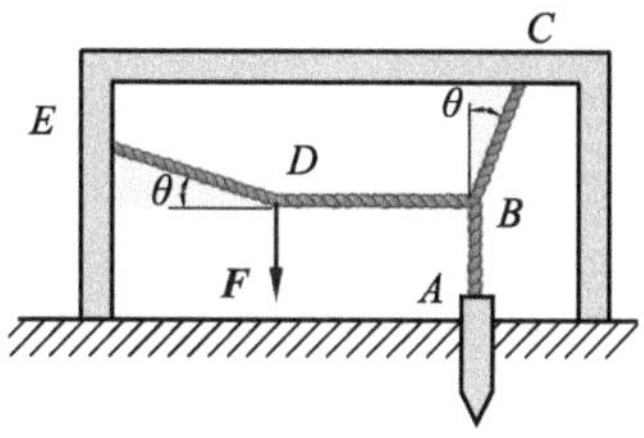

Fig. 2-52

2-7 The beam in Fig. 2-53 is subjected to a force of $F=200$ kN, determine the moment of the force about point O.

2-8 As shown in Fig. 2-54, a uniform wheel of weight G and radius r, rests against an obstacle of height h. Determine the least force $\boldsymbol{F}$ necessary to turn the wheel over the obstacle. Take all the surface to be smooth.

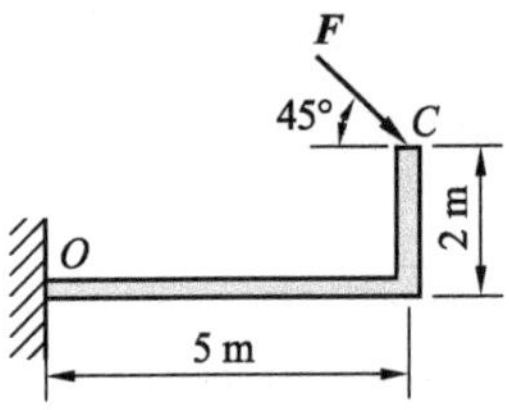

Fig. 2-53

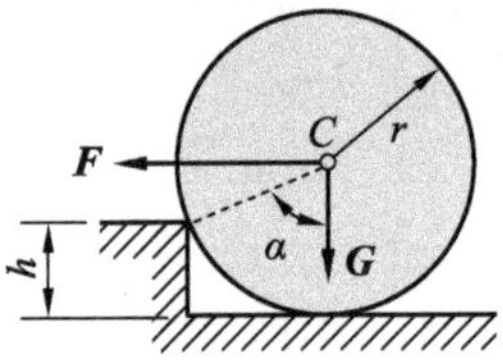

Fig. 2-54

2-9 In Fig. 2-55，three couples with forces $F_1=100$ N，$F_2=200$ N，$F_3=300$ N act on the triangular plate，determine the resultant couple moment.

2-10 As shown in Fig. 2-56，a couple with moment $\boldsymbol{M}$ acts on the beam AB，the length of beam is l，determine the reactions of the fixed pin supports A and roller B. Neglect the weight of the beam.

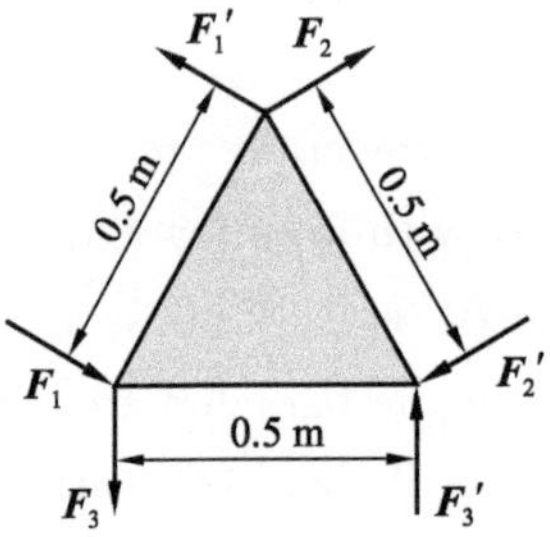

Fig. 2-55

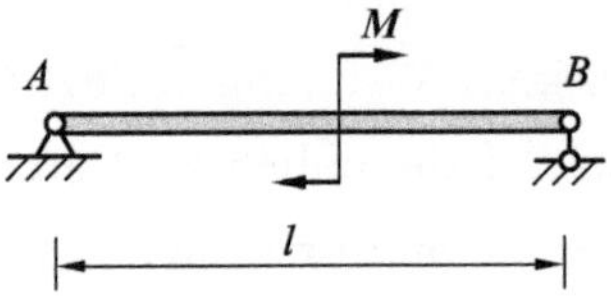

Fig. 2-56

2-11 In Fig. 2-57，the workpiece is fixed by two bolts A and B，and it is subjected to three couples of moment $M_1=M_2=10$ N • m，$M_3=20$ N • m，$l=200$ mm，determine the reactions of bolts A and B.

2-12 In Fig. 2-58，two angled parts are connected by the hinge B. Supports A and B are fixed pin. The system is loaded by couple $\boldsymbol{M}$. Neglect the weight of the parts，determine the reactions at A and C.

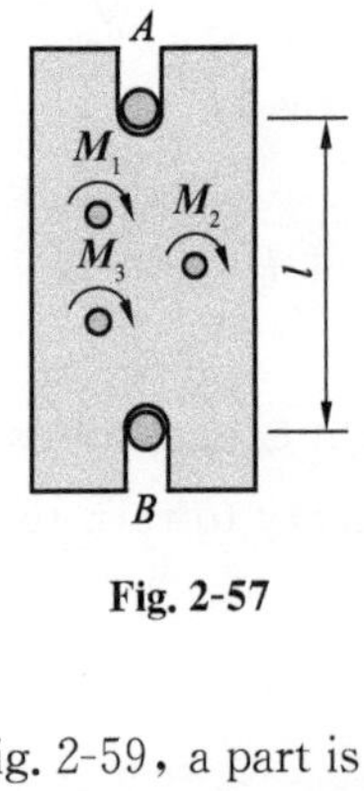

Fig. 2-57

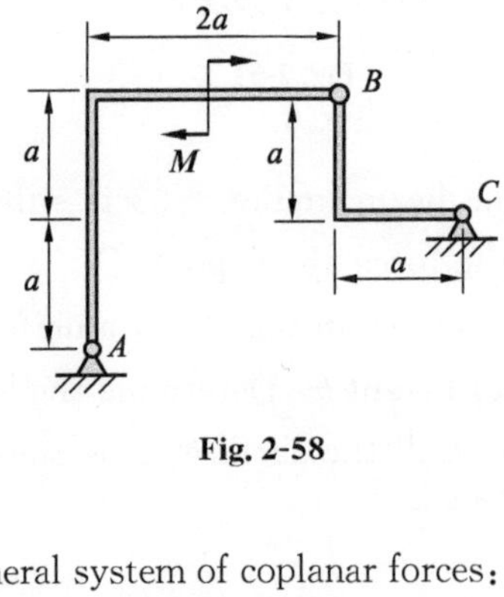

Fig. 2-58

2-13 In Fig. 2-59，a part is subjected to a general system of coplanar forces: $F_1=60$ kN，$F_2=80$ kN，$F_3=40$ kN，$F_4=110$ kN，$M=2\,000$ N • m. Determine the equivalent force and couple moment at point O.

2-14 As shown in Fig. 2-60, replace the force system acting on the beam AB by an equivalent force and couple moment at point A.

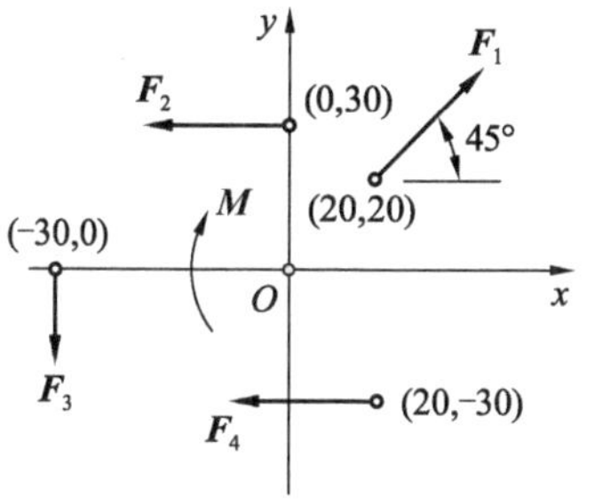

Fig. 2-59

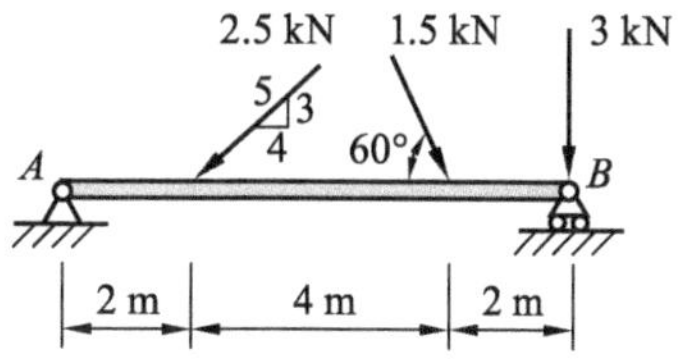

Fig. 2-60

2-15 In Fig. 2-61, the support beam is loaded by the uniformly distributed force q, the couple $\boldsymbol{M}$ and the force $\boldsymbol{F}$. Determine the reaction forces at the supports A and B.

2-16 In Fig. 2-62, the support beam is loaded by the uniformly distributed force $q=2$ N/mm, and the force $F=500$ N. Determine the reaction forces at the supports A and B. Set $a=500$ mm.

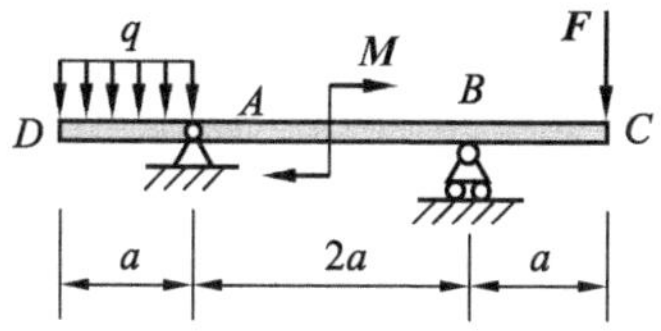

Fig. 2-61

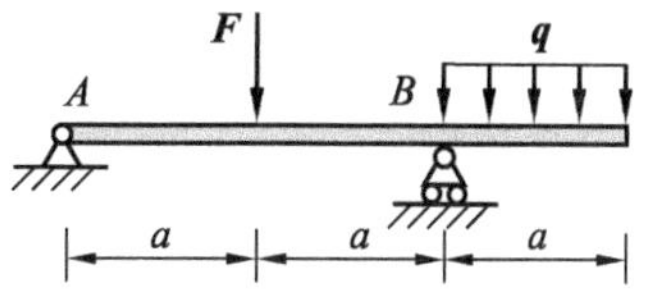

Fig. 2-62

2-17 In Fig. 2-63, the cantilever beam is loaded by the two forces F_1 and F_2. Determine the reactions at the fixed support A.

2-18 As shown in Fig. 2-64, the crane is supported by a pin at C and rope AB. If the load has a weight of $G=2$ kN, determine the horizontal and vertical components of reaction at the pin C and the force developed in rope AB on the crane.

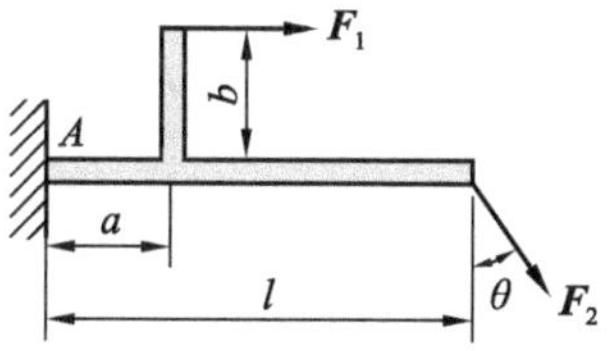

Fig. 2-63

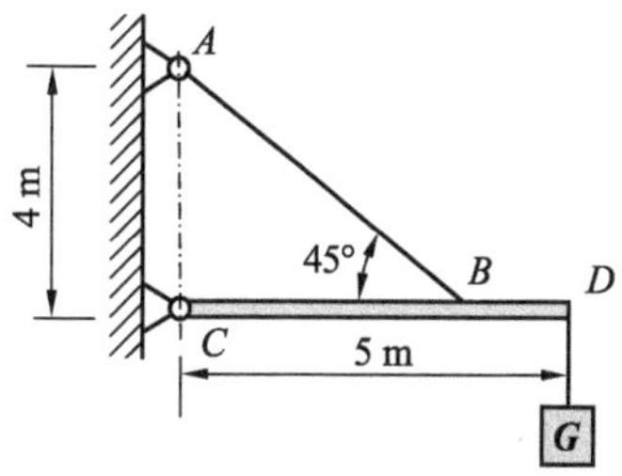

Fig. 2-64

2-19 In Fig. 2-65, the beam AB with a pulley is supported by fixed pin support A and rope BC. The end C of rope is fastened on the wall. Determine the reaction forces of support A and the force developed in rope BC. The pulley is frictionless, neglect weight of all members. Given: $G=1\ 800$ N, $AD=0.2$ m, $BD=0.4$ m, $r=0.1$ m.

2-20 In Fig. 2-66, the cable on a tow truck is subjected to a force of $T=6$ kN. Determine the magnitudes of the total brake frictional force F_s for the rear set of wheels B and the total normal forces at both front wheels A and both rear wheels B for equilibrium. The truck has a total mass of 4 000 kg and mass center at C.

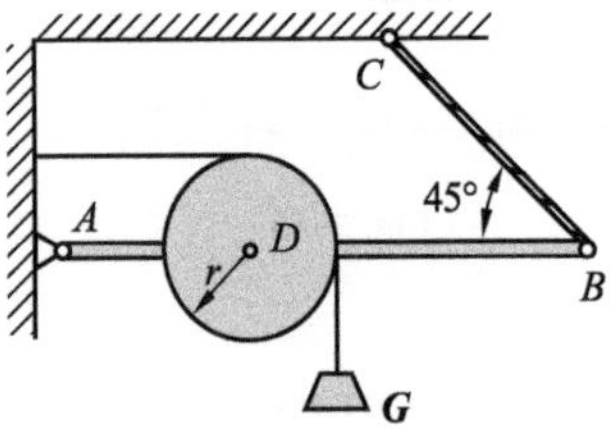

Fig. 2-65

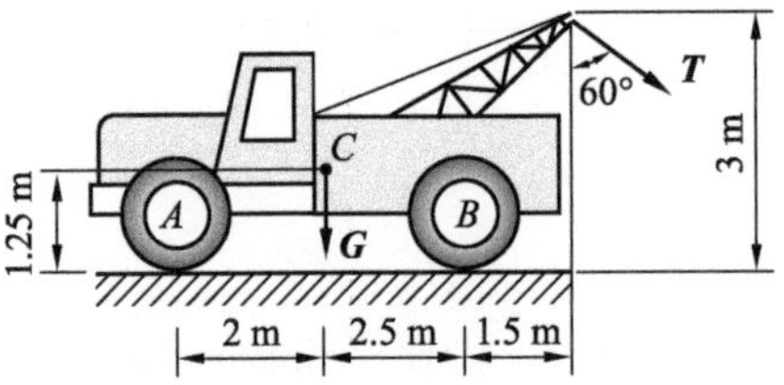

Fig. 2-66

2-21 As shown in Fig. 2-67, two angled member AC and ED are connected by the hinge D. The supports A and E are fixed pin. A force of $F=180$ N acts on the point C of member AC. Determine the reaction forces at A and E. Neglect the self-weight and friction.

2-22 The three-hinged arch structure consists of two arches, which are connected by the hinge C, Fig. 2-68. A force $\boldsymbol{F}$ acts on the arch AC, determine the reaction forces at the support A and B. Neglect the self-weight and friction.

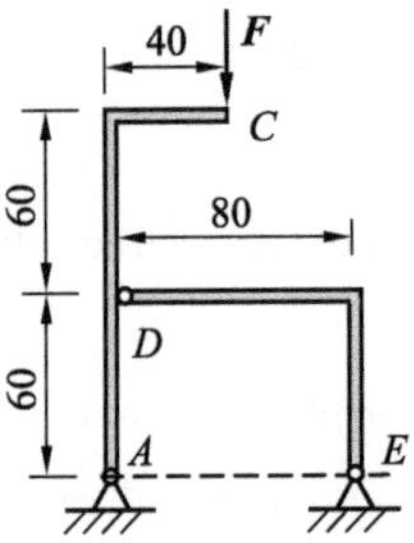

Fig. 2-67

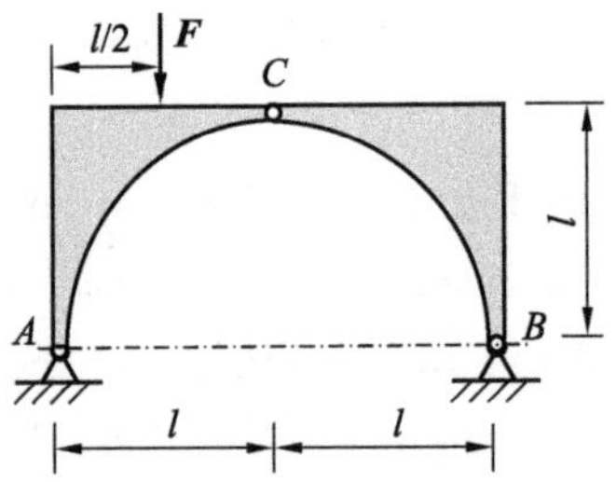

Fig. 2-68

2-23 The structure consists of two beams AD and BC, joined by the hinge C, Fig. 2-69. The supports in A and B are fixed pin. The system is loaded by the forces of $F_1=F$ and $F_2=2F$. Determine the reaction forces at the supports A, B and the hinge C. Neglect the weight of all members.

2-24 The beam BC and angled part AC are connected by the hinge C, Fig. 2-70. The part AC is fixed at A and the beam BC is roller support at B. The system is loaded by the force F. Determine the reaction forces at support A and B. Neglect the weight of all members.

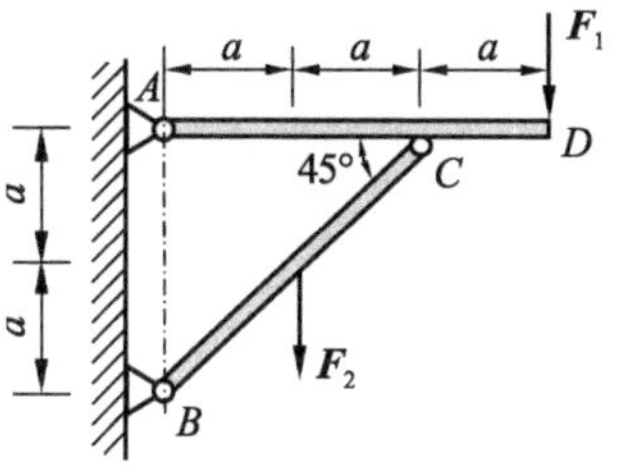

Fig. 2-69

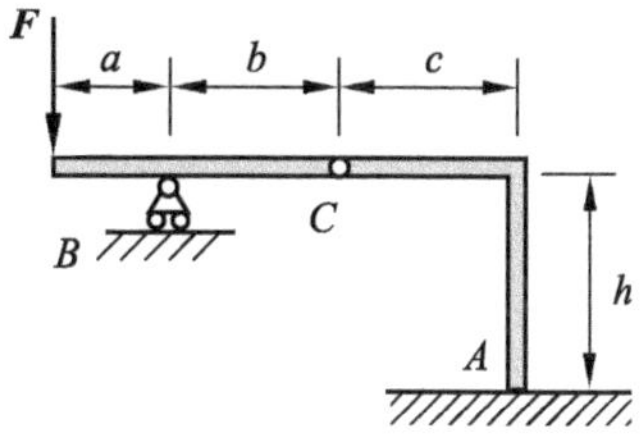

Fig. 2-70

2-25 In Fig. 2-71, two beams AB and BC are joined by the hinge B. The support A is fixed support and C is roller support with an inclined angle θ. Uniformly distributed force q acts on the beam BC, determine reaction forces at the support A, C and hinge B. Neglect the weight of all members.

2-26 In Fig. 2-72, two beams AB and BC are joined by the hinge B. The support A is fixed support and C is roller support. A couple of $\boldsymbol{M}=\boldsymbol{F}a$ acts on the beam BC and a force $\boldsymbol{F}$ acts on the beam AB, determine reaction forces at the support A, C and hinge B. Neglect the weight of all members.

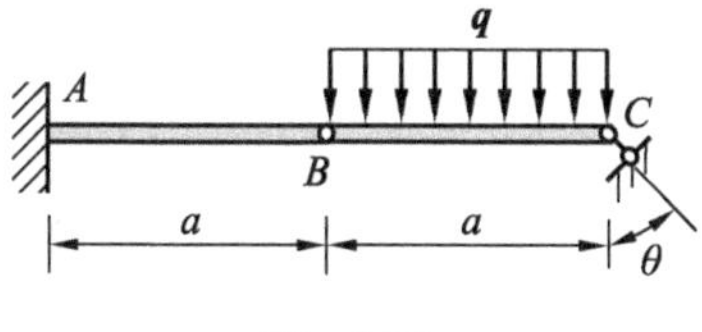

Fig. 2-71

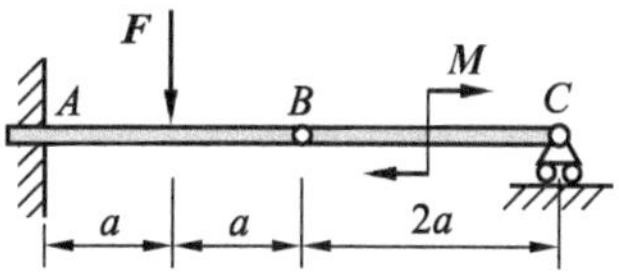

Fig. 2-72

2-27 As shown in Fig. 2-73, the holder consists of beams AD and EC, which are joined by the hinge B. The supports in A and E are fixed pin. The radius of pulley is $r=0.015$ m, one end of the rope is fastened at the end C of the beam EC. If the load is of $G=1\ 000$ N, determine reaction forces of support A and E. Neglect weight and friction of beams, pulley and rope.

2-28 The holder consists of three members, joined by the hinge B, C and D, in Fig. 2-74. The support in A is fixed pin and B is roller. One end of the rope is attached to the wall. If the load is of $G=1\ 200$ N, determine the reaction forces of support A, B and the internal force of member BC. Neglect weight and friction of beams, pulley and rope.

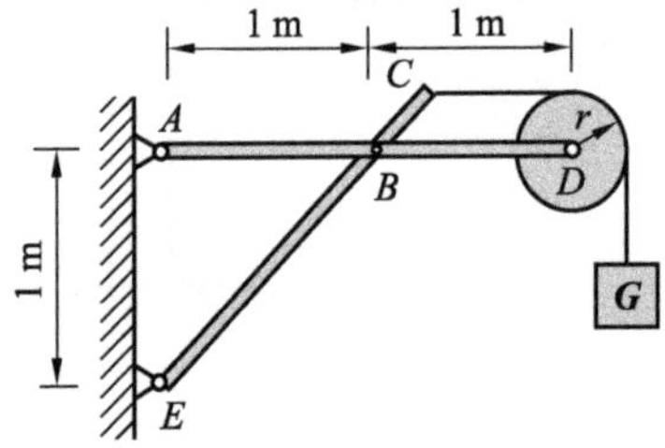

Fig. 2-73

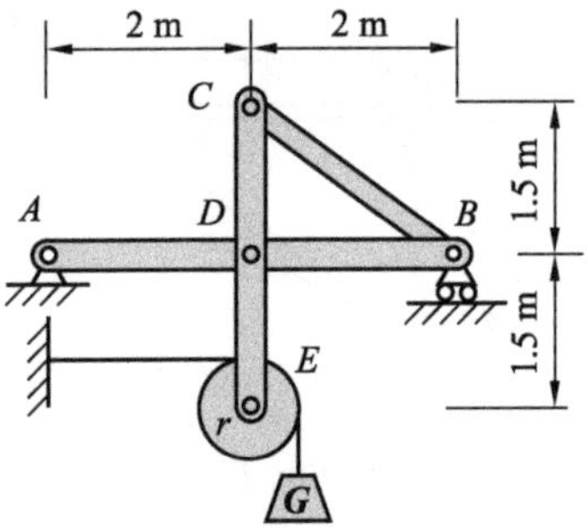

Fig. 2-74

2-29 The framework consists of three members, joined by the hinge A and D, in Fig. 2-75. The pin E can slide along the smooth chute of member AC. The supports B and C are fixed pin. If the force $\boldsymbol{F}$ acts on the end G, determine the reaction forces at the support A, B and D. Neglect the weight of all members.

2-30 As shown in Fig. 2-76, the truss is subjected to the force $F=450$ N at joint D. Determine the forces in each member of the truss.

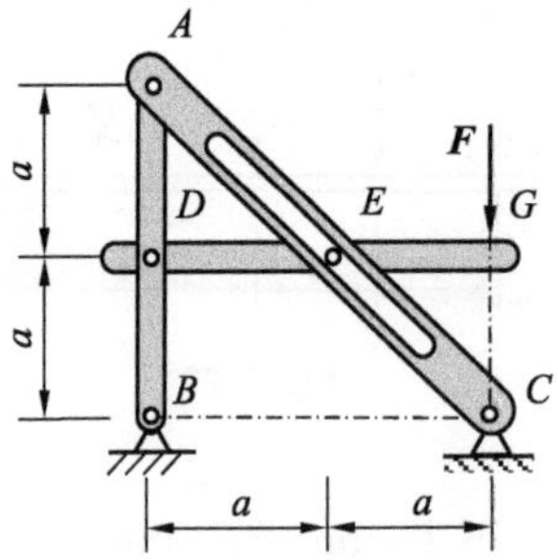

Fig. 2-75

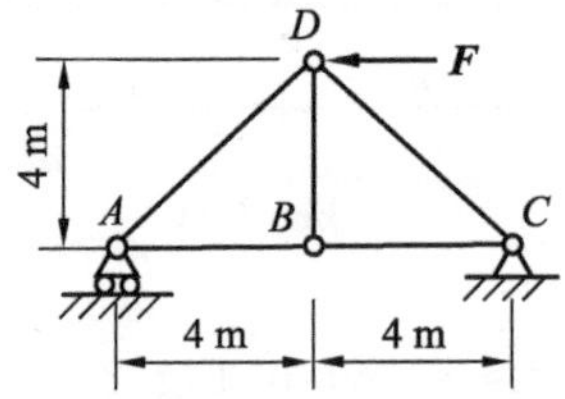

Fig. 2-76

2-31 The truss carries the force $F=300$ N at joint D, in Fig. 2-77. Determine the forces in each member of the truss.

2-32 A truss is loaded by force $\boldsymbol{F}$ at joint G, in Fig. 2-78. ABC is equilateral triangle and $AD=DB$, determine the forces in member CD.

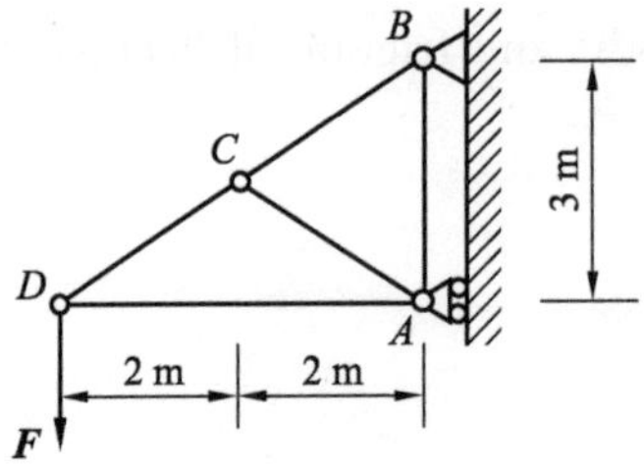

Fig. 2-77

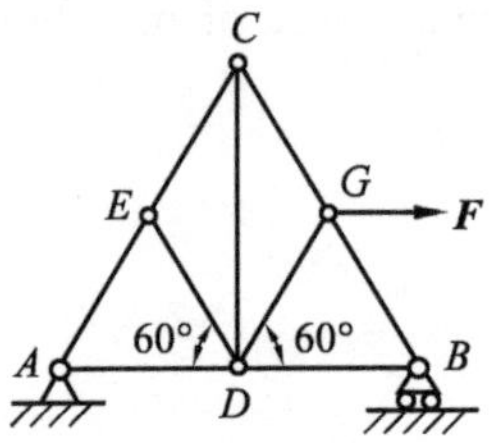

Fig. 2-78

2-33 The truss in Fig. 2-79 is loaded by four same forces $\boldsymbol{F}$ at the joint. Determine the forces in members 1, 2 and 3.

2-34 The truss in Fig. 2-80 is subjected to the forces $F_1 = 1\ 200$ N and $F_2 = 400$ N at joints E and C, respectively. Determine the forces in members GE, GC and BC.

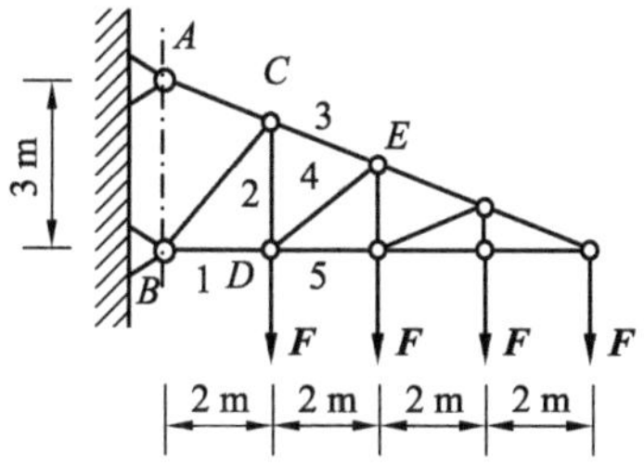

Fig. 2-79

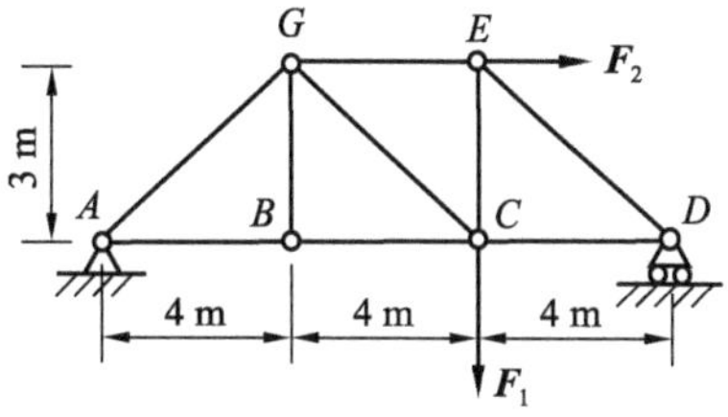

Fig. 2-80

Chapter 3

System of Forces in Space

Objectives

In this chapter, *spatial systems of forces* are studied. Students will learn how spatial systems of forces can be reduced and under which conditions they are in equilibrium. A correct free-body diagram and an appropriate application of the equilibrium conditions are the key to the solution of a spatial problem. In addition, the concept of the *center of gravity*, *center of mass*, and the *centroid* are discussed. It is shown how to determine the centroid of bodies, areas and lines.

3.1 Spacial System of Concurrent Forces

3.1.1 Addition of Spatial Concurrent Forces

It was shown in Section 2.2 that a force can be resolved into two components in a plane. Analogously, a force can be resolved uniquely into *three* components in *space*. Using a *rectangular coordinate system*, a force $\boldsymbol{F}$ may be represented by

$$\boldsymbol{F}=F_x\boldsymbol{i}+F_y\boldsymbol{j}+F_z\boldsymbol{k} \tag{3.1}$$

where the set of unit vectors, $\boldsymbol{i}$, $\boldsymbol{j}$, $\boldsymbol{k}$, is used to designate the directions of the x, y, z axes, respectively. The coordinates of three rectangular components along the x, y, z axes are given by Fig. 3-1a

$$\begin{aligned} F_x&=F\cos\alpha \\ F_y&=F\cos\beta \\ F_z&=F\cos\gamma \end{aligned} \tag{3.2}$$

The coordinate direction angles α, β, and γ measured between the *tail* of $\boldsymbol{F}$ and the *positive* x, y, z axes. Sometimes, the direction of $\boldsymbol{F}$ can be specified using two angles, such as shown in Fig. 3-1b, which yields

$$F_x = F\sin\gamma\cos\varphi$$
$$F_y = F\sin\gamma\sin\varphi \qquad (3.3)$$
$$F_z = F\cos\gamma$$

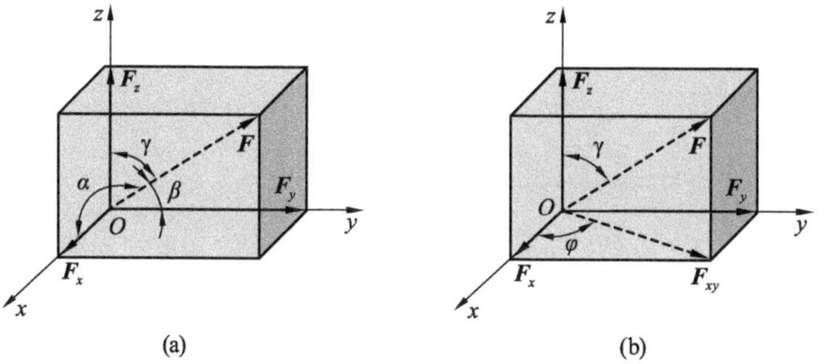

Fig. 3-1

In the case of a spatial system of n concurrent forces, the resultant is found through a successive application of the *parallelogram law* of forces in space. The resultant is the vector sum of all the force vectors. Mathematically, this is written as

$$\boldsymbol{F}_{\mathrm{R}} = \boldsymbol{F}_1 + \boldsymbol{F}_2 + \cdots + \boldsymbol{F}_n = \sum \boldsymbol{F} \qquad (3.4)$$

If each force $\boldsymbol{F}_i$ is represented by their rectangular components F_{ix}, F_{iy} and F_{iz}, according to Eq. (2.3), we obtain

$$\boldsymbol{F}_{\mathrm{R}} = F_{\mathrm{R}x}\boldsymbol{i} + F_{\mathrm{R}y}\boldsymbol{j} + F_{\mathrm{R}z}\boldsymbol{k}$$

Here coordinates of the resultant in space $F_{\mathrm{R}x}$, $F_{\mathrm{R}y}$ and $F_{\mathrm{R}z}$ equal to the algebraic sums of the respective x, y, z components of each force in the system.

$$F_{\mathrm{R}x} = F_{1x} + F_{2x} + \cdots + F_{nx} = \sum F_x$$
$$F_{\mathrm{R}y} = F_{1y} + F_{2y} + \cdots + F_{ny} = \sum F_y \qquad (3.5)$$
$$F_{\mathrm{R}z} = F_{1z} + F_{2z} + \cdots + F_{nz} = \sum F_z$$

The *magnitude* and *direction* of $\boldsymbol{F}_{\mathrm{R}}$ are given by

$$F_{\mathrm{R}} = \sqrt{F_{\mathrm{R}x}^2 + F_{\mathrm{R}y}^2 + F_{\mathrm{R}z}^2} \qquad (3.6)$$

$$\cos\alpha = \frac{F_{\mathrm{R}x}}{F_{\mathrm{R}}},\quad \cos\beta = \frac{F_{\mathrm{R}y}}{F_{\mathrm{R}}},\quad \cos\gamma = \frac{F_{\mathrm{R}z}}{F_{\mathrm{R}}} \qquad (3.7)$$

3.1.2 Equilibrium Conditions

For a spatial system of concurrent forces, the necessary and sufficient condition in *equilibrium* is that the resultant vector is the zero.

$$\boldsymbol{F}_{\mathrm{R}} = \sum \boldsymbol{F} = 0 \qquad (3.8)$$

This vector equation is equivalent to the *three* scalar equilibrium conditions. That is, the following three *scalar component equations* are satisfied.

$$\sum F_x = 0$$
$$\sum F_y = 0 \qquad (3.9)$$
$$\sum F_z = 0$$

These equations state that the algebraic sums of the x, y, z components of all the forces acting on the body must be zero. Using them we can solve for at most three unknowns.

Example 3-1

The 1 000 N block in Fig. 3-2a is suspended from the three bars that are pin connected at A. Determine the forces in the bars. Neglect the self-weight of all bars and rope. $CE=ED=12$ cm, $EA=24$ cm.

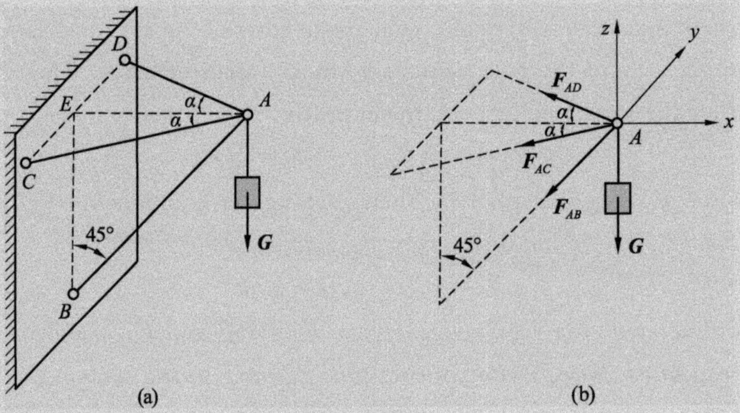

Fig. 3-2

Solution:

The connection at A is chosen for the analysis since the forces are concurrent at this point. The free-body diagram is shown in Fig. 3-2b. The unknown forces are assumed to be tensile forces. Therefore, the three scalar equations of equilibrium can be used, we have

$$\sum F_x = 0, \; -F_{AC}\cos\alpha - F_{AD}\cos\alpha - F_{AB}\sin\beta = 0$$
$$\sum F_y = 0, \; F_{AD}\sin\alpha - F_{AC}\sin\alpha = 0$$
$$\sum F_z = 0, \; -F_{AB}\sin\beta - G = 0$$

The angle α can be determined by applying the trigonometric, from Fig. 3-2b, we get

$$\sin\alpha = \frac{1}{\sqrt{5}}, \quad \cos\alpha = \frac{2}{\sqrt{5}}$$

Substituting the result of angle α into above equations, yields

$$F_{AB}=-1\ 414\ \text{kN},\ F_{AC}=F_{AD}=559\ \text{kN} \qquad \textit{Ans.}$$

The negative sign for the force F_{AB} indicates that the bar AB is actually in a state of compression.

3.2 The Moment Vector

3.2.1 Moment of a Force-Vector Formulation

If we establish x, y, z coordinate axes, then the moment of a force $\boldsymbol{F}$ about point O, or about the axis passing through O and perpendicular to the plane containing O and $\boldsymbol{F}$, as shown in Fig. 3-3, can be expressed using the vector cross product. Namely,

$$\boldsymbol{M}_O(\boldsymbol{F})=\boldsymbol{r}\times\boldsymbol{F}=\begin{vmatrix} \boldsymbol{i} & \boldsymbol{j} & \boldsymbol{k} \\ x & y & z \\ F_x & F_y & F_z \end{vmatrix}$$

$$=(yF_z-zF_y)\boldsymbol{i}+(zF_x-xF_z)\boldsymbol{j}+(xF_y-yF_x)\boldsymbol{k} \qquad (3.10)$$

where vector $\boldsymbol{r}$ is the *position vector* pointing from the reference point O to an arbitrary point on the action line of $\boldsymbol{F}$, Fig. 3-3. According to the properties of a cross-product, the moment vector $\boldsymbol{M}_O$ is perpendicular to the plane determined by $\boldsymbol{r}$ and $\boldsymbol{F}$, Fig. 3-3a. In order to distinguish between force vectors and moment vectors in the figures, a moment vector is represented with a *double head*.

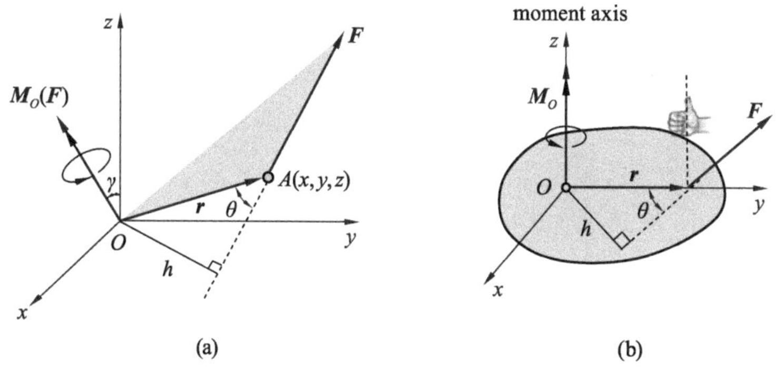

Fig. 3-3

The magnitude of the cross-product is numerically equal to the area of the parallelogram formed by $\boldsymbol{r}$ and $\boldsymbol{F}$:

$$M_O=rF\sin\theta=Fh \qquad (3.11)$$

When the moment of a force is computed about a point, the direction of $\boldsymbol{M}_O$ will be specified by using the "*right-hand rule*". The fingers of the right hand are curled such that they follow the sense of rotation. The *thumb* then *points* along the *moment axis* so that it gives the direction and sense of the moment vector, which is *upward* and *perpendicular* to the shaded plane containing F and h, as shown in Fig. 3-3b.

3.2.2 Moment of a Force about a Specified Axis

It is seen that the moment component M_z corresponds to the planar case considered. The magnitude of the moment of force about an axis z is equal to the scalar product of the moment of force about an arbitrary point O on that axis. If we consider magnitude and sense of rotation, then two quantities can be expressed by the moment vector, Fig. 3-4a.

$$\boldsymbol{M}_z = M_z \boldsymbol{k}$$

The vector $\boldsymbol{M}_z$ points in the direction of the z axis. The coplanar problem that was already treated in Section 2. 2 is reconsidered in Fig. 2-13. Force $\boldsymbol{F}$, which acts in the x, y-plane, has a moment M_O about point O, it is given by

$$M_z = Fh = xF_y - yF_x$$

The subscript z indicates that M_z exerts a moment about the z axis. The component $\boldsymbol{F}_x$ does *not* produce a moment or tendency to cause turning about the x axis since this force is *parallel* to the x axis. In coplanar problems, the body can only be rotated about the z axis. Therefore, the moment vector has only the component M_z, Fig. 3-4a. In spatial systems, there are three possibilities of rotation (about the three axes x, y and z). Hence, the moment vector $\boldsymbol{M}_O$ about point O has the three components M_x, M_y and M_z,

$$\boldsymbol{M}_O = M_x \boldsymbol{i} + M_y \boldsymbol{j} + M_z \boldsymbol{k} \tag{3.12}$$

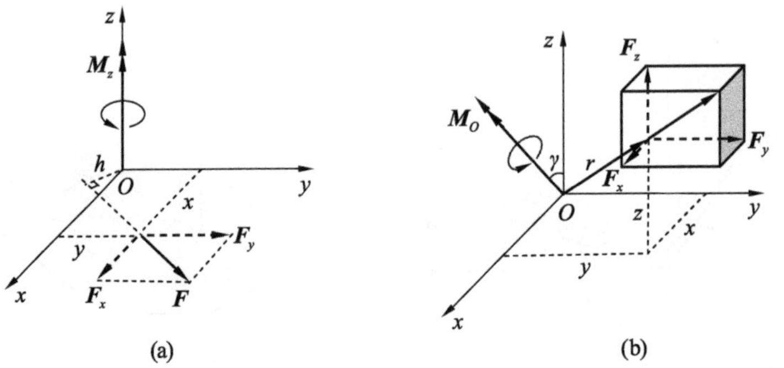

Fig. 3-4

Fig. 3-4b shows that the components, i. e. , the moments about the coordinate axes x, y, z, are obtained as follows:

$$\begin{aligned} [\boldsymbol{M}_O(\boldsymbol{F})]_x &= M_x(\boldsymbol{F}) = yF_z - zF_y \\ [\boldsymbol{M}_O(\boldsymbol{F})]_y &= M_y(\boldsymbol{F}) = zF_x - xF_z \\ [\boldsymbol{M}_O(\boldsymbol{F})]_z &= M_z(\boldsymbol{F}) = xF_y - yF_x \end{aligned} \tag{3.13}$$

This means, e. g. , the component M_x along the x axis is the *projection* of moment vector $\boldsymbol{M}_O$ onto the x axis. The magnitude and direction of the moment vector are given by

$$M_O = \sqrt{M_x^2 + M_y^2 + M_z^2} \tag{3.14}$$

$$\cos\alpha = \frac{M_x}{M},\ \cos\beta = \frac{M_y}{M},\ \cos\gamma = \frac{M_z}{M} \tag{3.15}$$

3.2.3 Moment of a Couple in Space

The moment of a couple in a plane has been discussed in Section 2. 3. Then the moment of a couple in *space* in Fig. 3-5 may be represented by the same vector production

$$\boldsymbol{M}(\boldsymbol{F}) = \boldsymbol{r} \times \boldsymbol{F} = \begin{vmatrix} \boldsymbol{i} & \boldsymbol{j} & \boldsymbol{k} \\ x & y & z \\ F_x & F_y & F_z \end{vmatrix}$$

$$= (yF_z - zF_y)\boldsymbol{i} + (zF_x - xF_z)\boldsymbol{j} + (xF_y - yF_x)\boldsymbol{k} \tag{3.16}$$

This result indicates that a couple moment is a free vector, i. e. , it can act at *any point*. The vector $\boldsymbol{r}$ points from an arbitrary point on the action line of $-\boldsymbol{F}$ to an arbitrary point on the action line of $\boldsymbol{F}$. As before, the moment vector $\boldsymbol{M}$ is orthogonal to the plane determined by $\boldsymbol{r}$ and $\boldsymbol{F}$. The moment of a couple, $\boldsymbol{M}$, Fig. 3-5, is defined as

$$M_O = Fh$$

where F is the magnitude of one of the forces and h is the perpendicular distance of moment arm between the forces. The *direction* and sense of the couple moment is determined by the *right-hand rule*, where the thumb indicates the direction when the fingers are curled with the sense of rotation caused by the two forces.

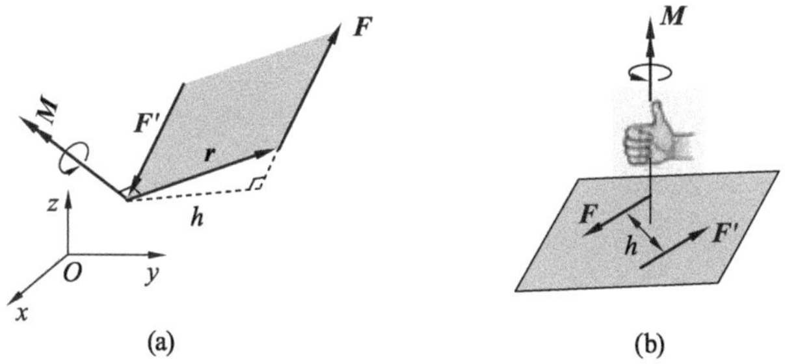

Fig. 3-5

If a body in space is subjected to several couple moments $\boldsymbol{M}_i$, the resultant moment $\boldsymbol{M}_R$ is obtained as the vector sum

$$\boldsymbol{M}_R = \sum \boldsymbol{M}_i(\boldsymbol{F}) \tag{3.17}$$

In general, problems projected in two dimensions should be solved using a scalar analysis since the moment arms and force components are easy to determine.

If the sum of the moments of all the forces in the system about point O is equal to zero, the rotational effect on the body vanishes. That is, the *equilibrium condition* for a spatial system of couple moments is

$$M_R = \sum M_i = 0 \tag{3.18}$$

Using the Eq. (3.14), three components M_x, M_y and M_z must be zero

$$\sum M_x = 0$$
$$\sum M_y = 0 \tag{3.19}$$
$$\sum M_z = 0$$

For a spatial couples' system, there are three equations being solved for three unknowns.

Example 3-2

The block is subjected to three forces of $F_1 = 400$ N, $F_2 = 500$ N, $F_3 = 600$ N, as shown in Fig. 3-6, determine the resultant moment of the forces about the x, y, and z.

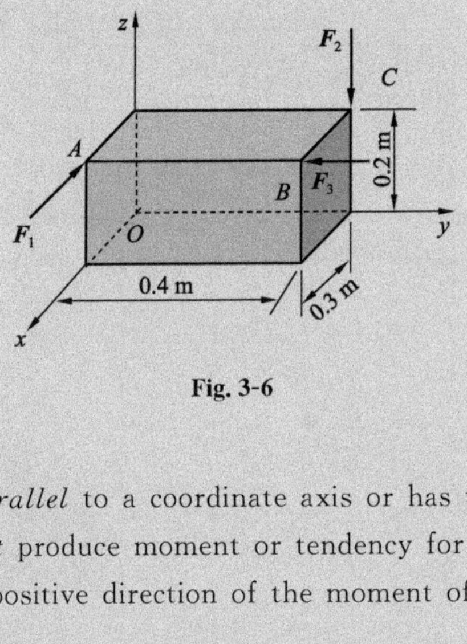

Fig. 3-6

Solution:

The force that is *parallel* to a coordinate axis or has the line of action passing through the axis does *not* produce moment or tendency for turning about that axis. Therefore, defining the positive direction of the moment of a force according to the right-hand rule, we have

$$M_x = -F_2 \times 0.4 + F_3 \times 0.2 = -500 \times 0.4 + 600 \times 0.2 = -80 \text{ N} \cdot \text{m} \qquad \textit{Ans.}$$

$$M_y = -F_1 \times 0.2 = -400 \times 0.2 = -80 \text{ N} \cdot \text{m} \qquad \textit{Ans.}$$

$$M_z = -F_3 \times 0.3 = -600 \times 0.3 = -180 \text{ N} \cdot \text{m} \qquad \textit{Ans.}$$

The negative signs indicate that resultant moment M_z, M_y, and M_z act in the $-x$, $-y$ and $-z$ direction, respectively.

Example 3-3

A force $\boldsymbol{F}$ acts on the rectangular box from point A to B as shown in Fig. 3-7. Determine the moment of the forces about point O.

Solution Ⅰ:

The force $\boldsymbol{F}$ is resolved into three components in the x, y, z direction,

$$F_x = F\sin\alpha = \frac{F \cdot a}{\sqrt{a^2+c^2}}$$

$$F_y = 0$$

$$F_z = -F\cos\alpha = -\frac{F \cdot c}{\sqrt{a^2+c^2}}$$

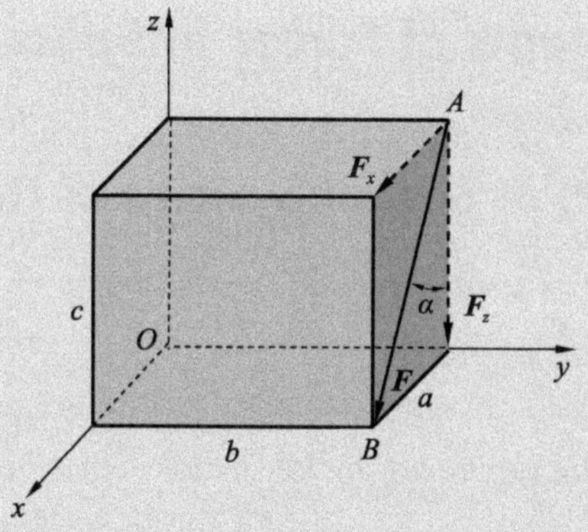

Fig. 3-7

Using the Eq. (3. 13), the moments of the force $\boldsymbol{F}$ about three axes are

$$M_x = yF_z - zF_y = -\frac{Fbc}{\sqrt{a^2+c^2}}$$

$$M_y = zF_x - xF_z = \frac{Fac}{\sqrt{a^2+c^2}}$$

$$M_z = xF_y - yF_x = -\frac{Fab}{\sqrt{a^2+c^2}}$$

Using the Eq. (3. 12), the moment vector of the force $\boldsymbol{F}$ about point O is

$$\boldsymbol{M}_O = M_x\boldsymbol{i} + M_y\boldsymbol{j} + M_z\boldsymbol{k}$$

$$= \left(-\frac{Fbc}{\sqrt{a^2+c^2}}\right)\boldsymbol{i} + \frac{Fac}{\sqrt{a^2+c^2}}\boldsymbol{j} + \left(-\frac{Fab}{\sqrt{a^2+c^2}}\right)\boldsymbol{k} \qquad Ans.$$

Solution Ⅱ:

Using the vector cross product approach, the force and position vectors are

$$\boldsymbol{r} = a\boldsymbol{i} + b\boldsymbol{j} + c\boldsymbol{k}$$

$$\boldsymbol{F} = \frac{F \cdot a}{\sqrt{a^2+c^2}}\boldsymbol{i} + 0\boldsymbol{j} - \frac{F \cdot c}{\sqrt{a^2+c^2}}\boldsymbol{k}$$

Using the Eq.(3.10), the moment vector of the force $\boldsymbol{F}$ about point O is therefore,

$$\boldsymbol{M}_O(\boldsymbol{F})=\boldsymbol{r}\times\boldsymbol{F}=\begin{vmatrix} \boldsymbol{i} & \boldsymbol{j} & \boldsymbol{k} \\ a & b & c \\ \dfrac{F\cdot a}{\sqrt{a^2+c^2}} & 0 & \dfrac{F\cdot c}{\sqrt{a^2+c^2}} \end{vmatrix}$$

$$=-\frac{Fbc}{\sqrt{a^2+c^2}}\boldsymbol{i}+\frac{Fac}{\sqrt{a^2+c^2}}\boldsymbol{j}-\frac{Fab}{\sqrt{a^2+c^2}}\boldsymbol{k} \qquad \textit{Ans.}$$

It is seen that two methods can obtain the same result. Sometimes the solution Ⅰ provides a more convenient method for analysis than solution Ⅱ, since the direction of the moment and the moment arm for each component force are easy to establish.

3.3 General Systems of Forces in Space

3.3.1 Simplification of Spatial Systems of General Forces

Consider a *general* system of forces in *space* shown in Fig. 3-8a. Similar to the procedure used in a coplanar problem (Section 2.4), an arbitrary reference point O is chosen in space. The forces $\boldsymbol{F}_i$ are moved to parallel lines of action that pass through this point. Since the effect of the forces on the body must not be changed, the corresponding moments $\boldsymbol{M}_i$ of the forces have to be added. We get an equivalent system that consists of a system of concurrent forces and a system of moments, Fig. 3-8b. Using the methods of the previous section, this system can be reduced to a *resultant force* $\boldsymbol{F}_R$ and a resultant moment $\boldsymbol{M}_O$, Fig. 3-8c, respectively,

$$\boldsymbol{F}_R=\boldsymbol{F}_1+\boldsymbol{F}_2+\cdots+\boldsymbol{F}_n=\sum\boldsymbol{F}$$

$$M_O=M_O(\boldsymbol{F}_1)+M_O(\boldsymbol{F}_2)+\cdots+M_O(\boldsymbol{F}_n)=\sum M_O(\boldsymbol{F}) \tag{3.20}$$

The resultant force $\boldsymbol{F}_R$ *is independent of the choice of point* O; the resultant moment $\boldsymbol{M}_O$, however, depends on this choice. Hence, there are many possible ways to reduce a given general system of forces to a resultant force and a resultant moment.

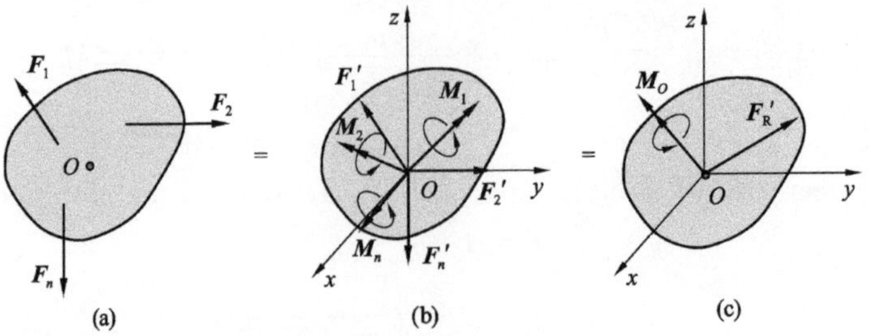

Fig. 3-8

In general, the equivalent resultant force $\boldsymbol{F}_R$ acting at point O and a resultant couple moment $\boldsymbol{M}_R$ are not perpendicular to one another, as shown in Fig. 3-9a and cannot be further reduced to an equivalent single resultant force. However, the resultant couple moment $\boldsymbol{M}_R$ can be resolved into components $\boldsymbol{M}_1$, $\boldsymbol{M}_2$ *parallel and perpendicular to the line of action of* $\boldsymbol{F}_R$, respectively, Fig. 3-9a.

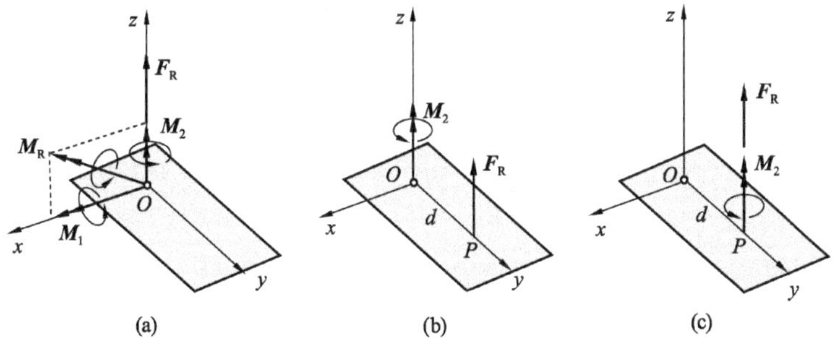

Fig. 3-9

(1) $M_1 \perp F_R$

The perpendicular component $\boldsymbol{M}_1$ and $\boldsymbol{F}_R$ can be replaced by a single force that acts at point P. A distance d from point O, Fig. 3-9b, is perpendicular to both the $\boldsymbol{M}_1$ and the line of action of $\boldsymbol{F}_R$. And this distance d can be determined from $d = \boldsymbol{M}_1 / \boldsymbol{F}_R$.

(2) $M_2 /\!/ F_R$

In this case, we cannot further simplify. But $\boldsymbol{M}_2$ can be moved to point P, Fig. 3-9c, since it is a free vector. This combination will tend to translate and rotate the body about its axis, and is referred to as a *wrench or screw* that is the simplest system to represent any general force and couple moment system acting on a body.

3.3.2 Equilibrium Conditions

The conditions for equilibrium of a rigid body subjected to a three-dimensional forces system require that both the *resultant force* $\boldsymbol{F}_R$ and *resultant couple moment* $\boldsymbol{M}$ about an arbitrary point O be equal to *zero*. The two conditions are expressed in vector form as

$$F_R' = 0, \quad M_O = 0 \tag{3.21}$$

In components, we have scalar equation of equilibrium as

$$\begin{gathered} \sum F_x = 0, \quad \sum F_y = 0 \\ \sum F_z = 0, \quad \sum M_x = 0 \\ \sum M_y = 0, \quad \sum M_z = 0 \end{gathered} \tag{3.22}$$

There are *six scalar equilibrium conditions* corresponds to the six degrees of freedom of a rigid body in space: translations in the x, y and z directions and rotations about the corresponding coordinate axes. These *six scalar equilibrium equations* may be used to solve for at most *six unknowns* shown on the free-body diagram. Eq.(3.22) expresses that the sum of the external force components acting in the x, y, and z directions must be zero, and the sum of the moment components about the x, y, and z axes must be zero.

3.3.3 Spatial Parallel Forces

Consider the special case of a system of parallel forces. For example, if all the forces act in the z-direction, Fig. 3-10, that is, the lines of action are parallel. Then components of all forces $F_{ix} = 0$ and $F_{iy} = 0$. In this case, the equilibrium conditions in the x and y directions for the forces and the moment equation about the axis which is parallel to the z-axis are identically satisfied. Therefore, the equilibrium conditions (3.22) reduce to

$$\sum F_z = 0,\ \sum M_x = 0,\ \sum M_y = 0 \tag{3.23}$$

We have only three independent equilibrium equations for three unknowns.

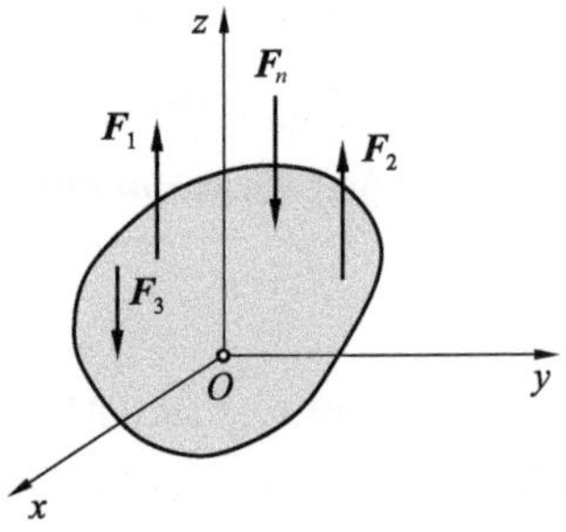

Fig. 3-10

Example 3-4

A rectangular block is subjected to six forces, F_1 to F_6 shown in Fig. 3-11. Determine the resultant $\boldsymbol{F}_R$ and the resultant moment $\boldsymbol{M}$ with respect to point A. Set $F_1 = F_2 = F$, $F_3 = F_4 = 2F$, $F_5 = F_6 = 3F$, $b = a$, $c = 2a$.

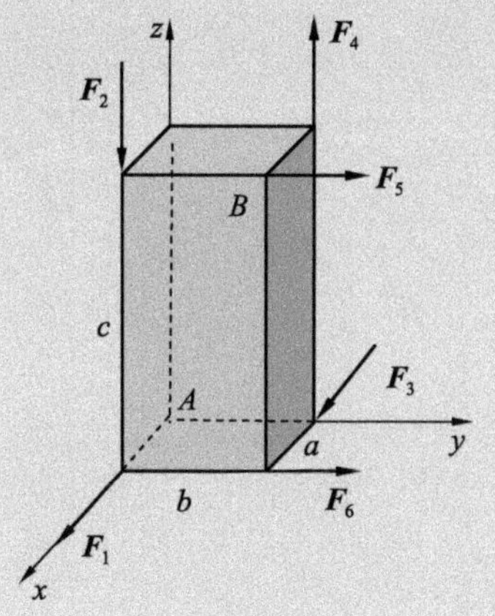

Fig. 3-11

Solution:

The coordinate system is chosen such that its origin is at A, Fig. 3-11. The components of the resultant force vector $\boldsymbol{F}_R$ are obtained as the sum of the given force components:

$$F_{Rx}=F_1+F_3=3F$$
$$F_{Ry}=F_5+F_6=6F$$
$$F_{Rz}=-F_2+F_4=F$$

The magnitude of $\boldsymbol{F}_R$ can be written as

$$F_R=\sqrt{F_{Rx}^2+F_{Ry}^2+F_{Rz}^2}=\sqrt{3^2+6^2+1^2}F$$
$$=\sqrt{46}F \qquad \textit{Ans.}$$

The components of the moment $\boldsymbol{M}$ are obtained as the sum of the moments about the x, y and z axes of all forces, respectively.

$$M_{Rx}=bF_4-cF_5=-4aF$$
$$M_{Ry}=aF_2=aF$$
$$M_{Rz}=aF_5+aF_6-bF_3=4aF$$

Hence, magnitude of $\boldsymbol{M}$ can be written as

$$M_R=\sqrt{4^2+1^2+4^2}Fa=\sqrt{33}Fa \qquad \textit{Ans.}$$

Example 3-5

Determine the reaction forces that the ball-and-socket joint at A, the smooth journal bearing at B, and the roller support at C exert on the rod shown in Fig. 3-12a.

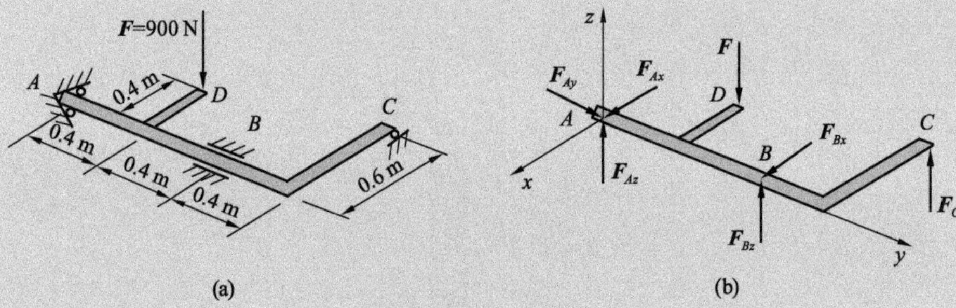

Fig. 3-12

Solution:

The free-body diagram of the rod is shown in Fig. 3-12b. Each of unknown reactions is assumed to acting a positive coordinate direction. A direct solution for F_{Ay} can be obtained by summing forces along the y axis.

$$\sum F_y=0,\ F_{Ay}=0 \qquad Ans.$$

The force F_{Bx} can be found by summing moments about the z axis.

$$\sum M_z = 0,\ -F_{Bx} \times 0.8 = 0$$

$$F_{Bx} = 0 \qquad Ans.$$

Then summing forces along the x axis, the force F_{Ax} can be determined,

$$\sum F_x = 0,\ F_{Ax} + F_{Bx} = 0$$

$$F_{Ax} = 0 \qquad Ans.$$

Summing moments about the y axis, the force F_C can be determined,

$$\sum M_y = 0,\ F_C \times 0.6 - 900 \times 0.4 = 0$$

$$F_C = 600\ \text{N} \qquad Ans.$$

Summing moments about the x axis, the force F_{Bz} can be determined,

$$\sum M_x = 0,\ F_{Bz} \times 0.8 + F_C \times 1.2 - 900 \times 0.4 = 0$$

$$F_{Bz} = -450\ \text{N} \qquad Ans.$$

The negative sign indicates that F_{Bz} acts downward. Finally, summing the forces along the z axis

$$\sum F_z = 0,\ F_{Az} + F_{Bz} + F_C - 900 = 0$$

$$F_{Az} = 750\ \text{N} \qquad Ans.$$

Example 3-6

A homogeneous plate of weight G is supported by six bars and loaded by a force $F=2G$ at joint A in Fig. 3-13a. Determine the reaction forces in each bar.

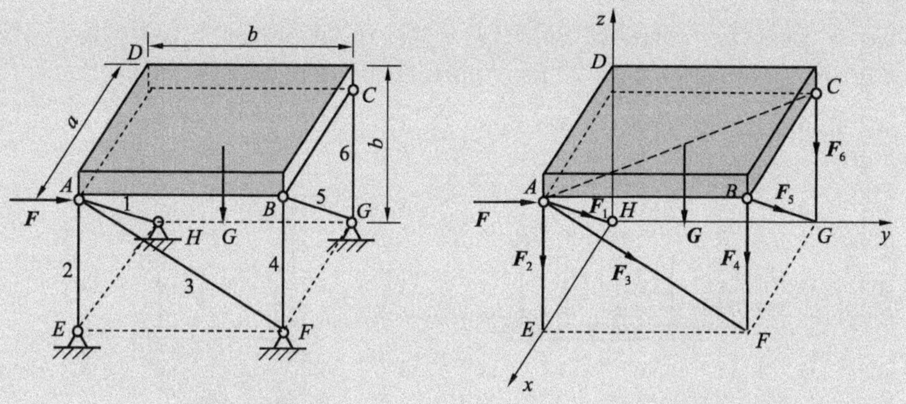

Fig. 3-13

Solution:

The free-body diagram of plate is shown in Fig. 3-13b, the forces of bars are assumed to be tensile forces. Choosing the suitable axes (not always x, y, z axes) such that as many moments as possible about its origin are zero, the following equilibrium conditions are obtained.

$$\sum M_{AE}(F) = 0,\ F_5 = 0 \qquad Ans.$$

$$\sum M_{EF}(F) = 0,\ F_1 = 0 \qquad Ans.$$

$$\sum M_{AC}(F) = 0,\ F_4 = 0 \qquad Ans.$$

$$\sum M_{AB}(F) = 0,\ -F_6 a - G\frac{a}{2} = 0$$

$$F_6 = -\frac{G}{2} \qquad Ans.$$

$$\sum M_{DH}(F) = 0,\ F_3 a\cos 45° + Fa = 0$$

$$F_3 = -2.824G \qquad Ans.$$

$$\sum M_{FG}(F) = 0,\ -F_2 b - G\frac{b}{2} + Fb = 0$$

$$F_2 = 1.5G \qquad Ans.$$

3.4 Center of Gravity and Centroid

3.4.1 Center of Parallel Forces

The forces, whose lines of action are parallel to each other, are known as *parallel forces*. The magnitude and position of the resultant force, of a system of parallel forces may be found out, the direction of the resultant $\boldsymbol{F}_R$ coincides with the direction of the individual forces. As a simple example, consider a beam (weight neglected) that is loaded by a system of parallel forces $\boldsymbol{F}_i$, the coordinate x is introduced with an arbitrarily chosen origin O, Fig. 3-14a. Applying the *Principle of Moments*.

$$x_C \cdot F_R = \sum x_i \cdot F_i$$

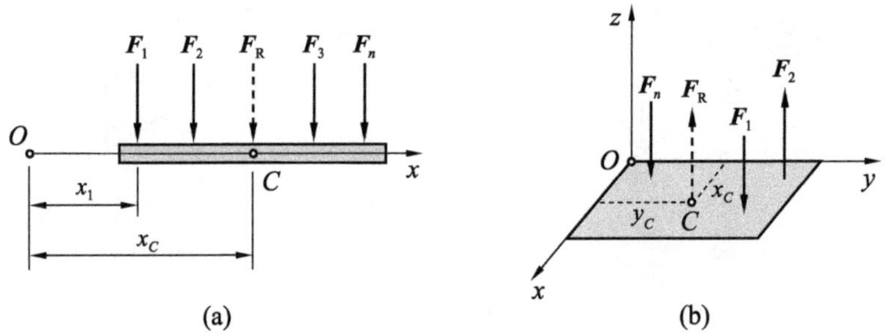

Fig. 3-14

Then the action line of the resultant $\boldsymbol{F}_R$ is found to be located at the distance from the origin O of the coordinate system.

$$x_C = \frac{\sum x_i \cdot F_i}{F_R} \tag{3.24}$$

The corresponding point C (an arbitrary point on the action line of $\boldsymbol{F}_R$) is called the *center of parallel forces*.

In the case of spatial parallel forces as shown in Fig. 3-14b, analogous equations are obtained for the coordinates y and z. Thus, the location of the center of the parallel forces is determined by

$$x_C = \frac{\sum x_i \cdot F_i}{F_R}, \quad y_C = \frac{\sum y_i \cdot F_i}{F_R}, \quad z_C = \frac{\sum z_i \cdot F_i}{F_R} \tag{3.25}$$

The considerations can also be applied to systems of continuously distributed loads and the sums are replaced by integrals.

3.4.2 Center of Gravity and Center of Mass

If a rigid body on the surface of the earth that is subjected to the action of the earth's gravitational, the center of gravity $\boldsymbol{G}$ is a point which locates the resultant weight of a system of particles. To show how to determine this point, consider the system of n particles as shown in Fig. 3-15a. The weights of each particle form a system of parallel forces. To find the x, y, z coordinates of $\boldsymbol{G}$, the resultant weight is equal to the total weight of all n particles; that is

$$\boldsymbol{G} = \sum \boldsymbol{G}_i$$

This resultant weight passes through a point called the *center of gravity*. Applying the *Principle of Moments*, the sum of the moments of the weights of all the particles about the y axis is then equal to the moment of the resultant weight about same axis. This yield

$$x_C \cdot \boldsymbol{G} = \sum x_i \cdot \boldsymbol{G}_i$$

Similarly, summing moments about the x axis, we can obtain the y_C coordinate, i. e. ,

$$-y_C \cdot \boldsymbol{G} = -\sum y_i \cdot \boldsymbol{G}_i$$

Although the weights do not produce a moment about the z axis, we can obtain the z_C coordinate of C by imagining the coordinate system, with the particles fixed in it, as being rotated 90° about the x (or y) axis, Fig. 3-15b. Summing moments about the x axis, we have

$$-z_C \cdot \boldsymbol{G} = -\sum z_i \cdot \boldsymbol{G}_i$$

We can generalize these formulas, and write the *center of gravity* of the particles system with respect to the x, y, z axes in the form

$$x_C = \frac{\sum x_i \cdot G_i}{G},\ y_C = \frac{\sum y_i \cdot G_i}{G},\ z_C = \frac{\sum z_i \cdot G_i}{G} \tag{3.26}$$

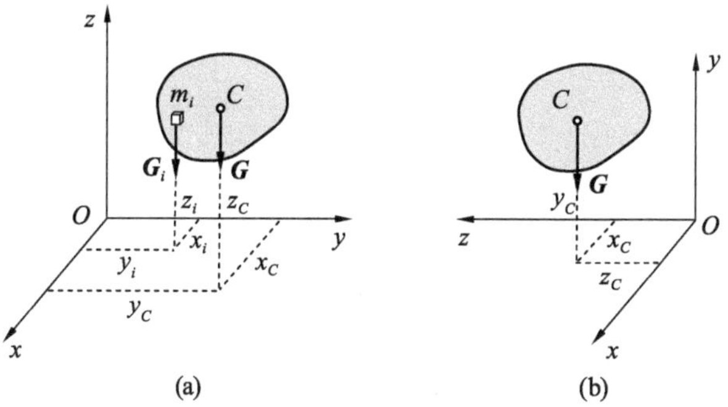

Fig. 3-15

If we want to locate the body's *center of mass*, this location can be determined by substituting $G_i = m_i g$ into Eq. (3. 26). Since g is constant, it can cancel out from both the numerator and denominator, then we have

$$x_C = \frac{\sum x_i \cdot m_i}{m},\ y_C = \frac{\sum y_i \cdot m_i}{m},\ z_C = \frac{\sum z_i \cdot m_i}{m} \tag{3.27}$$

where m is the resultant mass. The center of mass coincides with the center of gravity if the gravitational is assumed to be uniform and parallel.

A rigid body is composed of an infinite number of particles, and so it is necessary to use integration rather than a discrete summation of the term in which the arbitrary particle has an element weight dG or element mass dm. The resulting equations are

$$x_C = \frac{\int_V x\,\mathrm{d}G}{G},\ y_C = \frac{\int_V y\,\mathrm{d}G}{G},\ z_C = \frac{\int_V z\,\mathrm{d}G}{G} \tag{3.28}$$

or

$$x_C = \frac{\int_V x\,\mathrm{d}m}{m},\ y_C = \frac{\int_V y\,\mathrm{d}m}{m},\ z_C = \frac{\int_V z\,\mathrm{d}m}{m} \tag{3.29}$$

3.4.3 Centroid

The *centroid* is a point which defines the *geometric center* of an object. For the volume, the differential volume element $\mathrm{d}V$ is given by $\mathrm{d}m=\rho\mathrm{d}V$, and the mass of the whole body is the integral of the mass elements. If the body is made from a homogeneous material, the density ρ is a constant and cancel out from the Eq. (3.29). Thus, the *centroid of the volume* or geometric center of body can be obtained.

$$x_C = \frac{\int_V x\,\mathrm{d}V}{V},\ y_C = \frac{\int_V y\,\mathrm{d}V}{V},\ z_C = \frac{\int_V z\,\mathrm{d}V}{V} \tag{3.30}$$

If the body is considered to be a thin plate with a constant density and a constant thickness, as shown in Fig. 3-16a, then the location of the centroid of a homogeneous plane area can be determined from integral similar to Eq. (3.30), the coordinates of the *centroid of the area* is

$$x_C = \frac{\int_A x\,\mathrm{d}A}{A},\ y_C = \frac{\int_A y\,\mathrm{d}A}{A} \tag{3.31}$$

If a line that lies in a plane, its centroid C can be determined if the infinitesimal area $\mathrm{d}A$ is replaced by a line element $\mathrm{d}l$ and the area A is replaced by the length l of the line, Fig. 3-16b, then the coordinates of the *centroid of a line* is

$$x_C = \frac{\int_l x\,\mathrm{d}l}{l},\ y_C = \frac{\int_l y\,\mathrm{d}l}{l} \tag{3.32}$$

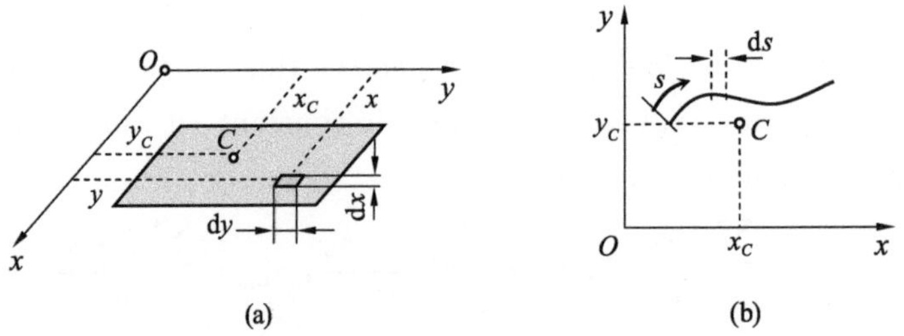

Fig. 3-16

3.4.4 Methods to Determine the Centroid

(1) Symmetry

The *centroids* of some shapes may be partially or completely specified by using conditions of *symmetry*.

① If a body possesses a plane of geometrical symmetry, the centroid will lie in this plane.

② If the area has *an axis of symmetry*, the centroid of the area lies on this axis, Fig. 3-17.

③ In the case of *two or three axes of symmetry*, the centroid is determined by the *point of intersection* of these axes, Fig. 3-17a, b and c.

In some cases, the centroid is located at a point that is not on the object.

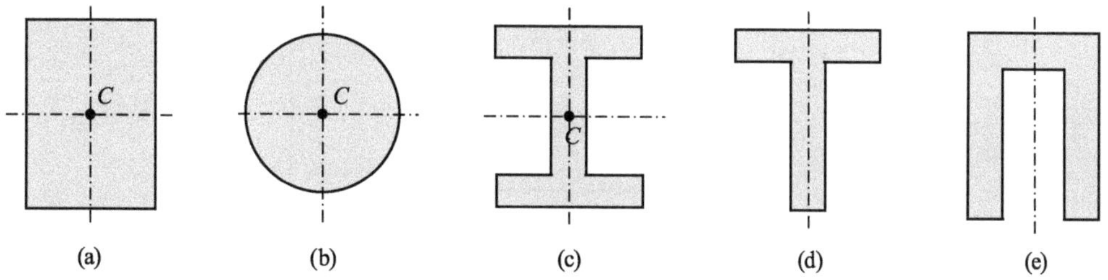

Fig. 3-17

(2) Composite Body

A *composite body* consists of several connected simpler shaped bodies, which may be rectangular, triangular, etc. Such a body can often be sectioned or divided into its composite parts and, provided the *weight* and location of the center of gravity of each of these parts are assumed to be known. Thus, the integrals are replaced by sum. We have

$$x_C = \frac{\sum x_i \cdot G_i}{\sum G_i}, \quad y_C = \frac{\sum y_i \cdot G_i}{\sum G_i}, \quad z_C = \frac{\sum z_i \cdot G_i}{\sum G_i} \tag{3.33}$$

The centroid for composite lines, areas, and volumes can be found using relations analogous to Eq. (3.33). These equations can also be used for areas with holes or cut-out sections. The holes are then considered to be "*negative*" areas.

Centroid for common shapes of lines, areas, shells, and volumes that often make up a composite body are given in the Tab. 3-1.

Tab. 3-1 Location of centroid

Type	Diagram	Location of centroid
Triangle	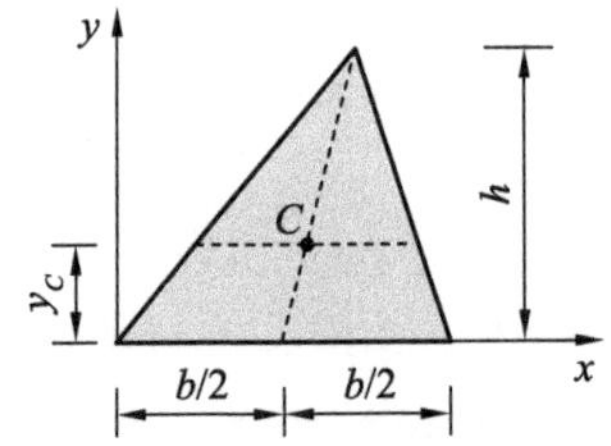	$y_C = \frac{1}{3}h$
Trapezium	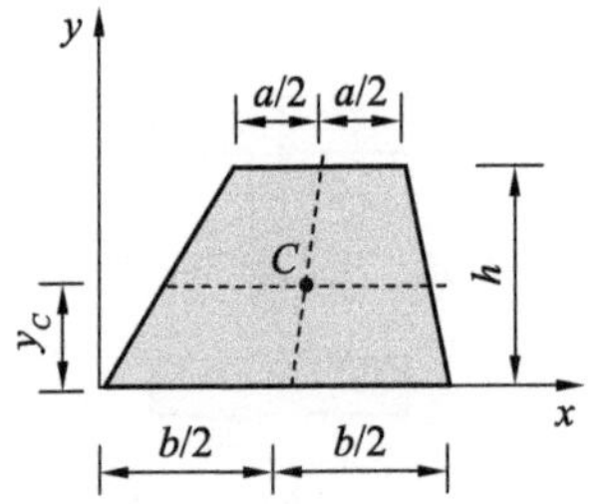	$y_C = \frac{h(2a+b)}{3(a+b)}$
Circular arc	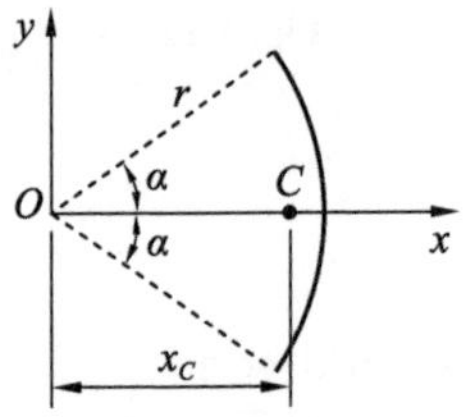	$x_C = \frac{r\sin\alpha}{\alpha}$ $\alpha = \frac{\pi}{2},\ x_C = \frac{2r}{\pi}$
Circular segment	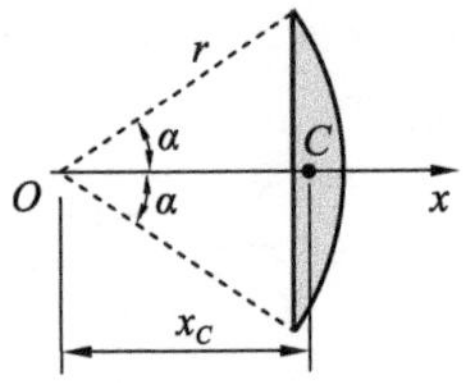	$x_C = \frac{2}{3}\frac{r^3\sin^3\alpha}{S}$ $S = \frac{r^2(2\alpha - \sin 2\alpha)}{2}$
Circular sector	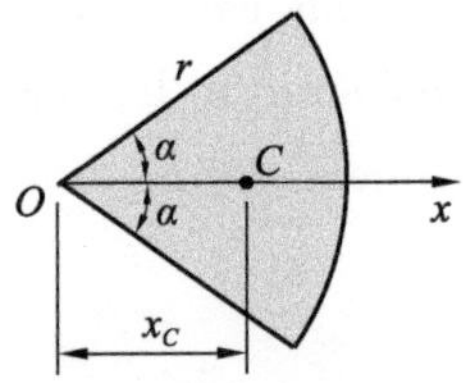	$x_C = \frac{2}{3}\frac{r\sin\alpha}{\alpha}$ a semicircular area $\alpha = \frac{\pi}{2},\ x_C = \frac{4r}{3\pi}$
Section of annulus	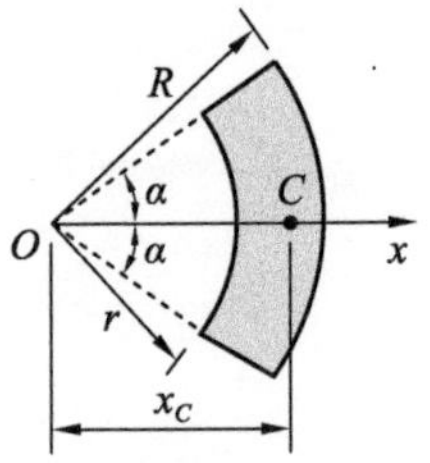	$x_C = \frac{2}{3}\frac{R^3 - r^3}{R^2 - r^2}\frac{\sin\alpha}{\alpha}$

Continued

Type	Diagram	Location of centroid
Quadratic parabola	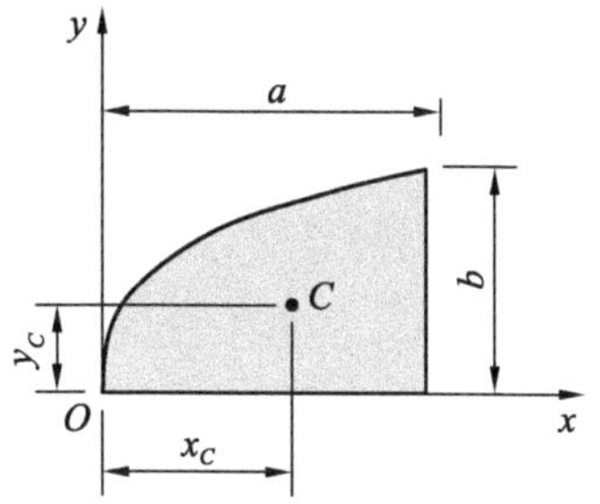	$x_C=\frac{3}{5}a$ $y_C=\frac{3}{8}b$
Half a sphere	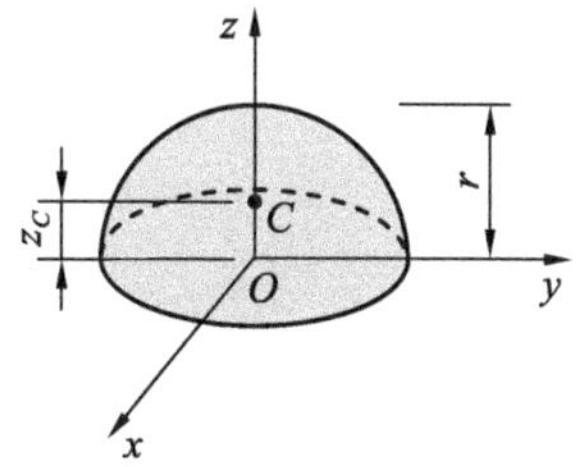	$z_C=\frac{3}{8}r$
Cone	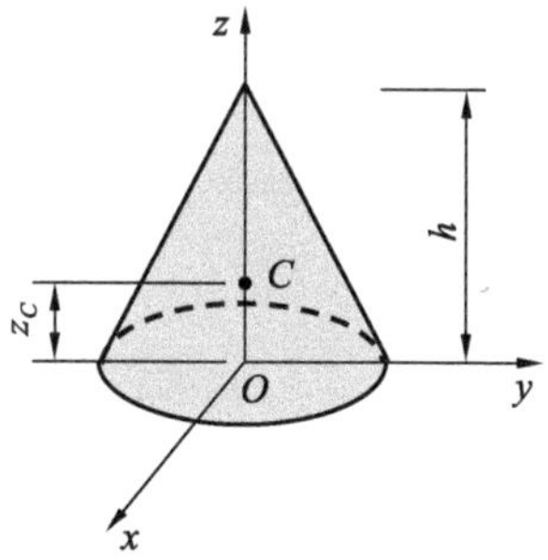	$z_C=\frac{1}{4}h$

Example 3-7

Determine the distance y_C measured from the x axis to the centroid of the area of the triangle shown in Fig. 3-18.

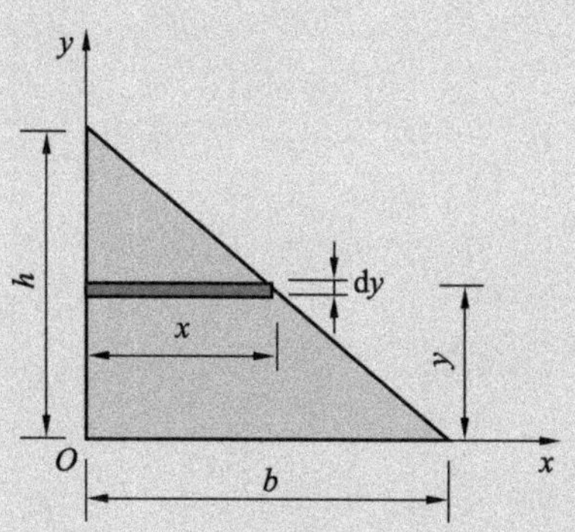

Fig. 3-18

Solution:

First, we introduce the coordinate system as shown in Fig. 3-18. Assume a rectangular element having a thickness dy. Here every point of the element has the same distance x from the y axis. The area of the element is

$$dA = x \cdot dy = \frac{b}{h}(h-y)dy$$

The coordinates of the centroid can be determined from Eq. (3.31)

$$y_C = \frac{\int_A y\,dA}{\int_A dA} = \frac{\int_0^h y\frac{b}{h}(h-y)dy}{\int_0^h \frac{b}{h}(h-y)dy} = \frac{\frac{1}{6}bh^2}{\frac{1}{2}bh} = \frac{1}{3}h \qquad Ans.$$

This result shows that the centroid is located at one-third the height, measured from the base of the triangle. If we want to determine the distance x_C measured from the y axis to the centroid of the triangle shown in Fig. 3-18, analogous equations are used for the coordinates x_C which students can try to do.

Example 3-8

Determine the centroid of the L-shaped area shown in Fig. 3-19a. The measurements are given in mm.

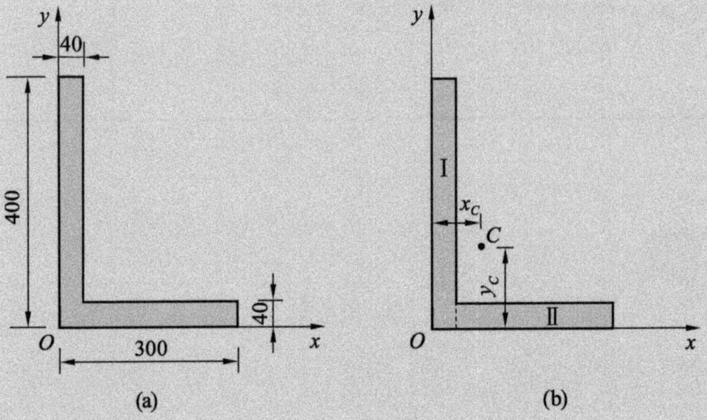

Fig. 3-19

Solution:

We choose a coordinate system and consider the plate is divided into two rectangles as shown in Fig. 3-19b. The centroid of each segment is located and area of each segment is calculated as follows:

$$A_1 = 400 \times 40 = 16\ 000\ \text{mm}^2,\ x_1 = 20\ \text{mm},\ y_1 = 200\ \text{mm}$$

$$A_2 = 260 \times 40 = 10\ 400\ \text{mm}^2,\ x_2 = 170\ \text{mm},\ y_2 = 20\ \text{mm}$$

Using the Eq. (3.33), we have

$$x_C=\frac{x_1\cdot A_1+x_2\cdot A_2}{A_1+A_2}=\frac{20\times16\ 000+170\times10\ 400}{16\ 000+10\ 400}=79.1\ \text{mm} \qquad Ans.$$

$$y_C=\frac{y_1\cdot A_1+y_2\cdot A_2}{A_1+A_2}=\frac{200\times16\ 000+20\times10\ 400}{16\ 000+10\ 400}=129.1\ \text{mm} \qquad Ans.$$

Example 3-9

A circle is removed from a homogeneous circular plate, as shown in Fig. 3-20a. Locate the centroid C of the remaining area.

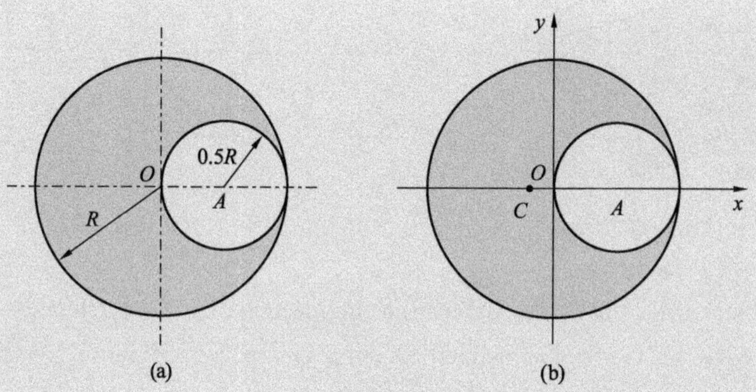

Fig. 3-20

Solution:

By symmetry, the centroid lies on the x-axis of the chosen coordinate system, i. e., $y_C=0$ as shown in Fig. 3-20b. The plate is considered as a composite area which consists of two parts: the big circle and small circle. Since the region of the small circle is void of material, this part has to be subtracted from the larger circle. That is, the area is *negative*. The centroid of each segment is located and area of each segment is calculated as follows:

$$A_1=\pi R^2,\ x_1=0$$

$$A_2=\pi\left(\frac{R}{2}\right)^2,\ x_2=0.5R$$

Using the Eq. (3.33), we have

$$x_C=\frac{x_1\cdot A_1+x_2\cdot A_2}{A_1+A_2}=\frac{0\times\pi R^2-0.5R\times\frac{\pi}{4}R^2}{\pi R^2-\frac{\pi}{4}R^2}$$

$$=-\frac{R}{6} \qquad Ans.$$

Exercises

3-1 In Fig. 3-21, three cables AE, BE and DE are used to support a homogeneous square plate of $G=18$ kN. Line CE of length 2.4 m is vertical to plate. Determine the tension in three cables.

3-2 Determine the tension developed in cables AB, AC, and AD, in Fig. 3-22.

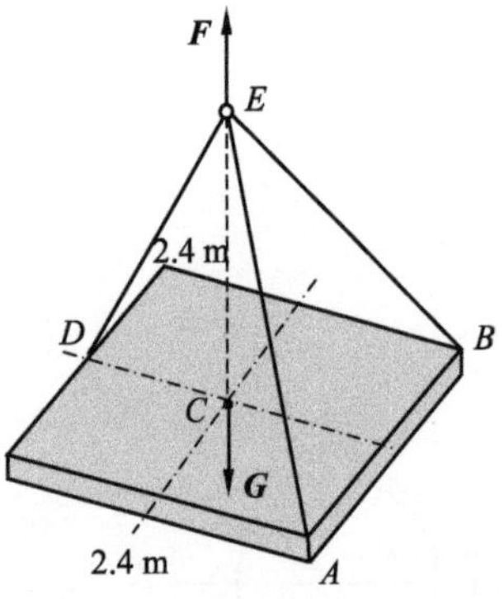

Fig. 3-21

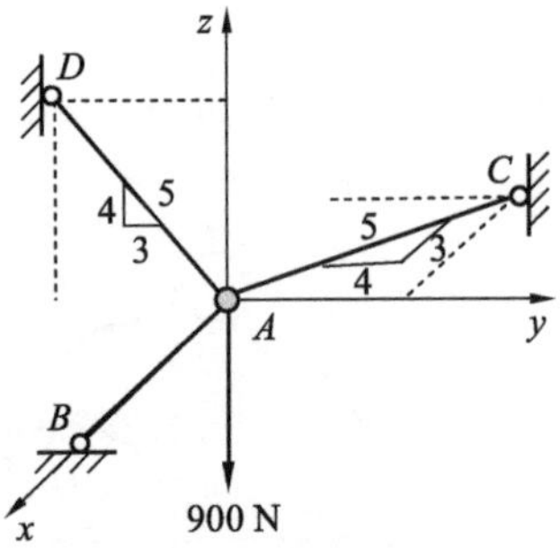

Fig. 3-22

3-3 As shown in Fig. 3-23, determine the force acting along the axis of each of the three struts AD, BD, and CD needed to support the 10 kN block. Neglect the self-weight of all bars and rope.

3-4 Determine the magnitude of the moment that the force $\boldsymbol{F}$ exerts about the z axis of the shaft, in Fig. 3-24.

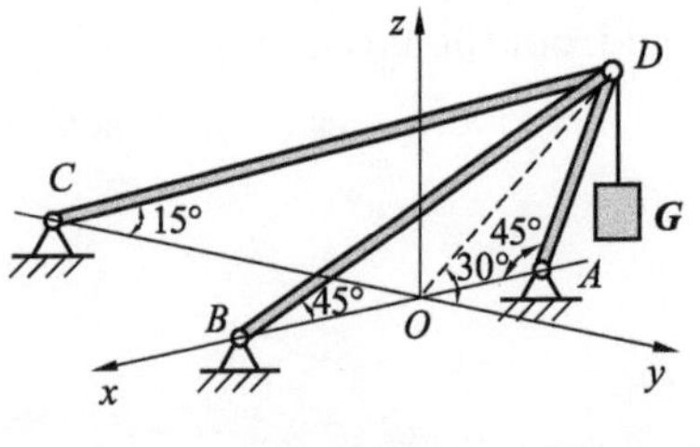

Fig. 3-23

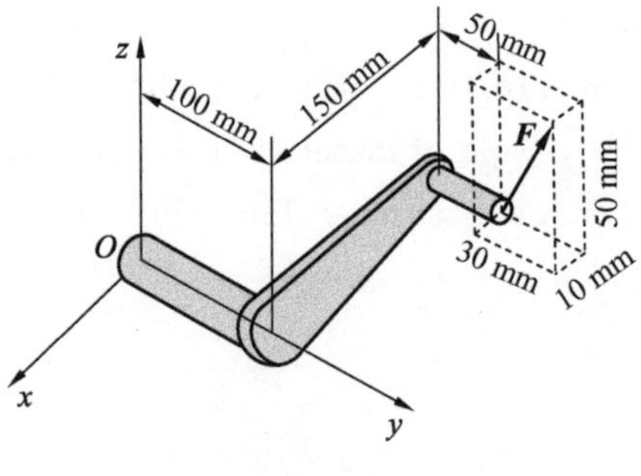

Fig. 3-24

3-5 Determine the magnitude of the moments of the force $\boldsymbol{F}$ about the x, y, and z axes in Fig. 3-25.

3-6 In Fig. 3-26, a force $\boldsymbol{F}$ acts on the square from A to B. Determine the resulting moment of the forces about point O.

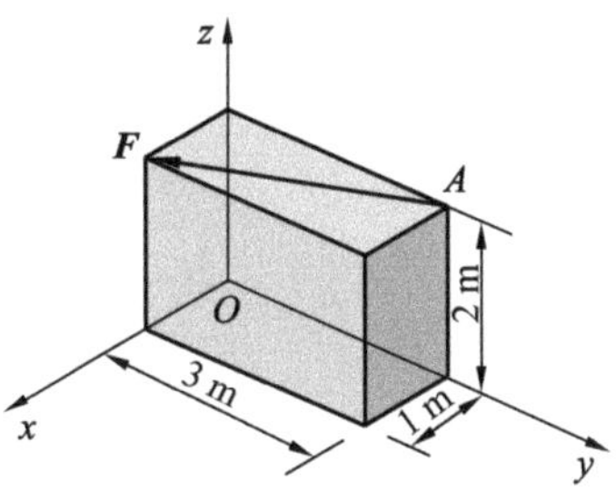

Fig. 3-25

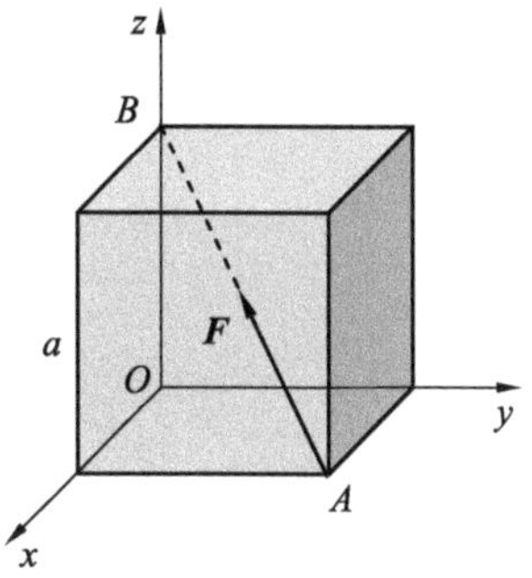

Fig. 3-26

3-7 A slab is subjected to four parallel forces in Fig. 3-27. Determine the magnitude and direction of a resultant force equivalent to the given forces system and locate its point of application on the slab.

3-8 A platform truck in Fig. 3-28 supports the uniform block having a weight of $G=400$ N. Determine the vertical reactions on the three wheels at A, B, and C. Neglect the mass of the platform truck.

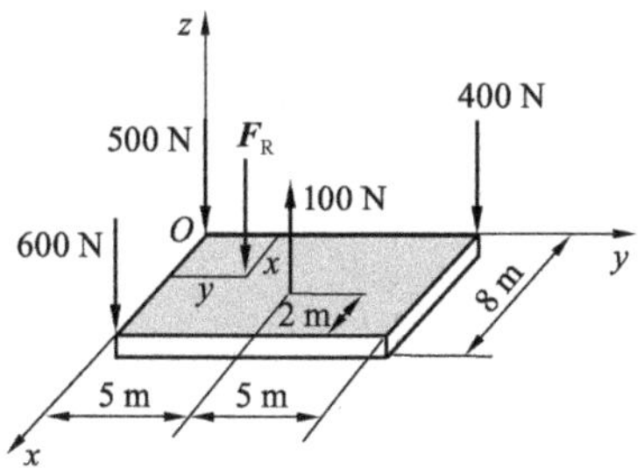

Fig. 3-27

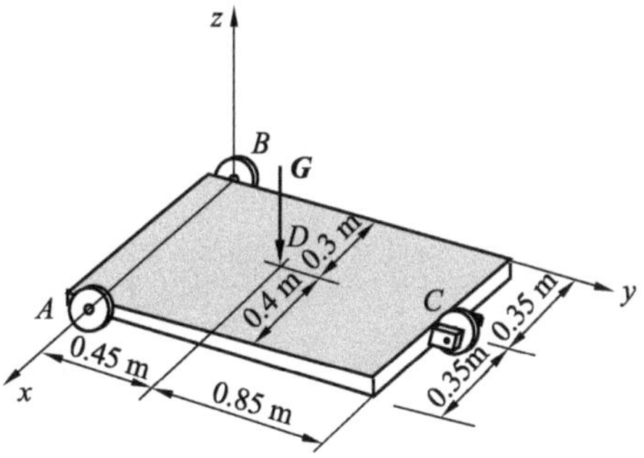

Fig. 3-28

3-9 As shown in Fig. 3-29, if force $F=6$ kN, $x=0.75$ m and $y=1$ m, determine the tension developed in cables AB, CD, and EF. Neglect the weight of the plate.

3-10 In Fig. 3-30, the 200 N homogeneous rectangular plate is supported by a ball-and-socket joint at A, a hinge B and a cable at C. Determine the tension in cable CE and reactions in support A and B.

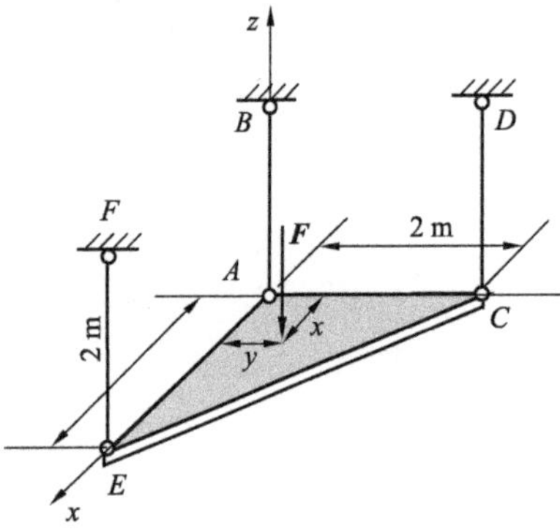

Fig. 3-29

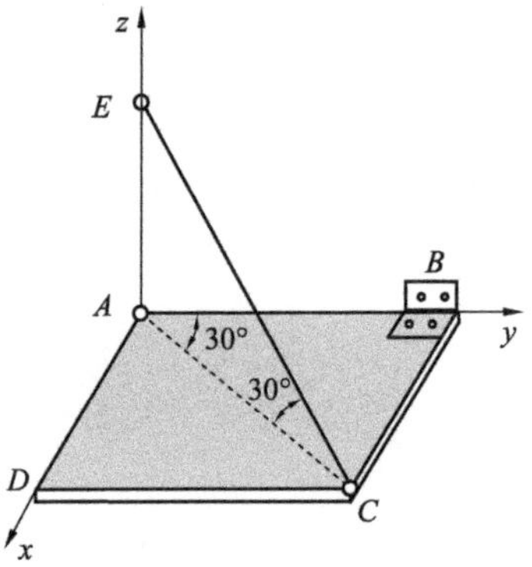

Fig. 3-30

3-11　In Fig. 3-31, the shaft is supported by three smooth journal bearings at A, B, and C. Determine the components of reaction at these bearings.

3-12　In Fig. 3-32, a homogeneous plate is supported by six bars and loaded by a force $\boldsymbol{F}$ at A. Calculate the forces in the bars. Neglect the weight of plate and bars.

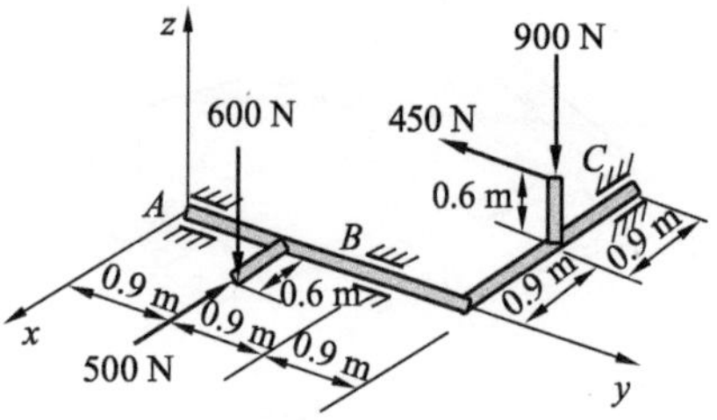

Fig. 3-31

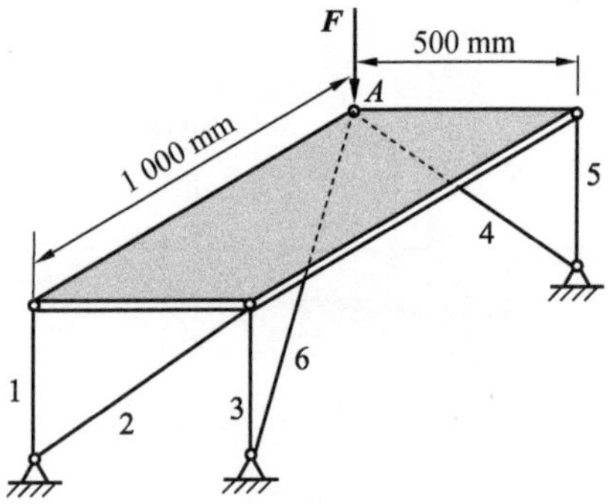

Fig. 3-32

3-13　In Fig. 3-33, determine the distance y_C measured from the x axis to the centroid of the area of the semicircle.

3-14　In Fig. 3-34, a square is removed from a homogeneous rectangle plate, locate the centroid C of the remaining area.

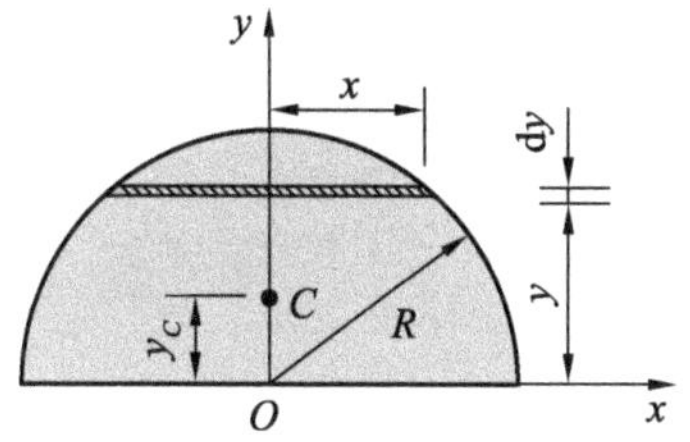

Fig. 3-33

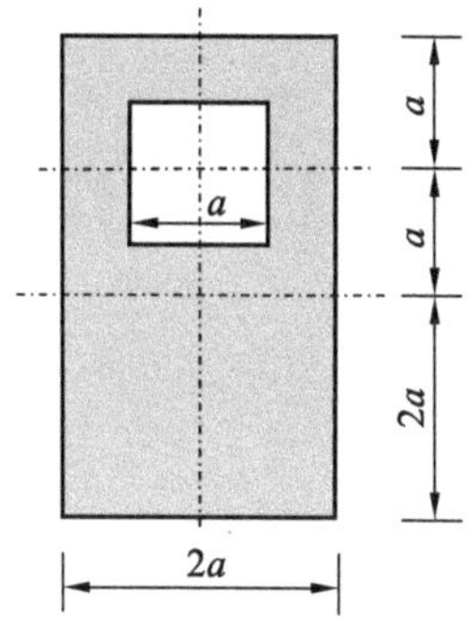

Fig. 3-34

3-15 In Fig. 3-35, determine the coordinates of the centroid of homogeneous I-shape structure. The measurements are given in mm.

3-16 In Fig. 3-36, determine the coordinates of the centroid of homogeneous channel structure. The measurements are given in mm.

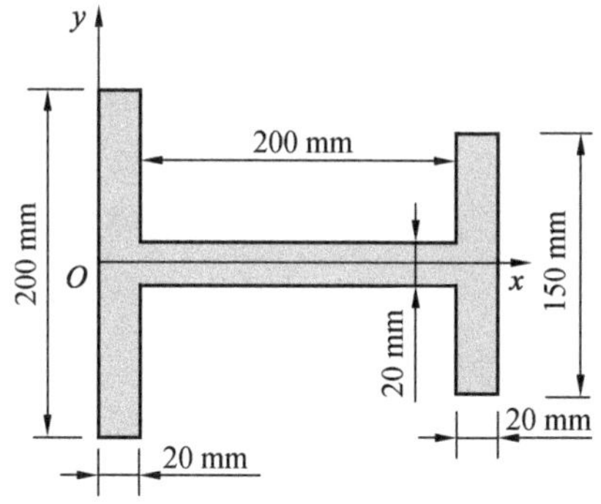

Fig. 3-35

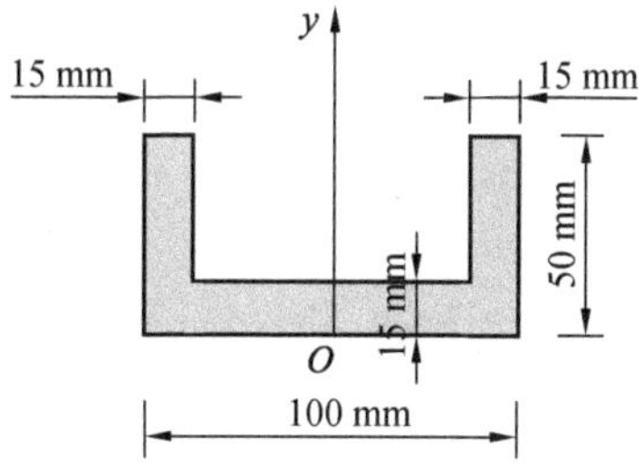

Fig. 3-36

3-17 As shown in Fig. 3-37, a square is removed from a circular plate. Locate the centroid C of the remaining area.

3-18 As shown in Fig. 3-38, a circle is removed from a triangle plate. Locate the centroid C of the remaining area.

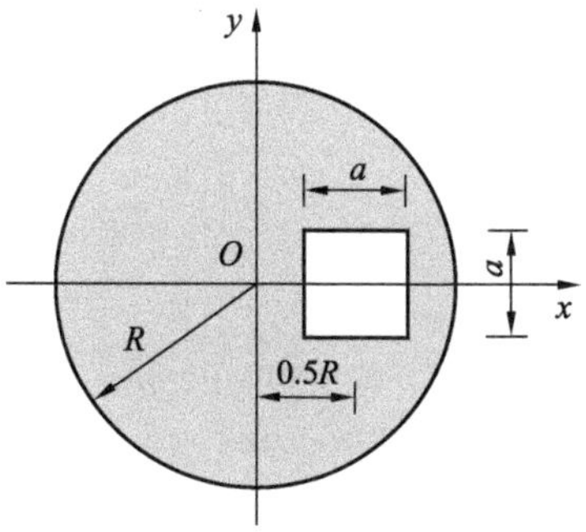

Fig. 3-37

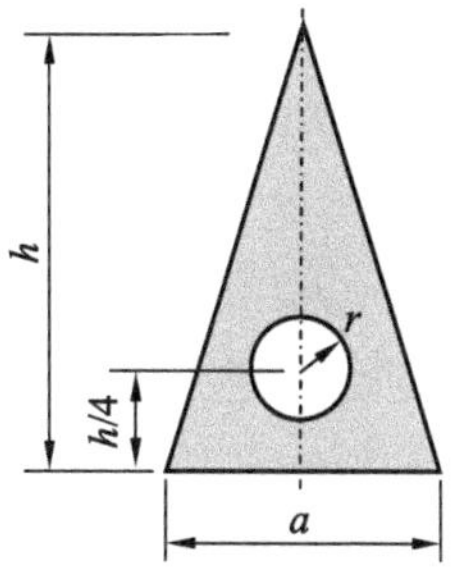

Fig. 3-38

Chapter 4

Friction

Objectives

In this chapter, the concept of friction and rolling resistance is introduced. The specific applications of frictional force on engineering are analyzed. After studying this chapter, students should be able to apply the Coulomb theory of friction to determine the forces in systems with contact and understand how to carry out equilibrium analysis of rigid bodies subjected to frictional force.

4.1 Sliding Friction

4.1.1 Theory of Static Friction

So far, it has been assumed that all bodies considered have *smooth* surfaces. But the friction is of great practical relevance. It is static friction that enables motion on solid ground. For instance, the car moves on the surface of the road, the frictional forces at the contact areas need to decelerate the car. This chapter, we will study the effects of friction. *Friction* may be defined as a force that resists the movement of two contacting surfaces that slide relative to one another. This force always acts *tangent* to the surface at point of contact with other bodies and is directed so as to oppose the possible or existing motion between the surfaces.

In general, two types of friction can occur between surfaces. *Fluid friction* exists when the contacting surfaces are separated by a film of fluid. Lubricants can significantly decrease friction in the case of moveable machine parts. But the fluid friction is studied in fluid mechanics. The following investigations are restricted to the case of *dry friction* occurring due to the roughness of any solid body's surface. This type of friction is often called *Coulomb friction*.

(1) Static Frictional Force

Fig. 4-1a shows a block of uniform weight $\boldsymbol{G}$ resting on a rough horizontal surface. A horizontal pulling force $\boldsymbol{F}$ applied to the block. As long as the force $\boldsymbol{F}$ is smaller than a certain *limit*, the block tends to slide but stays at rest and equilibrium. Draw the free-body diagram of the block as shown in Fig. 4-1b, for equilibrium, the normal forces $\boldsymbol{F}_N$ must act *upward* to balance the block's weight $\boldsymbol{G}$ (notice that $\boldsymbol{F}_N$ acts a distance x to the right of the line of action of $\boldsymbol{G}$), and a *tangential force* $\boldsymbol{F}_s$ between the surface and the block due to the surfaces' roughness acts to the left to prevent the block from moving. This tangential force is called *static frictional force*. Using the equilibrium conditions along the x axis, we have

$$\sum F_x = 0, \; F - F_s = 0$$

$$F_s = F$$

This states that the static frictional force $\boldsymbol{F}_s$ is a *reaction force* that can be determined from the equilibrium conditions without requiring additional assumptions. The direction of the static frictional force $\boldsymbol{F}_s$ *always opposes* to the direction of the motion or the tendency to motion. Using the moment equilibrium about point C, we can determine the distance x.

$$\sum M_C = 0, \; G \cdot x - F \cdot h = 0$$

$$x = \frac{Fh}{G}$$

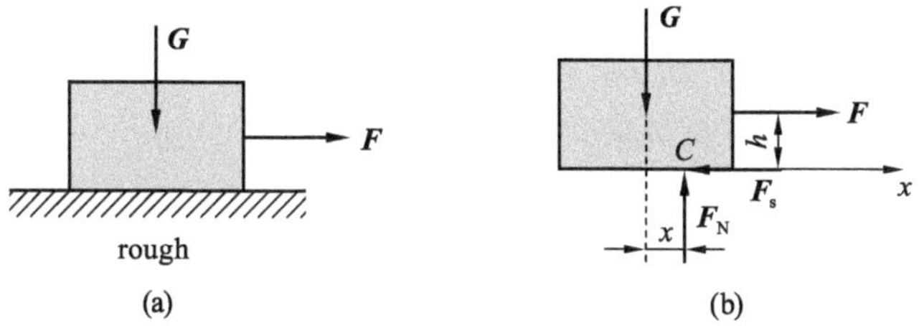

Fig. 4-1

(2) Coulomb's Theory of Friction

In Fig. 4-1, as the pulling force $\boldsymbol{F}$ is slowly increased, the static frictional force $\boldsymbol{F}_s$ increases correspondingly until it attains its *maximum value* $\boldsymbol{F}_{smax}$, called the *limiting static frictional force*. Then the block is on the *impending sliding* since any further increase in $\boldsymbol{F}$ cause the block to move. *Charles Augustine de Coulomb* showed in his experiments that this *limiting static frictional force* F_{smax} is approximation proportional to the normal force $\boldsymbol{F}_N$.

$$F_{smax} = \mu_s F_N \tag{4.1}$$

where the proportionality μ_s is commonly referred to the *coefficient of static friction*.

It is dimensionless and depends solely on the roughness of surfaces in contact, irrespective of their size. Typical values for μ_s of different configurations found in many engineering handbooks, are given in Tab. 4-1.

A body will be at rest as long as the *condition of static friction* is fulfilled

$$0 \leqslant F_s \leqslant F_{smax} \tag{4.2}$$

Tab. 4-1 Coefficients of sliding friction

Contact materials	Coefficient of static friction		Coefficient of kinetic friction	
	Dry	Oiled	Dry	Oiled
Steel on steel	0.15	0.10～0.12	0.10	0.05～0.10
Ski on snow	0.10～0.30		0.04～0.20	
Steel on cast iron	0.18	0.10	0.16	0.05～0.15
Steel on bronze	0.15	0.10～0.15	0.15	0.10～0.15
Cast iron on cast iron	0.45	0.25	0.15	0.07～0.12
Bronze on bronze	0.20	0.10	0.18	0.07～0.10
Leather on metal	0.30～0.50	0.15	0.60	0.15
Wood on wood	0.40～0.60	0.10	0.20～0.50	0.07～0.15
Steel on ice	0.03～0.05		0.015	

4.1.2 Kinetic Friction

If the force $\boldsymbol{F}$ acting on the block is increased so that it becomes slightly greater than *the limiting static frictional force* $\boldsymbol{F}_{smax}$, the block will begin to slide on the surface with increasing speed, as shown in Fig. 4-2a. Draw the free-body diagram of the block as shown in Fig. 4-2b, again, there is a frictional force transferred between the block and the base, called the *kinetic frictional force* F_d. Since it tends to prevent the movement of the block, its direction is opposed to the direction of the motion.

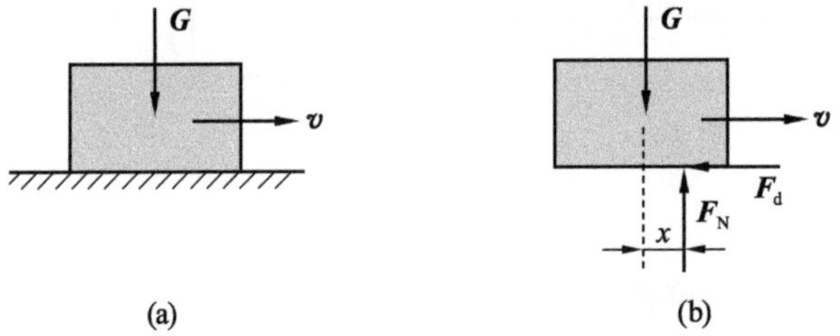

Fig. 4-2

Coulomb demonstrated experimentally that the *kinetic frictional force* $\boldsymbol{F}_d$ is directly proportional to the resultant normal force

$$F_d = \mu_d F_N \tag{4.3}$$

Here the proportionality μ_d is called the *coefficient of kinetic friction*. It is also dimensionless and depends solely on the roughness of surfaces in contact. In general, its value is smaller than the formerly introduced coefficient of static friction μ_s (compare Tab. 4-1).

$$\mu_d < \mu_s$$

In contrast to static frictional forces, the sense of direction of kinetic frictional forces cannot be assumed arbitrarily and its sense is opposite to that of slip velocity.

The above effects regarding friction can be summarized as follow:

① Static friction. $\boldsymbol{F}_s < \mu_s \boldsymbol{F}_N$. The body stays at rest, the static frictional force $\boldsymbol{F}_s$ can be calculated from the equilibrium conditions.

② Limiting friction. $\boldsymbol{F}_s = \mu_s \boldsymbol{F}_N$. The body is still at rest but on the verge of moving. *Limiting static frictional force* $\boldsymbol{F}_s$ reaches a maximum value.

③ Kinetic friction. $\boldsymbol{F}_d = \mu_d \boldsymbol{F}_N$. If the body slips, the kinetic frictional force $\boldsymbol{F}_d$ acts as an active force at the contacting surface.

Example 4-1

A box with weight $G=200$ N resting on a rough horizontal surface, in Fig. 4-3a. The coefficient of static friction is $\mu_s=0.5$, the coefficient of kinetic friction is $\mu_d=0.4$. If a force $\boldsymbol{F}$ is applied to the box, determine the frictional force if $F=90$ N and $F=110$ N, respectively.

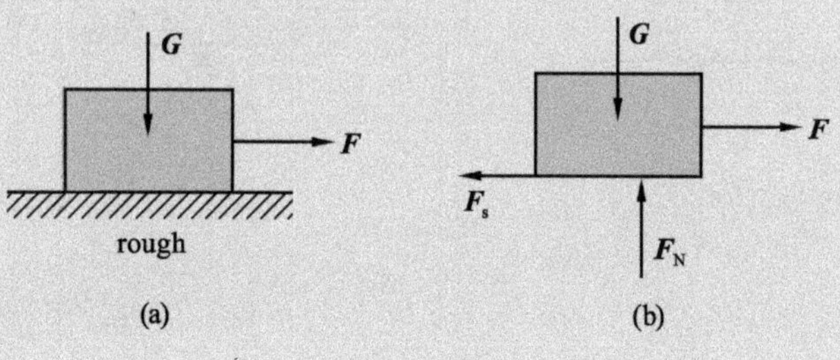

Fig. 4-3

Solution:

The free body diagram is shown in Fig. 4-3b, in order to calculate the frictional force, we need to judge that it is the force of static friction or kinetic friction.

(1) If F=90 N

$$F_N = G = 200 \text{ N}$$

$$F_{smax} = \mu_s F_N = 200 \times 0.5 = 100 \text{ N}$$

Since, $F < F_{smax}$, the box is at rest, using the *equation of equilibrium*

$$\sum F_x = 0,\ F - F_s = 0$$

$$F_s = 90 \text{ N} \qquad \textit{Ans.}$$

(2) If F=110 N

Since, $F > F_{smax}$, the box starts to move, we need to use the *kinetic frictional force* Eq. (4.3)

$$F_d = \mu_d F_N = 200 \times 0.4 = 80 \text{ N} \qquad \textit{Ans.}$$

4.2 Angle of Static Friction and Self-Locking

As shown in Fig. 4-4a, the normal force $\boldsymbol{F}_N$ and the static frictional force $\boldsymbol{F}_s$ can be combined into a *resultant force* $\boldsymbol{F}_{Rs}$. Its direction is defined by the angle φ, from the figure,

$$\tan \varphi = \frac{F_s}{F_N} \tag{4.4}$$

The angle φ increases as the static frictional force $\boldsymbol{F}_s$ is increased. Once the $\boldsymbol{F}_s$ attains its maximum value (*limiting static frictional force*), $\boldsymbol{F}_s = \boldsymbol{F}_{smax}$, then the limit angle φ_s, called *angle of static friction*, as shown in Fig. 4-4b, yields

$$\tan \varphi_s = \frac{F_{smax}}{F_N} = \frac{\mu_s F_N}{F_N} = \mu_s \tag{4.5}$$

In this case, the *angle of static friction* is related to the coefficient of static friction μ_s.

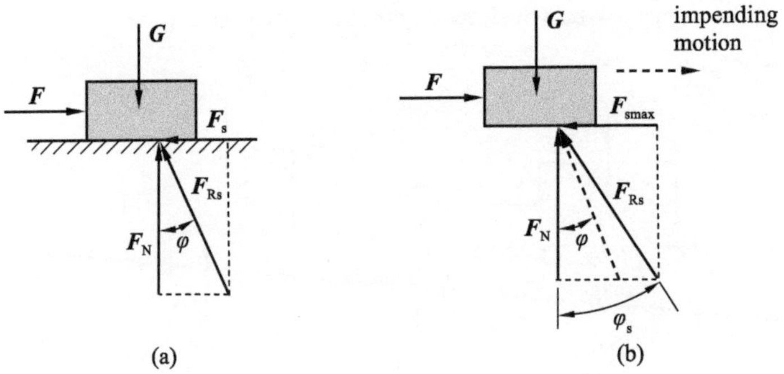

Fig. 4-4

In a plane, the static friction angle φ_s is drawn on both sides of the normal n, in Fig. 4-5a. In three-dimensional space, the static friction angle φ_s can also be interpreted by so-called "*static friction cone*" in Fig. 4-5b.

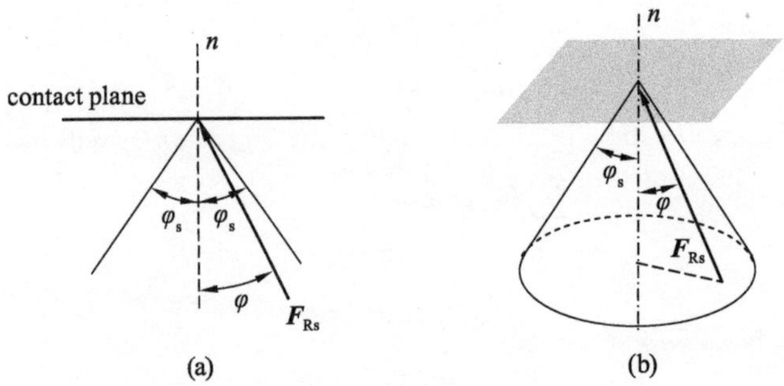

Fig. 4-5

If the resultant force $\boldsymbol{F}_{Rs}$ corresponding to an arbitrarily external load $\boldsymbol{F}_{RA}$ is located within this *angle of static friction*, that is, $\theta < \varphi_s$, a body stays at rest, Fig. 4-6a. If $\boldsymbol{F}_{Rs}$ lies outside of this *angle of static friction*, equilibrium is no longer possible, the body starts to move, Fig. 4-6b. A body can be at rest if the action line of the resultant of all applied forces lies within the *angle of static friction* (no matter how great the resultant is), this phenomenon is called *self-locking*. The conditions of self-locking is

$$\theta<\varphi_s$$

where θ is the angle between the *resultant* $\boldsymbol{F}_{RA}$ *of all applied forces* and the normal line n.

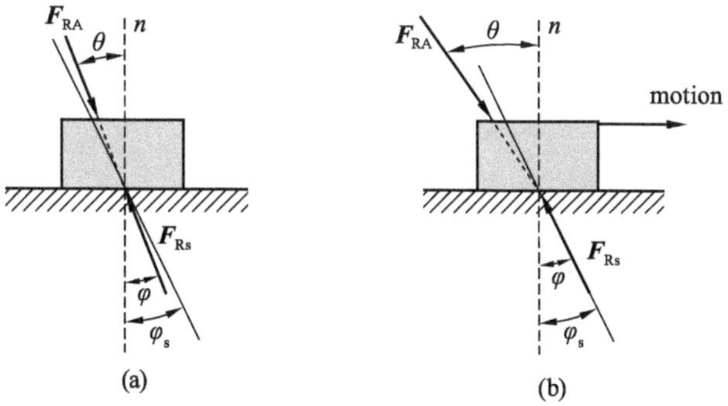

Fig. 4-6

In engineering, *self-locking* can be used to analyze and design as the wedges, jack, etc. For example, the square-threaded screws are often used as fasteners, e. g. , jack. If we unwind the thread by one revolution, we can obtain the slope with the *lead angle* θ, as shown in Fig. 4-7a. Let us consider the case of the screw that is subjected to upward impending motion. The *entire unraveled thread* can be represented as a block A whose free-body diagram is shown in Fig. 4-7b. The force $\boldsymbol{G}$ is the vertical force acting on the thread. If the angle of static friction φ_s is equal to θ, then the direction of the frictional force must be reversed so that resultant reaction $\boldsymbol{F}_{Rs}$ *acts on the other side of* $\boldsymbol{F}_N$ and vertically to balance $\boldsymbol{G}$. The screw will be on the verge of winding downward. If the angle of static friction φ_s is greater than θ, then the screw remains in place under any axial load $\boldsymbol{G}$. For this to occur, a screw is said to be *self-locking*.

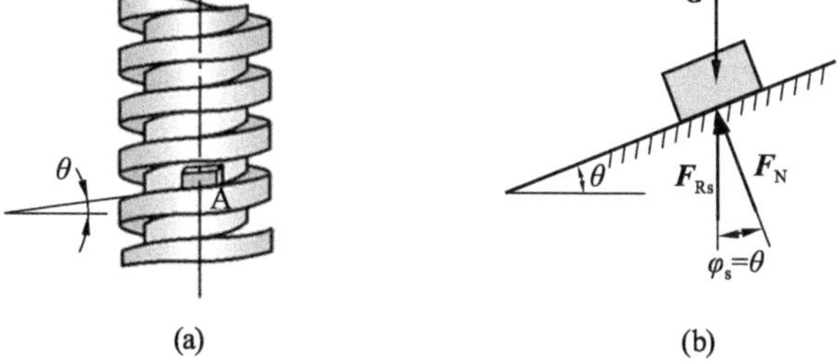

Fig. 4-7

Example 4-2

The jack shown in Fig. 4-8a has a square thread. Determine the largest *lead angle* θ so that the jack is self-locking. The coefficient of static friction is $\mu_s = 0.1$ in all contacting surfaces.

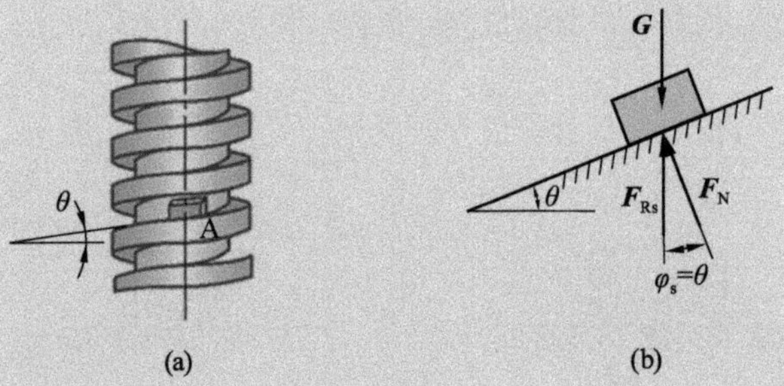

Fig. 4-8

Solution:

The free-body diagram is shown in Fig. 4-8b, if the jack is self-locking, the lead angle $\theta < \varphi_s$, the angle of static friction can be obtained by applying Eq. (4.5).

$$\tan \varphi_s = \mu_s = 0.1$$

$$\varphi_s = 5°43' \qquad \textit{Ans.}$$

The largest lead angle θ is $5°43'$, in engineering, the lead angle is usually $\theta = 4° \sim 4°30'$.

4.3 Equilibrium Problems Involving of Friction

If a rigid body is in equilibrium when it is subjected to a system of forces that includes the *effect* of friction, the forces system must satisfy not only the equations of equilibrium but also the laws that govern the frictional forces.

Determine the number of unknowns and compare this with the number of available equilibrium equations. If there are more unknowns than equations of equilibrium, it will be necessary to apply the frictional equation. If the equation $F_s = \mu_s F_N$ is to be used, it will be necessary to show $\boldsymbol{F}_s$ *acting in the proper direction on the free-body diagram*.

Example 4-3

A uniform block shown in Fig. 4-9a has a weight of $G = 200$ N. If a force of $F = 60$ N is applied to the block. Determine if it remains in equilibrium. The coefficient of static friction is $\mu_s = 0.3$.

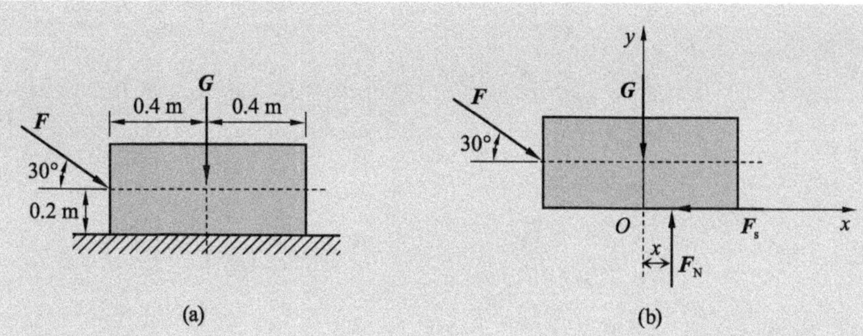

Fig. 4-9

Solution:

The free-body diagram of block is shown in Fig. 4-9b. The resultant normal force $\boldsymbol{F}_N$ act a distance x from the block's center line to counteract the tipping effect. There are *three* unknowns $\boldsymbol{F}_s$, $\boldsymbol{F}_N$ and x, which can be determined from the *three* equations of equilibrium.

$$\sum F_x = 0, \quad F\cos 30° - F_s = 0$$

$$\sum F_y = 0, \quad -F\sin 30° + F_N - G = 0$$

$$\sum M_O = 0, \quad F\sin 30° \times 0.4 - F\cos 30° \times 0.2 + F_N \cdot x = 0$$

Solving, we obtain

$$F_s = 52 \text{ N}, \quad F_N = 230 \text{ N}, \quad x = -7 \text{ mm}$$

Since x is negative, it indicates the *resultant* normal force $\boldsymbol{F}_N$ acts (slightly) to the *left* of the block's center line. Since $x < 0.4$ m, no tipping will occur. And the maximum frictional force is

$$F_{smax} = \mu_s F_N = 230 \times 0.3 = 69 \text{ N}$$

Since $F_s = 52$ N $<$ 69 N, no slip will occur, so the block remains in equilibrium.

Example 4-4

A block with weight $\boldsymbol{G}$ resting on a rough inclined plane, is subjected to an external horizontal force $\boldsymbol{F}$ as shown in Fig. 4-10a. Determine the range of $\boldsymbol{F}$ such that the block stays at rest. The coefficient of static friction μ_s, $\tan\theta > \mu_s$.

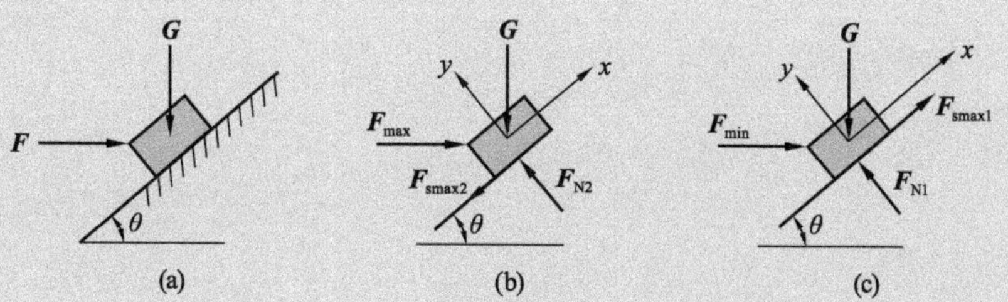

Fig. 4-10

Solution:

On the one hand, if $\boldsymbol{F}$ is a sufficiently large positive force, the block would move upwards. The static frictional force $\boldsymbol{F}_s$ then is oriented downwards, Fig. 4-10b. From the equilibrium conditions

$$\sum F_x = 0,\ F_{max}\cos\theta - F_{smax2} - G\sin\theta = 0$$

$$\sum F_y = 0,\ F_{N2} - F_{max}\sin\theta - G\cos\theta = 0$$

and the static friction condition

$$F_{smax2} = \mu_s F_{N2}$$

Solving, we get

$$F_{max} = \frac{\sin\theta + \mu_s\cos\theta}{\cos\theta - \mu_s\sin\theta}G \tag{a}$$

On the other hand, if $\boldsymbol{F}$ is too small, the block may slip downwards due to its weight. The static frictional force is then oriented upwards as shown in Fig. 4-10c. In this case, from the equilibrium equations

$$\sum F_x = 0,\ F_{min}\cos\theta + F_{smax1} - G\sin\theta = 0$$

$$\sum F_y = 0,\ F_{N1} - F_{min}\sin\theta - G\cos\theta = 0$$

and the static friction condition

$$F_{smax1} = \mu_s F_{N1}$$

Solving, we get

$$F_{min} = \frac{\sin\theta - \mu_s\cos\theta}{\cos\theta + \mu_s\sin\theta}G \tag{b}$$

Summarizing the results (a) and (b) yields the following admissible range for the force $\boldsymbol{F}$:

$$\frac{\sin\theta - \mu_s\cos\theta}{\cos\theta + \mu_s\sin\theta}G \leqslant F \leqslant \frac{\sin\theta + \mu_s\cos\theta}{\cos\theta - \mu_s\sin\theta}G \qquad \textit{Ans.}$$

Example 4-5

As shown in Fig. 4-11a, the drum can rotate without friction about point O, the coefficient of static friction between the brake bar and the drum is $\mu_s = 0.4$. Determine the magnitude of the force $\boldsymbol{F}$ needed to prevent the load of weight $G = 1\ 000$ N from falling downwards. Set $R = 0.5$ m, $r = 0.3$ m.

Solution:

Firstly, the free-body diagram for the drum is shown in Fig. 4-11b. The static frictional force F_s is opposite to the rotation sense. Summing the moment about the point O, we have

$$\sum M_O(F) = 0,\ -F_s \cdot R + G \cdot r = 0$$

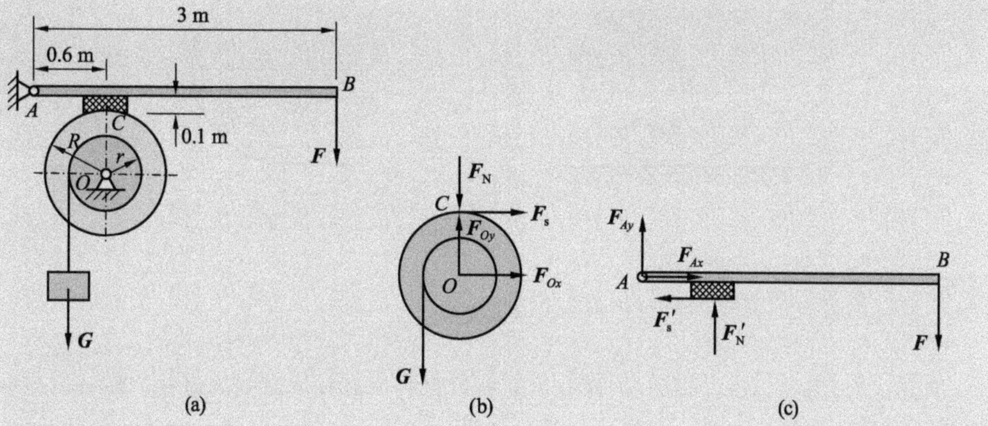

Fig. 4-11

Introducing them into the friction laws $F_s = \mu_s F_N$, we obtain

$$F_N = 1\ 500\ \text{N},\ F_s = 600\ \text{N} \qquad \textit{Ans.}$$

The free-body diagram for the lever bar is shown in Fig. 4-11c. Notice that the direction of the reaction friction is opposite. Summing the moment about the point A, we have

$$\sum M_A(F) = 0,\ F_N' \times 0.6 - F_s' \times 0.1 - F \times 3 = 0$$

Substituting the $F_N' = 1\ 500$ N, $F_s' = 600$ N into the equation, yields

$$F = 280\ \text{N} \qquad \textit{Ans.}$$

4.4 Rolling Resistance

Consider a *rigid* cylinder of weight $\boldsymbol{G}$ rolls at constant velocity along a *rigid* surface, a horizontal force $\boldsymbol{F}$ acts on the cylinder, as shown in Fig. 4-12a. The free-body diagram of the cylinder is shown in Fig. 4-12b, the normal force F_N and force of static friction F_s act at the point of contact, writing the equation of moments about point A we have

$$M_A = FR \tag{4.6}$$

An arbitrarily non-zero force $\boldsymbol{F}$ will initiate the cylinder motion. If the force $F > \mu_s F_N$, then the cylinder will slide on the plane. If $F < \mu_s F_N$, then the cylinder will roll on the plane.

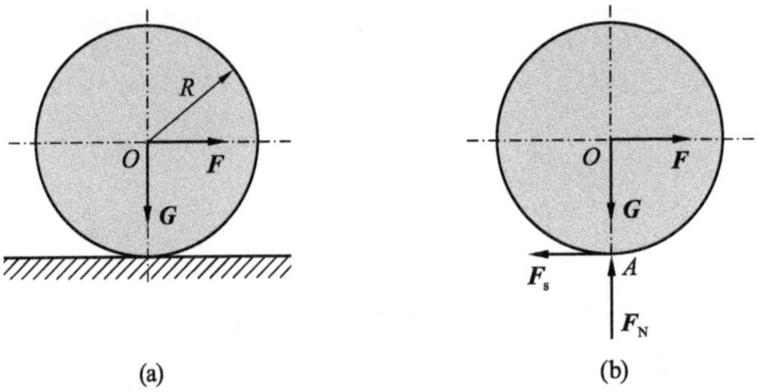

Fig. 4-12

Actually, however, no materials are perfectly rigid, and therefore the reaction of the surface on the cylinder consists of a distribution pressure. For example, consider the cylinder to be made of a very hard material, and the surface on which it rolls to be relatively soft, a horizontal driving force $\boldsymbol{F}$ is applied to the cylinder, Fig. 4-13a. Due to its weight, the cylinder compresses the surface underneath it and the surface is deformed. The free-body diagram of the cylinder is shown in Fig. 4-13a, the reaction distribution pressures acting on the cylinder are reduced to a resultant forces $\boldsymbol{F}_R$ and a couple moment $\boldsymbol{M}$ with respect to point A, Fig. 4-13b. The resultant force F_R can be resolved into a horizontal component $\boldsymbol{F}_s$ (static frictional force) and normal reaction $\boldsymbol{F}_N$, Fig. 4-13c. The static frictional force $\boldsymbol{F}_s$ opposite to the $\boldsymbol{F}$ since it retards the slide of the cylinder. The rotation direction of couple moment $\boldsymbol{M}$ is opposite to the rolling sense, so the couple is called *rolling resistance*.

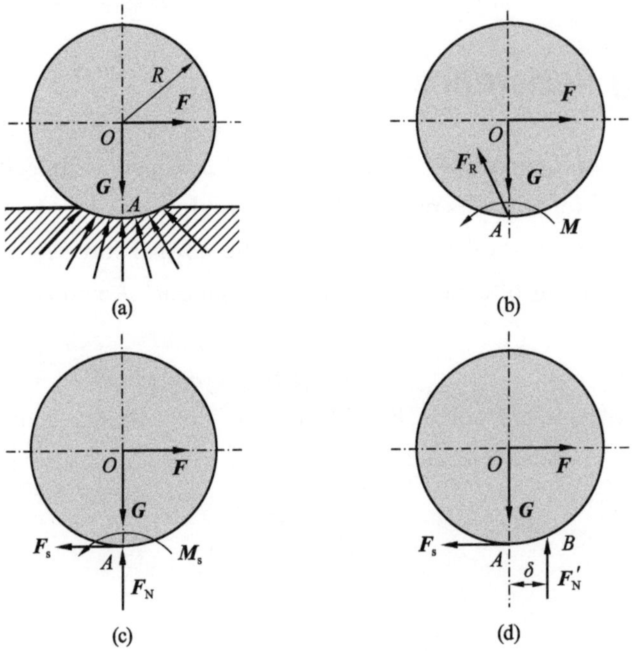

Fig. 4-13

Because the actual force $\boldsymbol{F}$ needed to overcome these effects is difficult to determine, we will consider that the normal pressure F_N and *rolling resistance* M acting on the cylinder may be replaced by a single force with a distance δ from point A, Fig. 4-13d, summing moments about point B gives

$$\delta=\frac{FR}{G} \tag{4.7}$$

Here the quantity δ we call a *coefficient of rolling resistance* or an *arm of rolling resistance*. Note that, as distinct from the dimensionless coefficient μ_s, the quantity δ possesses a dimension of length and characterizes the maximum distance at which the application point of reaction can move preserving the state of rest of the body. The motion will start when the moment FR overcomes the moment $M_{smax}=F_N\delta$, called a *moment of rolling resistance*.

A body would not roll as long as the *condition of rolling resistance* is fulfilled

$$0\leqslant M\leqslant M_{smax} \tag{4.8}$$

Several average values of the coefficient of rolling resistance δ for various materials of the cylinder and ground are given in Tab. 4-2.

Tab. 4-2 Coefficients of rolling resistance δ

Contact materials	δ/mm	Contact materials	δ/mm
Cast iron-Cast iron	0.5	Steel ball-Steel	0.01
Steel-Steel rail	0.05	Wooden-Wood	0.5~0.8
Wood-Steel	0.3~0.4	Tyre-Wood	2~10

Example 4-6

A homogeneous cylinder of weight $\boldsymbol{G}$ shown in Fig. 4-14a has a radius of $r=$ 25 mm and rests on an inclined plane of angle θ. If the coefficients of rolling resistance $\delta=0.5$ mm. Determine the maximum angle θ when the cylinder begins to roll down.

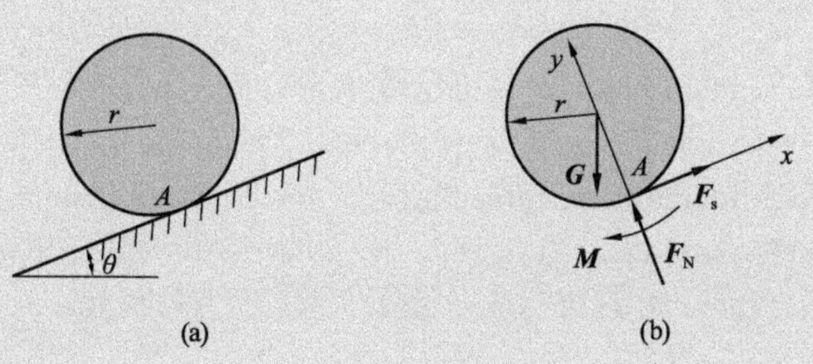

Fig. 4-14

Solution:

As shown on the free-body diagram, Fig. 4-14b, using the equation of equilibrium, we have

$$\sum F_y = 0,\ F_N - G\cos\alpha = 0$$

$$\sum M_A(\boldsymbol{F}) = 0,\ G\sin\alpha \cdot r - M = 0$$

And the *condition of rolling resistance*

$$M \leqslant M_{max} = \delta F_N$$

Solving, we obtain

$$\tan\alpha \leqslant \frac{\delta}{r}$$

$$\alpha \leqslant 1°9' \qquad \textit{Ans.}$$

Exercises

4-1 Two blocks have a weight of $G_A = 500$ N and $G_B = 200$ N, respectively, in Fig. 4-15. The coefficient of static friction between the surfaces of block A and B is $\mu_{s1} = 0.2$, the coefficient of static friction between the horizontal surfaces and block B is $\mu_{s2} = 0.10$, determine maximum force F_{max} that is applied to block A without causing either block to move.

4-2 If $F = 200$ N, determine the frictional force between the 500 N box and the ground, in Fig. 4-16. The coefficient of static friction between the box and the ground is $\mu_s = 0.3$.

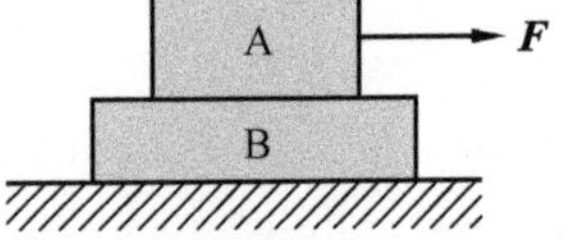

Fig. 4-15

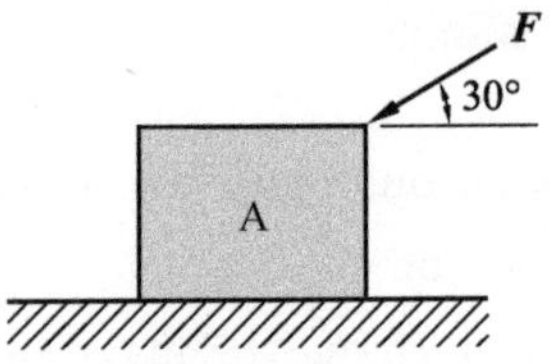

Fig. 4-16

4-3 A force $\boldsymbol{F}$ acts on block of weight $\boldsymbol{G}$ in a inclined angle $\theta = 25°$, in Fig. 4-17. If $F = G$, determine if the block is at rest or motion. The angle of static friction $\varphi_s = 20°$.

4-4 Determine the minimum force $\boldsymbol{F}$ to prevent the 30 kg rod AB from sliding, in Fig. 4-18. The contact surface at B is smooth, whereas the coefficient of static friction between the rod and the wall at A is $\mu_s = 0.2$.

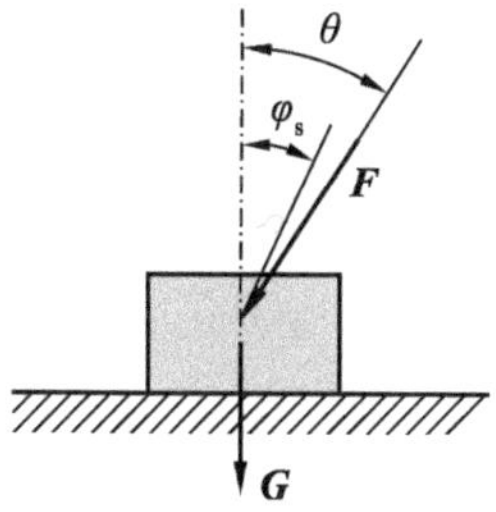

Fig. 4-17

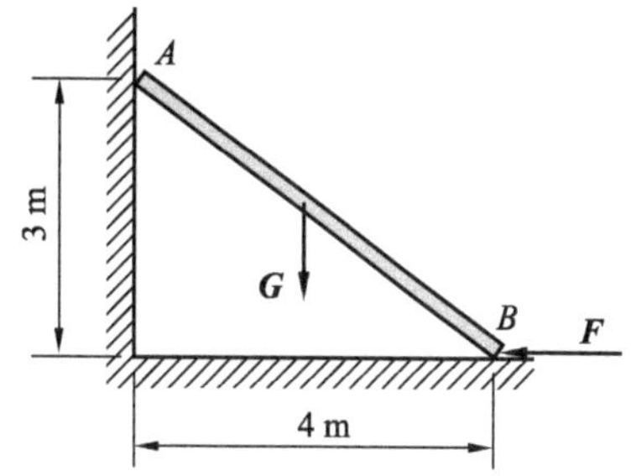

Fig. 4-18

4-5 In Fig. 4-19, the eccentric clamp device is used to exert a large normal force to compress the workpiece. If the coefficient of static friction is μ_s and the diameter of eccentric wheel is d, determine the required eccentricity e to keep self-lock.

4-6 In Fig. 4-20, the cylinder of weight $G=500$ N is put in corner of a smooth wall, determine the maximum force F to keep equilibrium. The coefficient of static friction between the cylinder and horizontal surfaces is 0.25. Set $R=200$ mm, $r=100$ mm.

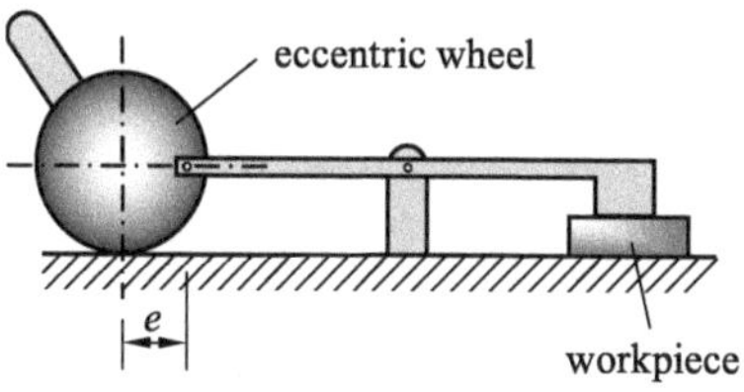

Fig. 4-19

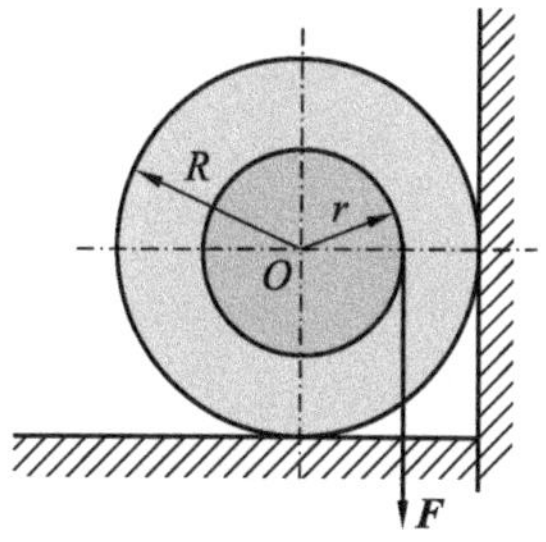

Fig. 4-20

4-7 In Fig. 4-21, the ladder of weight $G_1=200$ N is placed against a rough vertical wall with angle $\theta=60°$. A man with weight of $G_2=650$ N stands on the ladder. Determine the maximum position s that he can reach on the ladder. The coefficient of static friction between the ladder and wall or floor is $\mu_s=0.25$. Set $1=4$ m.

4-8 In Fig. 4-22, if the force $F=500$ N, determine the distance x of the force $\boldsymbol{F}$ needed to prevent the ball of weight $G=400$ N from falling. The coefficient of static friction between the ball and the lever is $\mu_s=0.20$. Neglect the weight of the angle lever.

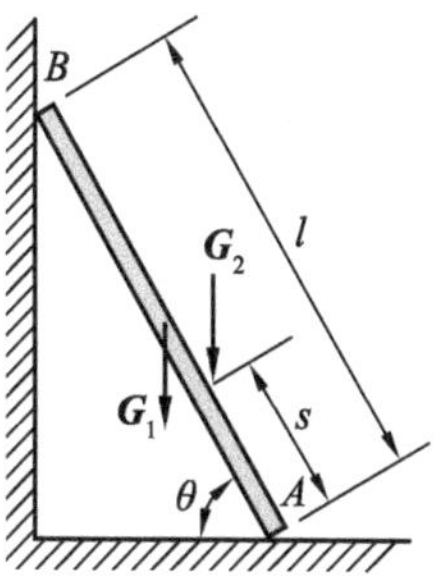

Fig. 4-21

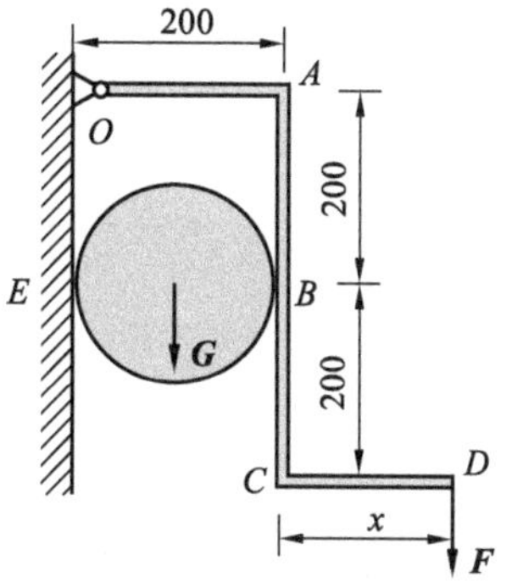

Fig. 4-22

4-9 In Fig. 4-23, the coefficient of static friction between the drum and brake bar is $\mu_s=0.4$. If the moment $M=35$ N · m, determine the smallest force $\boldsymbol{F}$ that needs to be applied to the brake bar in order to prevent the drum from rotating. Neglect the weight and thickness of the brake bar. The drum has a mass of 25 kg and radius of $r=0.125$ m.

4-10 In Fig. 4-24, a brick clamp is consisted of two angle bars that are pin connected at B. It is subjected a force $\boldsymbol{F}$ on point H to support a stack of bricks by applying a compressive force to the ends of the stack. If bricks have a total weight of $G=120$ N, determine the length b of brick clamp needed to prevent the bricks from falling. The coefficient of static friction between the bricks and clamp is $\mu_s=0.5$ and between any two brick is $\mu_s=0.5$.

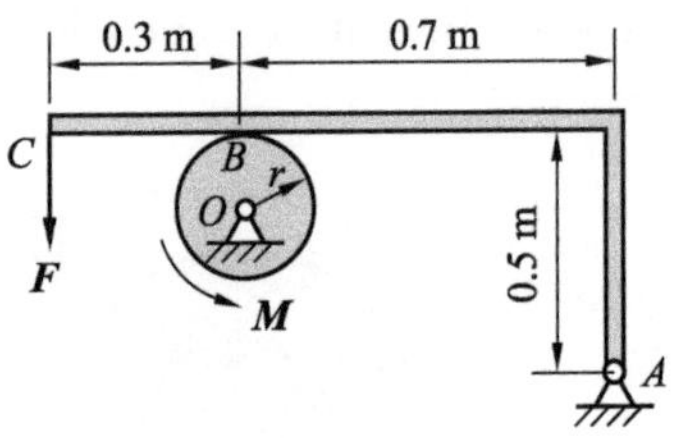

Fig. 4-23

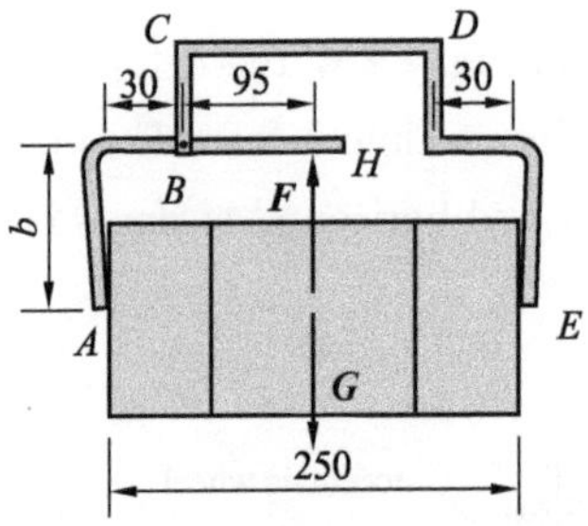

Fig. 4-24

4-11 In Fig. 4-25, the refrigerator has a weight of 180 N and rests on a floor for which $\mu_s=0.25$. If the man pushes horizontally on the refrigerator in the direction shown, determine the smallest magnitude of horizontal force needed to move it.

4-12 In Fig. 4-26, a force $\boldsymbol{F}$ is applied to the wheel. If the wheel has weight $\boldsymbol{G}$ and is rest on the rough ground, determine the moment of rolling resistance, static frictional force and reaction from the ground on the contact point A.

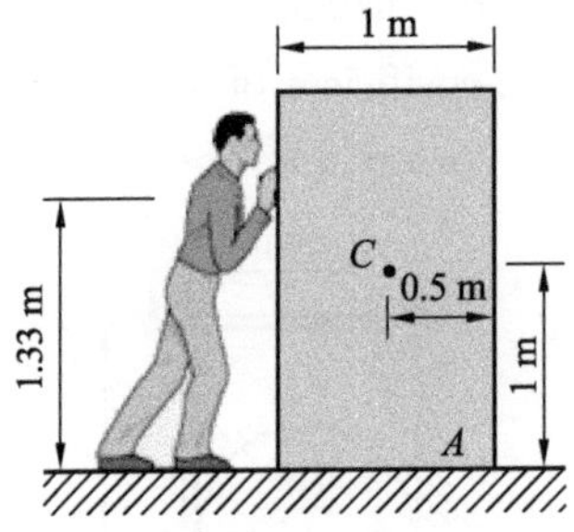

Fig. 4-25

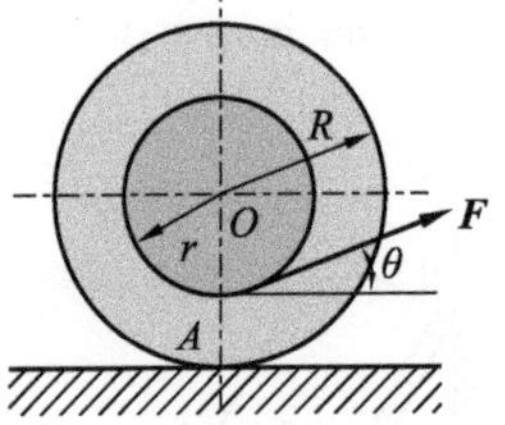

Fig. 4-26

Part 2 Kinematics

Engineering mechanics is divided into two areas of study, namely, *statics* and *dynamics*. Statics is concerned with the equilibrium of a body that is either at rest or moves with constant velocity. Here, we will consider dynamics which deals with the bodies in motion. It is further sub-divided into the subjects of *kinematics* and *kinetics*. *Kinematics* is the study of the geometric aspects of the motion, and kinetics is the analysis of the interplay between forces and motion. To develop these principles, the dynamics of a particle will be discussed first, followed by topics in rigid-body dynamics in two dimensions.

In the study of dynamics, we will use many concepts that we have already introduced in the study of statics, e. g. space, mass, force, moment. Fundamental concepts from statics such as section, the action-reaction law, and the force parallelogram law will also be employed. In the solution of concrete problems, the free-body diagrams will play a critical role, just as they did in the study of statics. For the study of motion, we will further introduce a new fundamental variable, time, and define new dynamical concepts and dynamical laws.

Before we study the interplay of forces and motion, we will first study the purely geometric aspects of motion, that is, "*kinematics*". In this regard, the notions of trajectory, velocity, and acceleration are introduced. Depending upon the type of motion (e. g. rectilinear, planar), we will describe these concepts using a variety of variables and coordinate systems. To develop these principles, the kinematics of a particle will be discussed first, followed by topics in the planar motion of a rigid body, in which the principles are applied first to simple, then to more complicated situations.

Chapter 5

Kinematics of a Particle

Objectives

In this chapter, we will first introduce the concepts of position, displacement, velocity, and acceleration. Further, the particle motions along a straight line or a curved path using different coordinate systems are discussed. After studying, students will learn how to describe the motion of a particle by its position, velocity, and acceleration in different coordinate systems and how such quantities can be determined.

5.1 General Curvilinear Motion

The kinematics of a particle is characterized by specifying the particle's *position*, *velocity*, and *acceleration* at any given instant.

5.1.1 Position

Consider a particle located at a point P along a space curve. The position of a particle in space is uniquely described by its *position vector* $\boldsymbol{r}$, Fig. 5-1a. This vector shows the instantaneous location of P relative to a fixed reference point in space O. If P changes location with time t, then $\boldsymbol{r}(t)$ describes the *trajectory* or *path* of P.

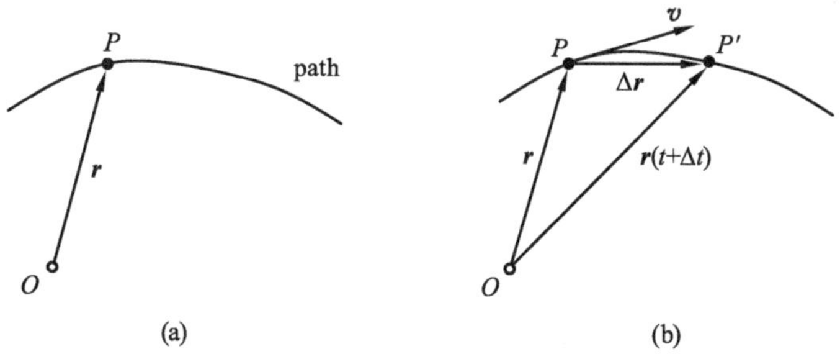

Fig. 5-1

$$\boldsymbol{r}=\boldsymbol{r}(t) \tag{5.1}$$

Notice that both the magnitude and direction of *position vector* $\boldsymbol{r}$ will change as the particle moves along the curve.

5.1.2 Velocity

Suppose that during a tiny time interval Δt, the particle moves from point P to the position P' along the curve, Fig. 5-1b. The change in the position vector $\Delta\boldsymbol{r}$ (*displacement*) over the time interval Δt is given by $\Delta\boldsymbol{r}=\boldsymbol{r}(t+\Delta t)-\boldsymbol{r}(t)$. The *velocity* of the particle is defined as the limit of the change in position vector $\Delta\boldsymbol{r}$ with respect to time,

$$\boldsymbol{v}=\lim_{\Delta t\to 0}\frac{\Delta\boldsymbol{r}}{\Delta t}=\frac{\mathrm{d}\boldsymbol{r}}{\mathrm{d}t}=\dot{\boldsymbol{r}} \tag{5.2}$$

Thus, the velocity $\boldsymbol{v}$ is the time derivative of the position vector $\boldsymbol{r}$, and time derivatives can be denoted by a superposed dot. *Velocity* is a vector, since $\mathrm{d}\boldsymbol{r}$ points in the direction of the tangent to the curve, the velocity is always *tangent* to this curve, Fig. 5-1b. The *magnitude* of the velocity vector $\boldsymbol{v}$, which is called the *speed*, will be discussed in the future course by introducing the arc-length to represent the space curve. The velocity has dimensions of distance/time and is often measured in units of m/s or km/h.

5.1.3 Acceleration

If in two neighboring positions P and P', at two different times t and $t+\Delta t$, the particle has velocities $\boldsymbol{v}(t)$ and $\boldsymbol{v}(t+\Delta t)$, Fig. 5-2a. Thus, the change in the velocity is given by $\Delta\boldsymbol{v}=\boldsymbol{v}(t+\Delta t)-\boldsymbol{v}(t)$. The *instantaneous acceleration* is defined as the limit of this change with respect to time,

$$\boldsymbol{a}=\lim_{\Delta t\to 0}\frac{\boldsymbol{v}(t+\Delta t)-\boldsymbol{v}(t)}{\Delta t}=\frac{\mathrm{d}\boldsymbol{v}}{\mathrm{d}t}=\frac{\mathrm{d}^2\boldsymbol{r}}{\mathrm{d}t^2}=\dot{\boldsymbol{v}}=\ddot{\boldsymbol{r}} \tag{5.3}$$

Therefore the acceleration $\boldsymbol{a}$ is the first derivative of velocity $\boldsymbol{v}$ and the second derivative of position vector $\boldsymbol{r}$. Acceleration is also a vector, but since $\Delta\boldsymbol{v}$, Fig. 5-2b does not have an obvious relation to the curve, in general it is not tangent to the path of motion. Acceleration is often measured in units of $\mathrm{m/s^2}$.

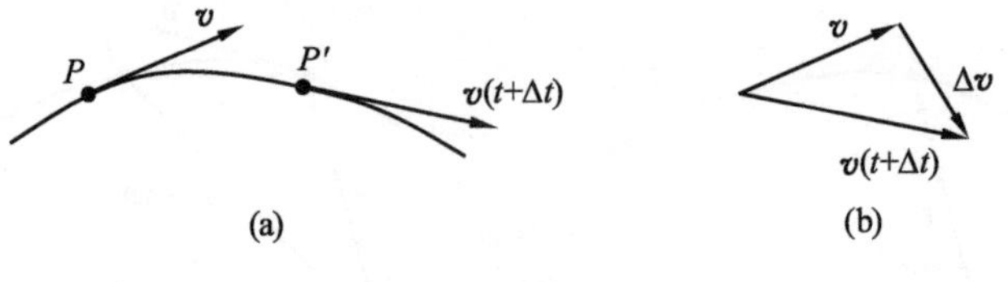

Fig. 5-2

Velocity and acceleration discussed is independent of a coordinate system. To solve specific problems, usually we need particular coordinates. In what follows, we will introduce two important coordinate systems.

5.2 Curvilinear Motion in Rectangular Coordinates

5.2.1 Position

A particle is at point P on the curve, if we want to describe motion in rectangular coordinates, we choose O as the origin of a fixed (in space) system x, y, z, with unit vectors $\boldsymbol{i}$, $\boldsymbol{j}$, $\boldsymbol{k}$ in the three coordinate directions in Fig. 5-3a, then its location is defined by the position vector

$$\boldsymbol{r}=x\boldsymbol{i}+y\boldsymbol{j}+z\boldsymbol{k} \tag{5.4}$$

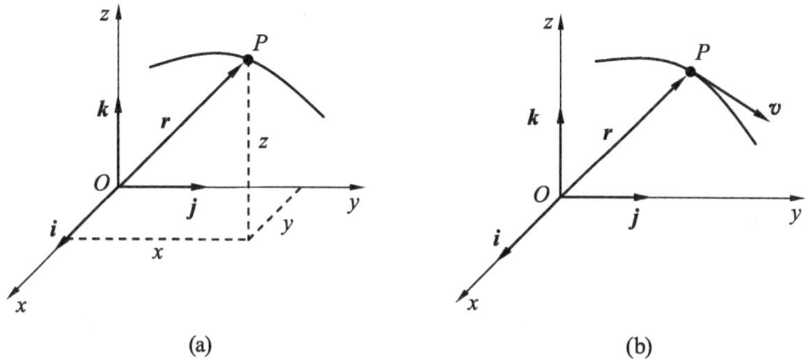

Fig. 5-3

When the particle moves, the x, y, z components of $\boldsymbol{r}=\boldsymbol{r}(t)$ will be functions of time,

$$x=f_1(t),\ y=f_2(t),\ z=f_3(t) \tag{5.5}$$

This is a parametric description of the trajectory with t as the parameter. If we eliminate time from the three component relations in Eq. (5.5), then we can obtain a time independent geometric description of the trajectory.

$$f(x,y,z)=0 \tag{5.6}$$

At any instant the magnitude of position vector $\boldsymbol{r}$ is defined as

$$r=\sqrt{x^2+y^2+z^2} \tag{5.7}$$

5.2.2 Velocity

The first time derivative of position vector $\boldsymbol{r}$ yields the velocity of the particle. If the x, y, z reference frame is fixed, then the direction and the magnitude of $\boldsymbol{i}$, $\boldsymbol{j}$, $\boldsymbol{k}$ do not change with time, which yields the final result,

$$\boldsymbol{v}=\frac{\mathrm{d}\boldsymbol{r}}{\mathrm{d}t}=\frac{\mathrm{d}x}{\mathrm{d}t}\boldsymbol{i}+\frac{\mathrm{d}y}{\mathrm{d}t}\boldsymbol{j}+\frac{\mathrm{d}z}{\mathrm{d}t}\boldsymbol{k} \tag{5.8}$$

where

$$v_x=\frac{dx}{dt}=\dot{x},\ v_y=\frac{dy}{dt}=\dot{y},\ v_z=\frac{dz}{dt}=\dot{z} \tag{5.9}$$

The "dot" notation represents the first-time derivatives of $x=x(t)$, $y=y(t)$, $z=z(t)$, respectively. Then the velocity has a magnitude

$$v=\sqrt{v_x^2+v_y^2+v_z^2} \tag{5.10}$$

The direction that is specified by

$$\cos(\boldsymbol{v},\boldsymbol{i})=\frac{v_x}{v},\ \cos(\boldsymbol{v},\boldsymbol{j})=\frac{v_y}{v},\ \cos(\boldsymbol{v},\boldsymbol{k})=\frac{v_z}{v} \tag{5.11}$$

This direction is always tangent to the path, as shown in Fig. 5-3b.

5.2.3 Acceleration

The acceleration of the particle is obtained by taking the first time derivative of Eq. (5.8) (or the second time derivative of Eq. (5.4)). We have

$$\boldsymbol{a}=\frac{d\boldsymbol{v}}{dt}=\frac{d^2\boldsymbol{r}}{dt^2}=\frac{d^2x}{dt^2}\boldsymbol{i}+\frac{d^2y}{dt^2}\boldsymbol{j}+\frac{d^2z}{dt^2}\boldsymbol{k}=a_x\boldsymbol{i}+a_y\boldsymbol{j}+a_z\boldsymbol{k} \tag{5.12}$$

where

$$a_x=\frac{dv_x}{dt}=\frac{d^2x}{dt^2},\ a_y=\frac{dv_y}{dt}=\frac{d^2y}{dt^2},\ a_z=\frac{dv_z}{dt}=\frac{d^2z}{dt^2} \tag{5.13}$$

The acceleration has a magnitude

$$a=\sqrt{a_x^2+a_y^2+a_z^2} \tag{5.14}$$

The direction specified by the unit vector. Since $\boldsymbol{a}$ represents the time rate of change in both the magnitude and direction of the velocity, in general, $\boldsymbol{a}$ will not be tangent to the path.

$$\cos(\boldsymbol{a},\boldsymbol{i})=\frac{a_x}{a},\ \cos(\boldsymbol{a},\boldsymbol{j})=\frac{a_y}{a},\ \cos(\boldsymbol{a},\boldsymbol{k})=\frac{a_z}{a} \tag{5.15}$$

5.2.4 Rectilinear Motion

Rectilinear motion is the simplest form of motion and it has many practical uses. For example, the *free fall* of a body in the earth's gravitational field or the travel of a train over a bridge is rectilinear motions.

If a particle P moves along a straight line, then we can assume without loss of generality that the x-axis is coincident with this line as shown in Fig. 5-4. The origin O on the path is a fixed point, and from this point the position vector $\boldsymbol{r}$ only has an x-component to specify the location of the particle at any given instant, likewise for the velocity $\boldsymbol{v}$ and the acceleration $\boldsymbol{a}$. Thus, we can dispense with the vector character of the position, velocity, and acceleration and obtain from

$$v=\frac{dx}{dt},\ a=\frac{dv}{dt}=\frac{d^2x}{dt^2} \tag{5.16}$$

This states that in a case of rectilinear motion, if the position x is known as a function of time t, then the instantaneous velocity and acceleration can be found via differentiation by this equation. In the case that $\boldsymbol{v}$ or $\boldsymbol{a}$ is negative, this means that the velocity and the acceleration is in the negative x-direction. An acceleration that decreases the magnitude of the velocity is known as a "deceleration".

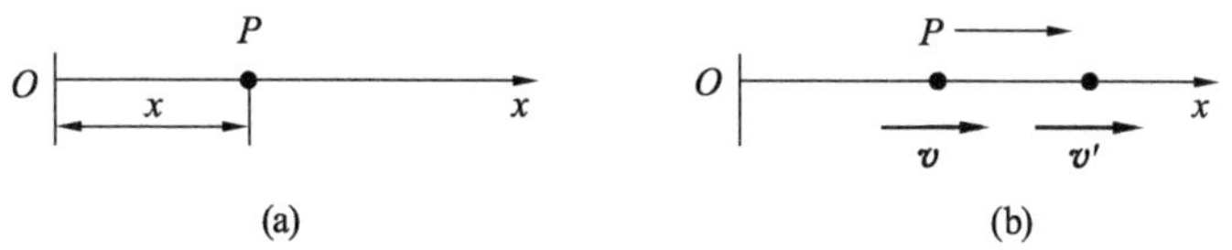

Fig. 5-4

① If the acceleration $\boldsymbol{a}=0$, then according to Eq. (5.16), $a=\mathrm{d}v/\mathrm{d}t=0$. Integration says that the velocity is constant.

$$v=\text{const.}$$

A motion with a constant velocity is known as a *uniform motion*. The position x can be found from $v=\mathrm{d}x/\mathrm{d}t$ via integration

$$x=x_0+vt \tag{5.17}$$

② If $a=\text{const.}$, a motion with a *constant acceleration* is called a *uniform acceleration*. Assume that $t_0=0$, the initial conditions for the velocity and position are

$$v(0)=v_0,\ x(0)=x_0$$

Then integration of Eq. (5.16), the velocity and position follow as

$$v=v_0+at \tag{5.18}$$

$$x=x_0+v_0t+\frac{1}{2}at^2 \tag{5.19}$$

So a constant acceleration a leads to a linear velocity $at+\boldsymbol{v}_0$ and to a quadratic position-time dependency x. In nature, the examples of *uniform acceleration* are *free fall* and other vertical motions in the earth's gravitational field. This acceleration is called the *earth's gravitational acceleration* $\boldsymbol{g}$. At the earth's surface, it has the value $g=9.81\ \mathrm{m/s^2}$, where small variations with geographical latitude are neglected.

③ If $a=a(t)$, in this case, the velocity v and the position x can be found via two successive integrations of Eq. (5.16) with respect to time. With initial conditions $v(0)=v_0$, $x(0)=x_0$ one has

$$v=v_0+\int_{t_0}^{t}a(t)\,\mathrm{d}t \tag{5.20}$$

$$x=x_0+\int_{t_0}^{t}v(t)\,\mathrm{d}t \tag{5.21}$$

Example 5-1

The car in Fig. 5-5 moves in a straight line such that for a short time, its velocity is defined by $\boldsymbol{v}=(3t^2+2t)$ m/s, where t is in seconds. Determine its position and acceleration when $t=3$ s. When $t=0$, $x=0$.

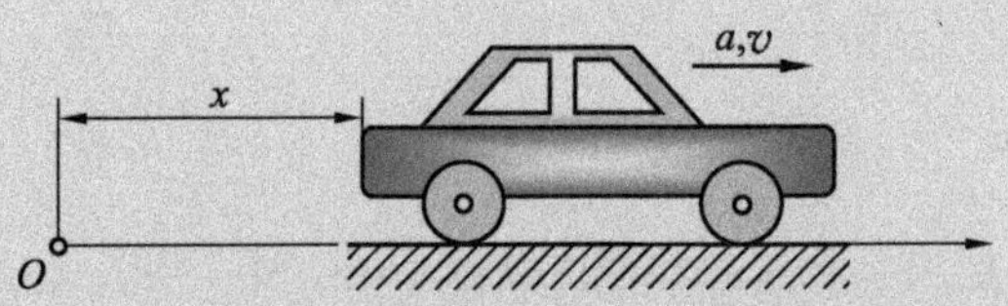

Fig. 5-5

Solution:

The position coordinate x extends from the fixed origin O to the car, positive to the right. When $t=0$, $x=0$, we have

$$v=\frac{\mathrm{d}x}{\mathrm{d}t}=3t^2+2t$$

$$\int_0^s \mathrm{d}x=\int_0^t (3t^2+2t)\mathrm{d}t$$

$$x=t^3+t^2$$

When $t=3$ s,

$$x=3^3+3^2=36 \text{ m} \qquad \textit{Ans.}$$

The acceleration is determined from $a=\mathrm{d}\boldsymbol{v}/\mathrm{d}t$, we have

$$a=\frac{\mathrm{d}\boldsymbol{v}}{\mathrm{d}t}=\frac{\mathrm{d}}{\mathrm{d}t}(3t^2+2t)=6t+2$$

When $t=3$ s,

$$a=6\times 3+2=20 \text{ m/s}^2 \qquad \textit{Ans.}$$

Note: The formulas for constant acceleration cannot be used to solve this problem, because the acceleration is a function of time.

Example 5-2

The crank OC in Fig. 5-6 rotates with constant angular velocity ω about fixed axis O and its another end C is pin connected with the middle of the link AB. The link is guided by two blocks at A and B, which can slide along two mutually perpendicular slots. Determine the trajectory, velocity and acceleration of point M. Set $OC=AC=BC=l$, $MC=a$, $\varphi=\omega t$.

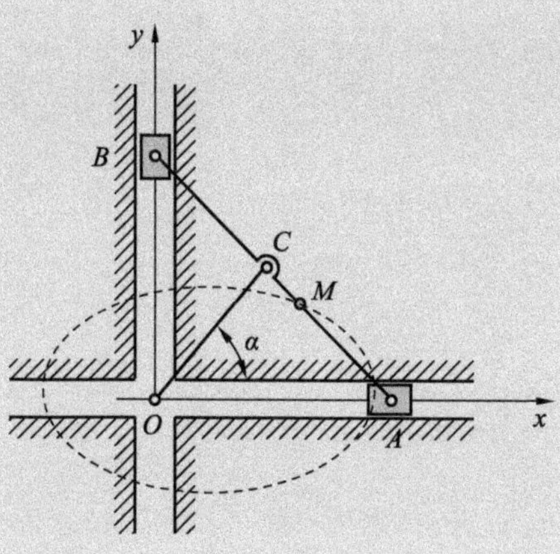

Fig. 5-6

Solution:

As shown in Fig. 5-6, the origin of the coordinates is set at point O. First, we need to describe the position of the point M.

$$x=(OC+CM)\cos\varphi=(l+a)\cos\omega t$$

$$y=AM\sin\varphi=(l-a)\sin\omega t$$

These are two parametric equations. We can eliminate time t from two component relations, and get geometric description of the trajectory.

$$\frac{x^2}{(l+a)^2}+\frac{y^2}{(l-a)^2}=1 \qquad \textit{Ans.}$$

It can be seen that the trajectory of point M is ellipse, as shown in Fig. 5-6.

The velocity component in the x, y direction is

$$v_x=\frac{\mathrm{d}x}{\mathrm{d}t}=-\omega(l+a)\sin\omega t$$

$$v_y=\frac{\mathrm{d}y}{\mathrm{d}t}=\omega(l-a)\cos\omega t$$

The magnitude of velocity is therefore

$$v=\sqrt{v_x^2+v_y^2}=\sqrt{\omega^2(l+a)^2\sin^2\omega t+\omega^2(l-a)^2\cos^2\omega t}$$

$$=\omega\sqrt{l^2+a^2-2al\cos 2\omega t} \qquad \textit{Ans.}$$

The direction is tangent to the path

$$\cos(\boldsymbol{v},\boldsymbol{i})=\frac{v_x}{v}=\frac{-(l+a)\sin\omega t}{\sqrt{l^2+a^2-2al\cos 2\omega t}}$$

$$\cos(\boldsymbol{v},\boldsymbol{j})=\frac{v_y}{v}=\frac{(l-a)\cos\omega t}{\sqrt{l^2+a^2-2al\cos 2\omega t}}$$

The relationship between the acceleration components is determined using the successive time derivatives, Eq. (5.13). We have

$$a_x=\frac{\mathrm{d}v_x}{\mathrm{d}t}=\frac{\mathrm{d}^2x}{\mathrm{d}t^2}=-\omega^2(l+a)\cos\omega t$$

$$a_y = \frac{dv_y}{dt} = \frac{d^2 y}{dt^2} = -\omega^2 (l-a)\sin \omega t$$

Thus, the magnitude of acceleration is

$$a = \sqrt{a_x^2 + a_y^2} = \sqrt{\omega^4 (l+a)^2 \cos^2 \omega t + \omega^4 (l-a)^2 \sin^2 \omega t}$$

$$= \omega^2 \sqrt{l^2 + a^2 + 2al\cos 2\omega t} \qquad \textit{Ans.}$$

The direction of $\boldsymbol{a}$ is

$$\cos(\boldsymbol{a}, \boldsymbol{i}) = \frac{a_x}{a} = \frac{-(l+a)\cos \omega t}{\sqrt{l^2 + a^2 + 2al\cos 2\omega t}}$$

$$\cos(\boldsymbol{a}, \boldsymbol{j}) = \frac{a_y}{a} = \frac{-(l-a)\sin \omega t}{\sqrt{l^2 + a^2 + 2al\cos 2\omega t}}$$

5.3 Normal and Tangential Coordinates

When the path of a particle is known, it is often convenient to describe the motion using *normal* and *tangential* coordinates which act normal and tangent to the path, respectively.

5.3.1 Normal and Tangential Components

Firstly, we will introduce a way of describing motion, consider the particle P shown in Fig. 5-7a, which moves in a plane along a fixed curve. The origin O on the path is a fixed point, and from this point the position coordinate s is used to specify the location of the particle at any given instant. The magnitude of s is the distance from O to the particle. Since the particle moves, s is a function of time

$$s = f(t) \tag{5.22}$$

Since the position of the particle can be designated by the position vector $\boldsymbol{r}$, the length of straight line segment $\Delta \boldsymbol{r}$ approaches the arc length Δs as $\Delta t \to 0$.

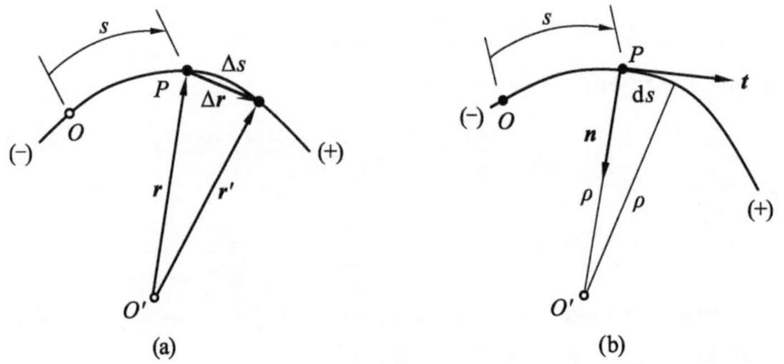

Fig. 5-7

Consider the particle shown in Fig. 5-7b, at a given instant it is at position s, measured from point O. We now consider a coordinate system that has its origin at the location of the particle P. The $\boldsymbol{t}$ axis is tangent to the curve at the point and is positive in the direction of increasing s. In this instant, the differential segments ds can be approximated by an arc that have a *radius of curvature* ρ and *center of curvature* O'. The normal axis $\boldsymbol{n}$ is perpendicular to the $\boldsymbol{t}$ axis with its positive sense directed *toward* the center of curvature O', Fig. 5-7b. This positive direction, which is always on the concave side of the curve, will be designated by the unit vector $\boldsymbol{n}$. The plane which contains the $\boldsymbol{n}$ and $\boldsymbol{t}$ axes is the so-called *osculating plane*.

If the particle P moves along a space curve, Fig. 5-8, then at a given instant, the $\boldsymbol{\tau}$ axis and $\boldsymbol{n}$ axis are uniquely specified in the osculating plane and are always perpendicular to one another. For spatial motion, a third unit vector, $\boldsymbol{b}$, defines the *binormal axis* which is perpendicular to $\boldsymbol{\tau}$ and $\boldsymbol{n}$, Fig. 5-8. The three unit vectors are related to one another by the vector cross product, e. g.,

$$\boldsymbol{b}=\boldsymbol{\tau}\times\boldsymbol{n} \tag{5.23}$$

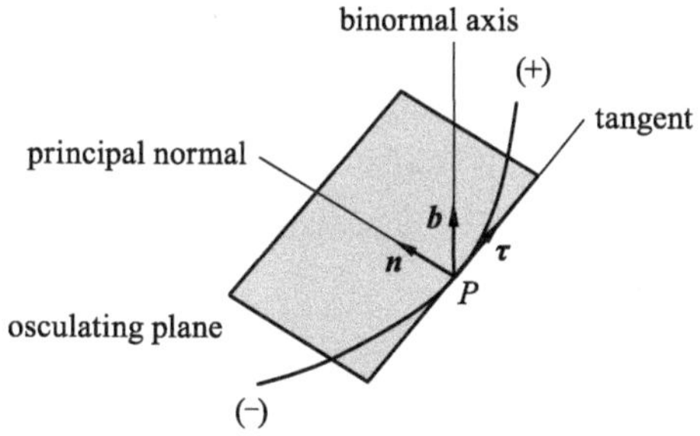

Fig. 5-8

5.3.2 Velocity

Using the arc-length $s(t)$, it follows from the expression for the vector, that the velocity is given by

$$\boldsymbol{v}=\frac{\mathrm{d}\boldsymbol{r}}{\mathrm{d}t}=\frac{\mathrm{d}\boldsymbol{r}}{\mathrm{d}s}\frac{\mathrm{d}s}{\mathrm{d}t} \tag{5.24}$$

Since d$\boldsymbol{r}$ points in the direction of the tangent and $|\mathrm{d}\boldsymbol{r}|=\mathrm{d}s$, one has $\mathrm{d}\boldsymbol{r}=\mathrm{d}s\boldsymbol{\tau}$. Noting that the *speed* $\boldsymbol{v}$ is determined by taking the time derivative of the path function $s(t)$.

$$v=\lim_{\Delta t\to 0}\frac{\Delta r}{\Delta t}=\lim_{\Delta t\to 0}\frac{\Delta s}{\Delta t}=\frac{\mathrm{d}s}{\mathrm{d}t}=\dot{s} \tag{5.25}$$

We get

$$\boldsymbol{v}=\frac{\mathrm{d}s}{\mathrm{d}t}\boldsymbol{\tau}=v\boldsymbol{\tau} \tag{5.26}$$

The velocity $\boldsymbol{v}$ has a direction that is always tangent to the path, Fig. 5-9.

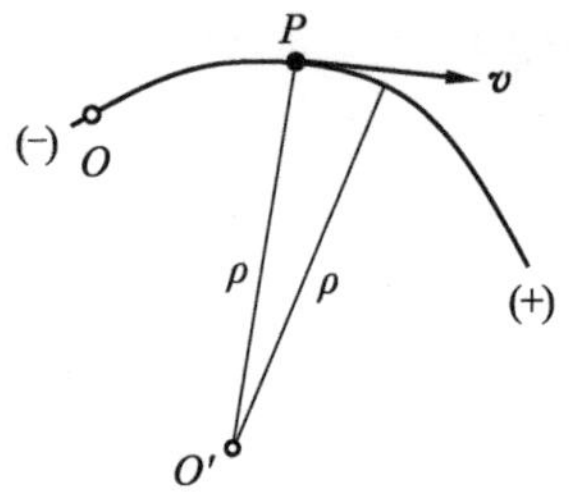

Fig. 5-9

5.3.3 Acceleration

The acceleration of the particle is the time derivative of the velocity Eq.(5.26), yields

$$\boldsymbol{a}=\frac{\mathrm{d}\boldsymbol{v}}{\mathrm{d}t}=\frac{\mathrm{d}}{\mathrm{d}t}(v\boldsymbol{\tau})=\frac{\mathrm{d}v}{\mathrm{d}t}\boldsymbol{\tau}+v\frac{\mathrm{d}\boldsymbol{\tau}}{\mathrm{d}t} \tag{5.27}$$

The unit vector $\boldsymbol{\tau}$ changes its direction by an angle $\mathrm{d}\varphi$ between two neighboring points P and P' on the path, Fig. 5-10a. The change $\mathrm{d}\boldsymbol{\tau}$ has a magnitude of $1 \cdot \mathrm{d}\varphi$ and points towards the center of curvature O', namely its direction is defined by $\boldsymbol{n}$, as shown in Fig. 5-10b. As the change of arc-length $\mathrm{d}s$ between P and P' can be expressed in terms of the angle $\mathrm{d}\varphi$ and the radius of curvature ρ ($\mathrm{d}s=\rho\mathrm{d}\varphi$), it follows that

$$\frac{\mathrm{d}\boldsymbol{\tau}}{\mathrm{d}t}=\frac{\mathrm{d}\varphi}{\mathrm{d}t}\boldsymbol{n}=\frac{1}{\rho}\frac{\mathrm{d}s}{\mathrm{d}t}\boldsymbol{n}=\frac{v}{\rho}\boldsymbol{n} \tag{5.28}$$

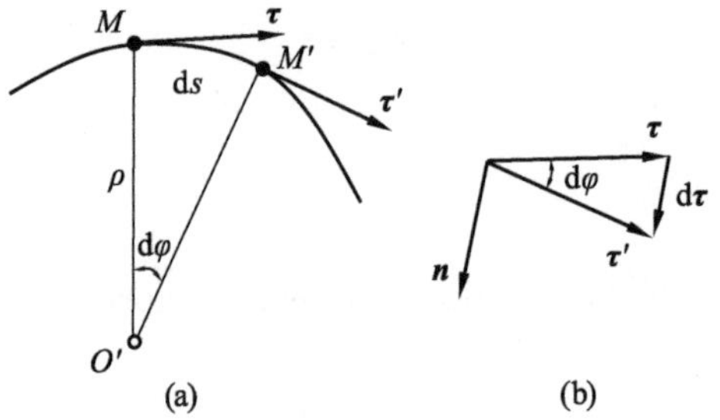

Fig. 5-10

Substituting into Eq.(5.27), gives an expression for the acceleration vector in terms of the *Normal* and *Tangential* frame

$$\boldsymbol{a}=a_{\mathrm{t}}\boldsymbol{\tau}+a_{\mathrm{n}}\boldsymbol{n}=\frac{\mathrm{d}v}{\mathrm{d}t}\boldsymbol{\tau}+\frac{v^2}{\rho}\boldsymbol{n} \tag{5.29}$$

where

$$a_{\mathrm{t}}=\frac{\mathrm{d}v}{\mathrm{d}t}=\dot{v},\quad a_{\mathrm{n}}=\frac{v^2}{\rho} \tag{5.30}$$

The acceleration is composed of two components: one is in the direction of the tangent to the path, namely the *tangential acceleration*, it is the result of the time rate of change in the *magnitude* of velocity, and one in the direction of the principal normal, namely the *normal acceleration*, it is the result of the time rate of change in the *direction* of the velocity. Note that the entire vector lies in the osculating plane. These two mutually perpendicular components are shown in Fig. 5-11. Therefore, the magnitude of acceleration is the positive value of

$$a=\sqrt{a_t^2+a_n^2}$$

$$\tan\alpha=\frac{|a_t|}{a_n} \tag{5.31}$$

In the special case of circular motion, $\rho=r=\text{const.}$, $s=r\varphi$ and $d\varphi/dt=\omega$. This gives the following velocity and acceleration components:

$$v=\dot{s}=r\omega$$

$$a_t=\dot{v}=r\dot{\omega}=\alpha r$$

$$a_n=\frac{v^2}{r}=r\omega^2 \tag{5.32}$$

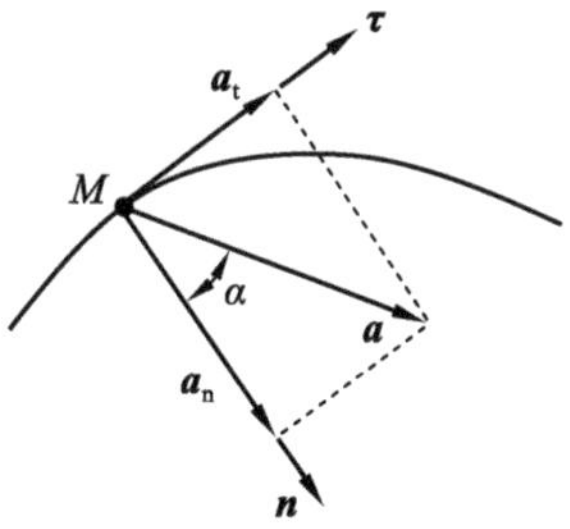

Fig. 5-11

Example 5-3

A train moves along a semicircular track that has a radius $R=800$ m, Fig. 5-12, the velocity of the train is $v_0=0$ when it enters the track, and the velocity $v=54$ km/h when it leaves the track after 2 min. If the train increases its speed at uniformly acceleration, determine its accelerations when it enters and leaves the semicircular track, respectively.

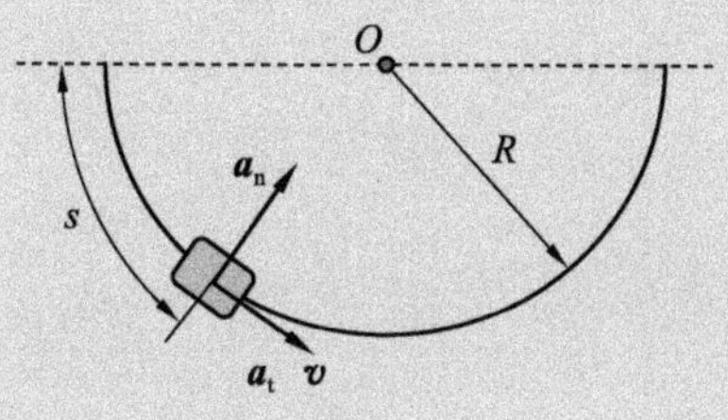

Fig. 5-12

Solution:

Since the train moves with a uniform acceleration along the track, so the tangential acceleration a_t is constant. Then we have

$$\frac{dv}{dt}=a_t=\text{const.}$$

Integrating,
$$\int_0^v dv=\int_0^t a_t dt$$

We get

$$v=a_t t$$

When $t=2$ min$=120$ s, $v=54$ km/h$=15$ m/s, we get

$$a_t=\frac{v}{t}=\frac{15}{120}=0.125\ \text{m/s}^2$$

At the enter point, $v=0$, so the normal acceleration a_n is zero, the acceleration of train is

$$a=a_t=0.125\ \text{m/s}^2 \qquad \textit{Ans.}$$

At the end point, the normal acceleration and tangent acceleration are not zero, we have

$$a_t=0.125\ \text{m/s}^2$$

$$a_n=\frac{v^2}{R}=\frac{15^2}{800}=0.281\ \text{m/s}^2$$

Therefore, the magnitude of acceleration at the end of track is

$$a=\sqrt{a_t^2+a_n^2}=\sqrt{0.125^2+0.281^2}=0.308\ \text{m/s}^2 \qquad \textit{Ans.}$$

Example 5-4

A race car C travels around the horizontal circular track that has a radius of 300 m, Fig. 5-13. If the car increases its speed at a constant tangent acceleration of 7 m/s², starting from rest, determine the time needed when it reaches an acceleration of 8 m/s² and its speed at this instant.

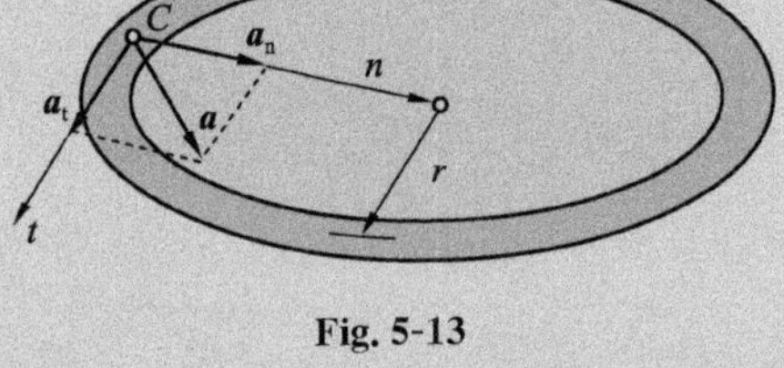

Fig. 5-13

Solution:

The origin of the $\boldsymbol{n}$ and $\boldsymbol{t}$ axes is coincident with the car at the instant considered as shown in Fig. 5-13. The $\boldsymbol{t}$ axis is in the direction of motion, and the positive $\boldsymbol{n}$ axis is directed toward the center of the circle. The velocity as a function of time must be determined first.

$$\boldsymbol{v}=\boldsymbol{v}_0+a_t t=7t \text{ m/s}$$

Then the normal acceleration is

$$a_n=\frac{\boldsymbol{v}^2}{r}=\frac{(7t)^2}{300}=0.163t^2 \text{ m/s}^2$$

The magnitude of acceleration can be related to its components. The time needed for the acceleration to reach 8 m/s² is therefore

$$a=\sqrt{a_t^2+a_n^2}$$

$$8=\sqrt{7^2+(0.163t^2)^2}$$

$$t=4.87 \text{ s} \qquad \textit{Ans.}$$

The speed at time $t=4.87$ s is

$$\boldsymbol{v}=7t=7\times 4.87=34.1 \text{ m/s} \qquad \textit{Ans.}$$

Exercises

5-1 In Fig. 5-14, a car travels along a straight line with a speed $\boldsymbol{v}=(0.5t^3-8t)$ m/s, where t is in seconds. Determine the acceleration of the car when $t=2$ s.

5-2 In Fig. 5-15, a truck travels along a straight line with a velocity of $\boldsymbol{v}=(4t-3t^2)$ m/s, where t is in seconds. Determine the position of the truck when $t=4$ s. $s=0$ when $t=0$.

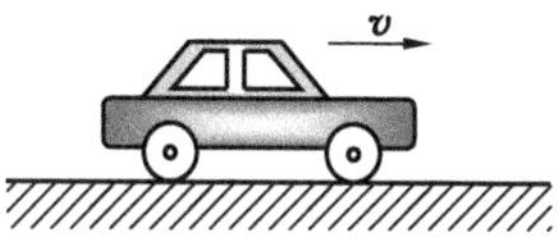

Fig. 5-14

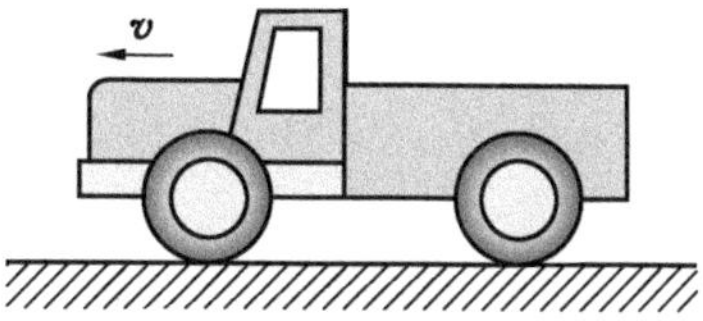

Fig. 5-15

5-3 In Fig. 5-16, a particle M travels along a straight-line path $y=0.5x$. If the x component of the particle's velocity is $\boldsymbol{v}_x=2t^2$ m/s, where t is in seconds, determine the magnitude of the particle's velocity and acceleration when $t=4$ s.

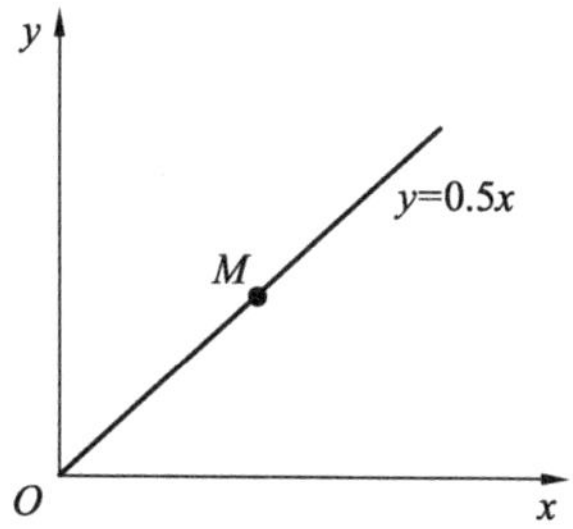

Fig. 5-16

5-4 The position of a particle is $\boldsymbol{r}=[(3t^3-2t)\boldsymbol{i}-(4t^{1/2}+t)\boldsymbol{j}+(3t^2-2)\boldsymbol{k}]$ m, where t is in seconds. Determine the magnitude of the particle's velocity and acceleration when $t=2$ s.

5-5 As shown in Fig. 5-17, the eccentric cam with radium R and eccentricity e rotates around axis O with constant angular speed ω, the bar AB is in reciprocationg motion upward and downward. Determine the equation of motion and velocity of the bar AB.

5-6 In Fig. 5-18, the rim M slides in both the rotating arm OA and fixed circular ring. If OA rotates with the constant angular speed ω. Determine the magnitude of equation, velocity and acceleration of rim M.

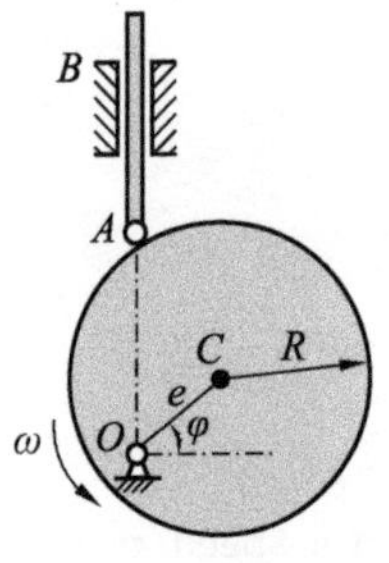

Fig. 5-17

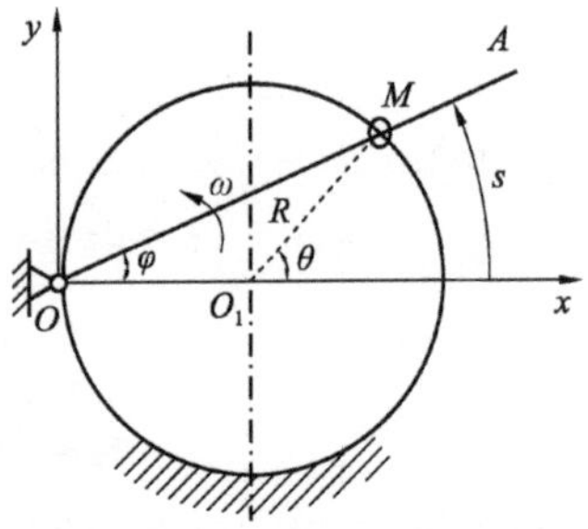

Fig. 5-18

5-7 In Fig.5-19, the automobile has a speed of 80 m/s at point A and an acceleration $\boldsymbol{a}$ having a magnitude of 10 m/s^2, acting in the direction shown. Determine the radius of curvature of the path at point A and the tangential component of acceleration.

5-8 In Fig. 5-20, starting from rest, the car travels around the circular path, $\rho=$ 50 m, at a speed $\boldsymbol{v}=0.2t^2$ m/s, where t is in seconds. Determine the magnitudes of the car's velocity and acceleration at the instant $t=3$ s.

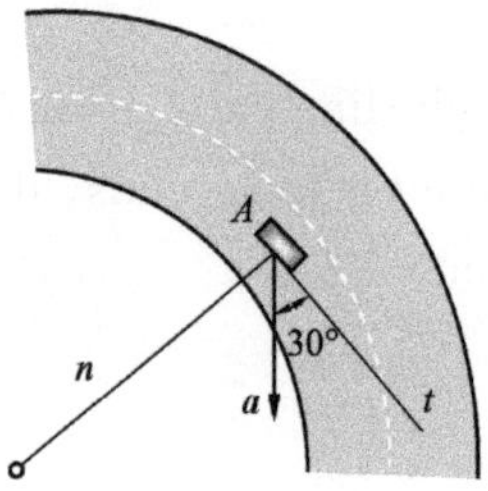

Fig. 5-19

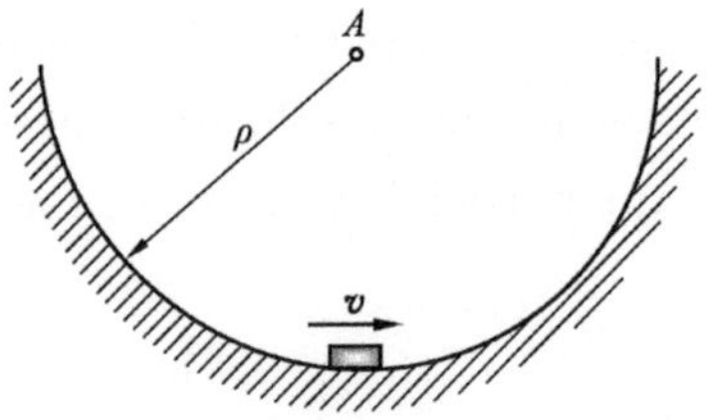

Fig. 5-20

5-9 At a given instant, a car travels along a circular curved road with a speed of 20 m/s while decreasing its speed at the rate of 3 m/s^2. If the magnitude of the car's acceleration is 5 m/s^2, determine the radius of curvature of the road.

5-10 Determine the maximum constant speed a race car can have if the acceleration of the car cannot exceed 7.5 m/s while rounding a track having a radius of curvature of 200 m.

5-11 As shown in Fig. 5-21, if the motorcycle has a deceleration of $a_t = -0.001s$ m/s^2 and its speed at position A is 25 m/s, determine the magnitude of its acceleration when it passes point B.

5-12 In Fig. 5-22, the truck travels along a circular road that has a radius of 50 m at a speed of 4 m/s. For a short distance when $t=0$ s, its speed is then increased by $a_t = 0.4t$ m/s^2, where t is in seconds. Determine the speed and the magnitude of the truck's acceleration when $t=4$ s.

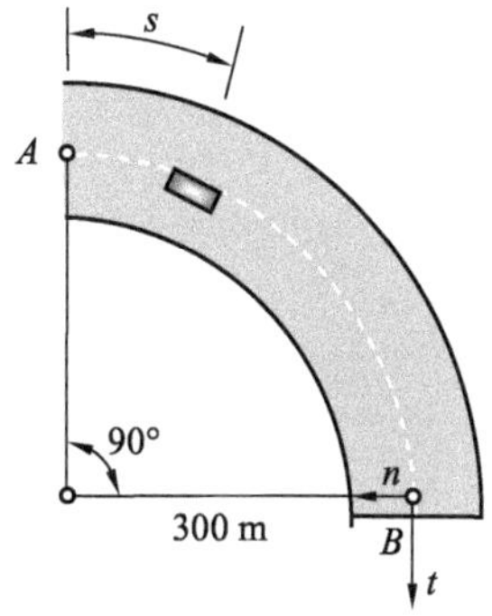

Fig. 5-21

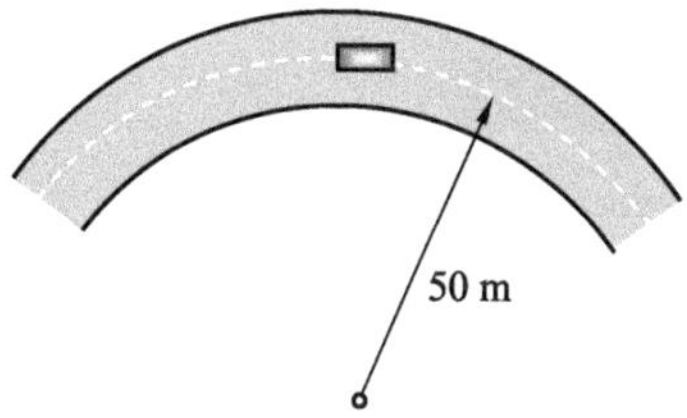

Fig. 5-22

Chapter 6

Translation and Rotation of Rigid Bodies

Objectives

In this chapter, we will study the kinematics of a rigid body, of particular interest will be translation and rotation about a fixed axis. We will derive the equations which describe the motion of rigid bodies. After studying, students can apply these equations of motion to solve the specific problems.

6.1 Translation

A *rigid body* may be considered to be a system of an infinite number of particles whose relative distances remain unchanged when the body is loaded. A motion that leaves the direction of the straight line between any two arbitrary points A and B of a rigid body unchanged is called a *translation*. When the paths of motion for any two points on the body are parallel lines, the motion is called *rectilinear translation*, e. g., the motion of the train in Fig. 6-1a. If the paths of motion are along curved lines which are equidistant, the motion is called *curvilinear translation*, e. g., the motion of the parallel bar AB in Fig. 6-1b.

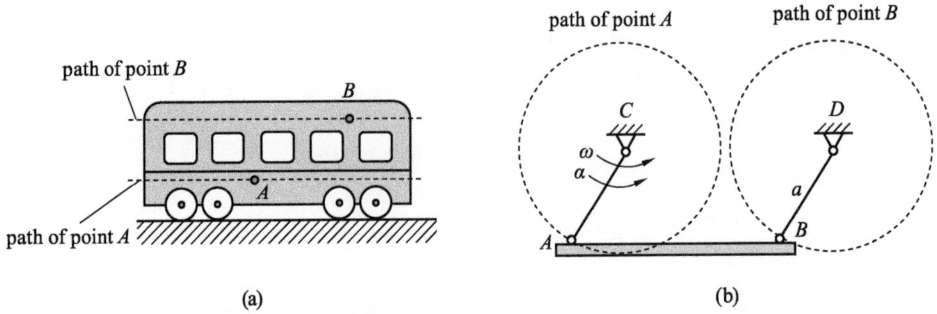

Fig. 6-1

6.1.1 Position

A rigid body is subjected to translation in the x-y plane, Fig. 6-2. The locations of arbitrary points A and B on the body are defined using position vectors $\boldsymbol{r}_A$ and $\boldsymbol{r}_B$. The position of B with respect to A is denoted by the relative-position vector $\boldsymbol{r}_{AB}$. By vector addition,

$$\boldsymbol{r}_A = \boldsymbol{r}_B + \boldsymbol{r}_{AB} \tag{6.1}$$

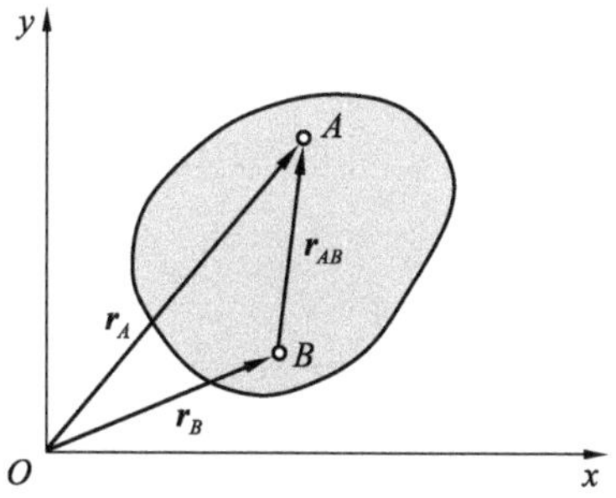

Fig. 6-2

6.1.2 Velocity

A relation between the instantaneous velocities of arbitrary point A and B is obtained by taking the time derivative of the Eq. (6.1), which yields

$$\boldsymbol{v}_A = \boldsymbol{v}_B + \frac{\mathrm{d}\boldsymbol{r}_{AB}}{\mathrm{d}t} \tag{6.2}$$

Because the body is translating, the direction of $\boldsymbol{r}_{AB}$ is *constant*. Since the magnitude of $\boldsymbol{r}_{AB}$ is also *constant* by the definition of a rigid body, the term $\frac{\mathrm{d}\boldsymbol{r}_{AB}}{\mathrm{d}t}=0$, therefore,

$$\boldsymbol{v}_A = \boldsymbol{v}_B \tag{6.3}$$

6.1.3 Acceleration

Taking the time derivative of the velocity Eq.(6.3) yields a similar relationship between the instantaneous accelerations of point A and B

$$\boldsymbol{a}_A = \boldsymbol{a}_B \tag{6.4}$$

The above two equations indicate that *all points* in a rigid body subjected to translation move with the *same velocity and acceleration*. The paths of different points all have the *same shape*. Thus, the motion of a single point of the rigid body arbitrarily chosen represents the motion of the complete body. As a result, the kinematics of particle motion can also be used to specify the kinematics of points located in a translating rigid body. This propriety makes that one body in translation motion will be considered as a particle.

6.2 Rotation about a Fixed Axis

If a motion all the particles of a rigid body move about a common axis that is fixed in space, the motion is called a *rotation about a fixed axis*. In this case, any point of the rigid body travels along a *circular path* whose plane is perpendicular to the axis.

6.2.1 Angular Position

The position of point P is defined by the position vector $\boldsymbol{r}$, which extends from origin point O to P. During the same time interval, the radius vectors sweep out the same angle. At the instant shown in Fig. 6-3a, the *angular position* is defined by the angle φ.

$$\varphi = f(t) \tag{6.5}$$

It is expressed in degrees, radians, or revolutions. We can refer to the sense of rotation as clockwise or counterclockwise. Here, we *arbitrarily* chose counterclockwise rotations as positive.

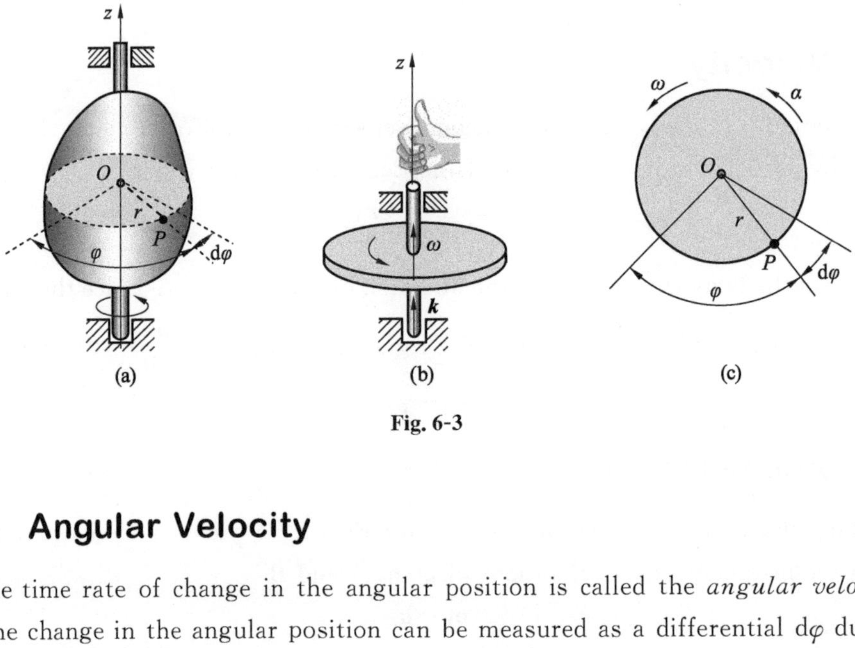

Fig. 6-3

6.2.2 Angular Velocity

The time rate of change in the angular position is called the *angular velocity* ω. Since the change in the angular position can be measured as a differential $\mathrm{d}\varphi$ during an instant of time $\mathrm{d}t$, then,

$$\omega = \frac{\mathrm{d}\varphi}{\mathrm{d}t} \tag{6.6}$$

The magnitude of angular velocity is often measured in rad/s. The direction is determined by the right-hand rule; that is, the fingers of the right hand are curled with the sense of rotation, in this case the thumb points upward, Fig. 6-3b. From the top view of the shaded plane, Fig. 6-3c, both φ and ω are counterclockwise, and the directional sense of $\boldsymbol{\omega}$ points outward from the page.

6.2.3 Angular Acceleration

The *angular acceleration* $\boldsymbol{\alpha}$ measures the time rate of change of the angular velocity. The magnitude is obtained through differentiation of Eq. (6. 6).

$$\alpha=\frac{\mathrm{d}\omega}{\mathrm{d}t}=\frac{\mathrm{d}^2\varphi}{\mathrm{d}t^2}=\dot{\omega}=\ddot{\varphi} \tag{6. 7}$$

The line of action of $\boldsymbol{\alpha}$ is the same as that for $\boldsymbol{\omega}$, Fig. 6-3b, however, its sense of direction depends on whether $\boldsymbol{\omega}$ is increasing or decreasing. If $\boldsymbol{\omega}$ is decreasing, then $\boldsymbol{\alpha}$ is called an *angular deceleration* and therefore has a sense of direction which is opposite to $\boldsymbol{\omega}$. If the angular acceleration of the body is constant, $\alpha=\alpha_c$, then

$$\omega=\omega_0+\alpha_c t \tag{6. 8}$$

$$\varphi=\varphi_0+\omega_0 t+\frac{1}{2}\alpha_c t^2 \tag{6. 9}$$

Here φ_0 and ω_0 are the initial values of the body's angular position and angular velocity, respectively.

In engineering, the *revolutions* n (r/min) is used to indicate the angular velocity, we can convert the revolutions to radians. Since there are 2π rad in one revolution, then

$$\omega=\frac{2\pi n}{60}=\frac{\pi n}{30} \tag{6. 10}$$

6. 3 Motion of Point in Rotational Rigid Body

6.3.1 Position

As the rigid body in Fig. 6-3a rotates about a fixed axis, point P in the body travels along a circular path of radius r with center at point O. From the top view, this path is contained within the shaded plane, Fig. 6-3c. Then the relation between the angular φ and the arch-length that the point P moves is obtained.

$$s=r\varphi \tag{6. 11}$$

6.3.2 Velocity

The magnitude of velocity (*speed*) can be found by time derivative of the Eq. (6. 11)

$$v=\frac{\mathrm{d}s}{\mathrm{d}t}=r\frac{\mathrm{d}\varphi}{\mathrm{d}t}=r\omega \tag{6. 12}$$

The direction of $\boldsymbol{v}$ is *tangent* to the circular path, Fig. 6-4a. The velocity $\boldsymbol{v}$ of point P can also be obtained by using the cross product of ω and $\boldsymbol{r}_p$. Here, $\boldsymbol{r}_p$ is directed from *any point* on the axis of rotation to point P, Fig. 6-4a. We have

$$\boldsymbol{v}=\frac{\mathrm{d}\boldsymbol{r}}{\mathrm{d}t}=\boldsymbol{\omega}\times\boldsymbol{r} \tag{6. 13}$$

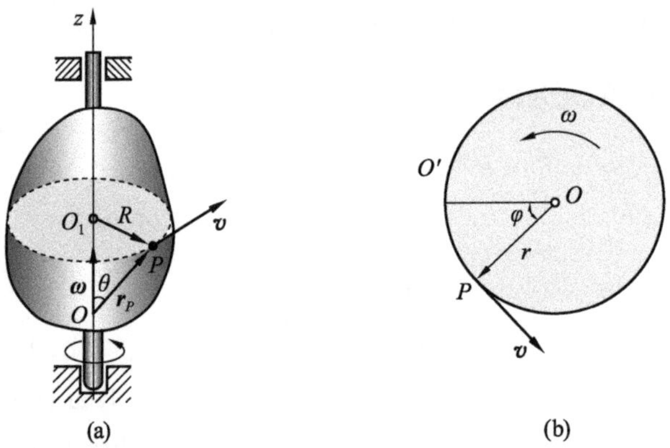

Fig. 6-4

6.3.3 Acceleration

The acceleration of point P can be expressed in terms of its normal and tangential components. Since $a_t = dv/dt$ and $a_n = v^2/\rho$, where $\rho = r$, $v = \omega r$, and $\alpha = d\omega/dt$, we have

$$a_t = \frac{dv}{dt} = r\frac{d\omega}{dt} = r\alpha \tag{6.14}$$

$$a_n = \frac{v^2}{\rho} = \frac{(r\omega)^2}{r} = r\omega^2 \tag{6.15}$$

The *tangential component of acceleration*, Fig. 6-5, represents the time rate of change in the velocity's magnitude. If the speed of P is increasing, then $\boldsymbol{a}_t$ acts in the same direction as $\boldsymbol{v}$, if the speed is decreasing, $\boldsymbol{a}_t$ acts in the opposite direction of $\boldsymbol{v}$, and if the speed is constant, $\boldsymbol{a}_t$ is zero.

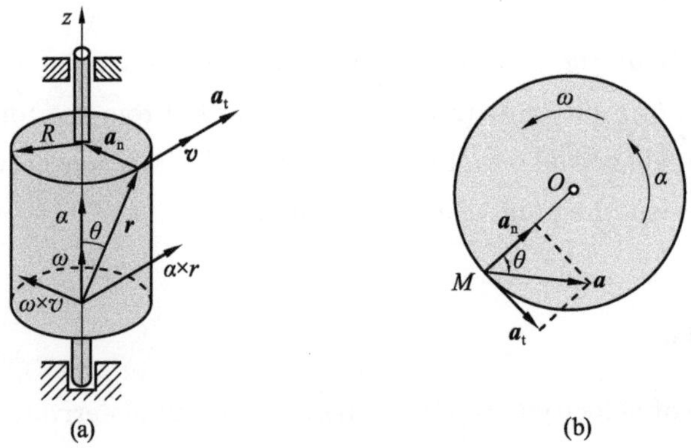

Fig. 6-5

The *normal component of acceleration* represents the time rate of change in the velocity's direction. The *direction* of $\boldsymbol{a}_n$ is always toward O, the center of the circular path, Fig. 6-5.

Like the velocity, the acceleration of point P can be expressed in terms of the vector cross product. Taking the time derivative of Eq. (6. 13), we have

$$\boldsymbol{a}=\frac{\mathrm{d}\boldsymbol{v}}{\mathrm{d}t}=\frac{\mathrm{d}\boldsymbol{\omega}}{\mathrm{d}t}\times\boldsymbol{r}+\boldsymbol{\omega}\times\frac{\mathrm{d}\boldsymbol{r}}{\mathrm{d}t} \tag{6.16}$$

Using $\alpha=\mathrm{d}\boldsymbol{\omega}/\mathrm{d}t$, and $\mathrm{d}\boldsymbol{r}/\mathrm{d}t=\boldsymbol{v}=\boldsymbol{\omega}\times\boldsymbol{r}$, yields

$$\boldsymbol{a}=\boldsymbol{\alpha}\times\boldsymbol{r}+\boldsymbol{\omega}\times(\boldsymbol{\omega}\times\boldsymbol{r})=\boldsymbol{a}_{\mathrm{t}}+\boldsymbol{a}_{\mathrm{n}} \tag{6.17}$$

Since $\boldsymbol{a}_{\mathrm{t}}$ and $\boldsymbol{a}_{\mathrm{n}}$ are perpendicular to one another, the magnitude of acceleration can be determined, Fig. 6-5b.

$$a=\sqrt{a_{\mathrm{t}}^2+a_{\mathrm{n}}^2}=r\sqrt{\alpha^2+\omega^4} \tag{6.18}$$

$$\tan\theta=\frac{|a_{\mathrm{t}}|}{a_{\mathrm{n}}}=\frac{|\alpha|}{\omega^2} \tag{6.19}$$

At an instant, the speed and acceleration of each point in body is directly proportional to its rotational radius, on a straight line passing through the rotation center, the distribution of velocities and accelerations is linear, the velocities being perpendicular on the line, as shown in Fig. 6-6a.

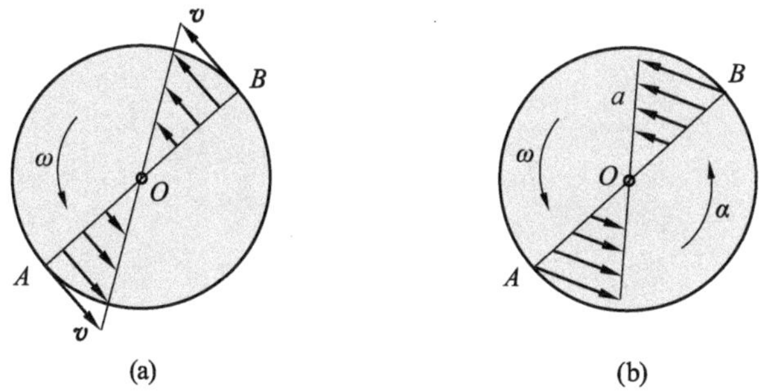

Fig. 6-6

Example 6-1

A transfer plate AB is pin connected with two cranks O_1A and O_2B, as shown in Fig. 6-7. If the crank O_1A rotates with $\varphi=15\pi t$ (where φ is in rad and t is in seconds). Determine the velocity and acceleration of the block C on the plate AB. $O_1A=O_2B=R=0.2$ m, $O_1O_2=AB$.

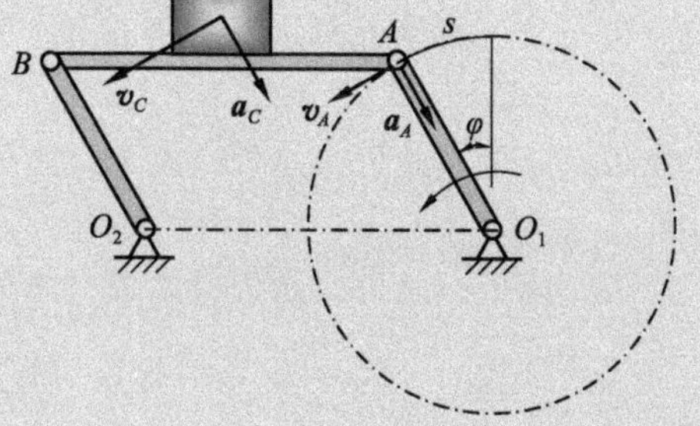

Fig. 6-7

Solution:

Since $O_1A=O_2B=R$, $O_1O_2=AB$, when the mechanism is in motion, plate AB is always parallel to O_1O_2, the plate AB is subjected to translation. Thus, all the particles (include block) in plate AB have the same velocity and same acceleration. Therefore, the velocity and acceleration of the point C are same as the velocity and acceleration of the point A. The crank O_1A is subjected to rotation about a fixed axis passing through point O_1. Point A on the crank has a circular path, Fig. 6-7, the motion of point A is

$$s=R\varphi=0.2\times15\pi t=3\pi t$$

The speed of point A is

$$v_A=\frac{ds}{dt}=3\pi=9.42\ \text{m/s}$$

The direction is shown in Fig. 6-7.

Since v_A is constant, the tangential acceleration of point A is

$$a_t=\frac{dv_A}{dt}=0$$

The normal component of acceleration of point A is

$$a_n=\frac{v_A^2}{R}=\frac{(3\pi)^2}{0.2}=444\ \text{m/s}^2$$

The direction of acceleration is shown in Fig. 6-7.

According to the properties of translation, we can obtain the velocity $\boldsymbol{v}_C$ and acceleration a_C of block C.

$$v_C=v_A=9.42\ \text{m/s} \qquad \textit{Ans.}$$

$$a_C=a_A=a_n=444\ \text{m/s}^2 \qquad \textit{Ans.}$$

The direction of velocity and acceleration is shown in Fig. 6-7.

Example 6-2

As shown in Fig. 6-8, when the gear rotates with angular displacement of $\varphi=(2t^2+3t)$ rad, where t is in seconds, determine its revolutions, angular velocity and angular acceleration when $t=10$ s.

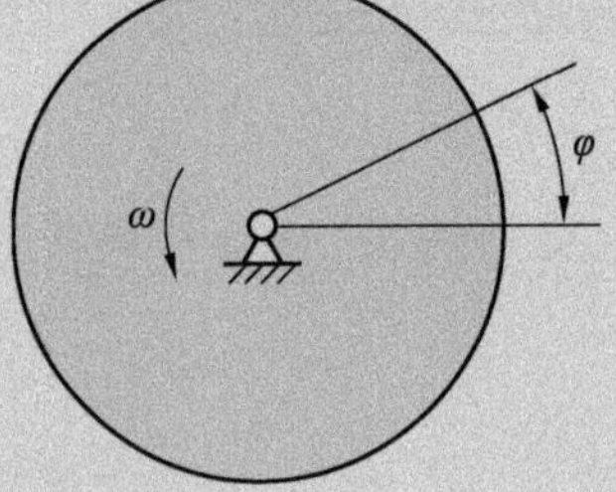

Fig. 6-8

Solution:

Since the equation of angular displacement is known, the angular position when $t=10$ s is

$$\varphi|_{t=10}=2\times10^2+3\times10=230 \text{ rad}$$

Using this result, the turning revolutions is

$$n=\frac{\varphi}{2\pi}=\frac{230}{2\pi}=36.6 \text{ r/min} \qquad \textit{Ans.}$$

The angular velocity of gear can be determined from

$$\omega=\frac{\mathrm{d}\varphi}{\mathrm{d}t}=4t+3$$

$$\omega|_{t=10}=4\times10+3=43 \text{ rad/s} \qquad \textit{Ans.}$$

The angular acceleration of gear can be determined from

$$\alpha=\frac{\mathrm{d}\omega}{\mathrm{d}t}=4 \text{ rad/s}^2 \qquad \textit{Ans.}$$

Thus the angular acceleration α is constant.

Example 6-3

The steel rope wraps around a drum wheel of a radius $r=0.2$ m, as shown in Fig. 6-9. If the angular displacement of the drum wheel is $\varphi=(4t-t^2)$ rad (t is in seconds), determine, when $t=1$ s, (1) the velocity and acceleration of a point M located on its rim, and (2) the velocity and acceleration of a weight A. Assume the steel rope does not slip on the pulley.

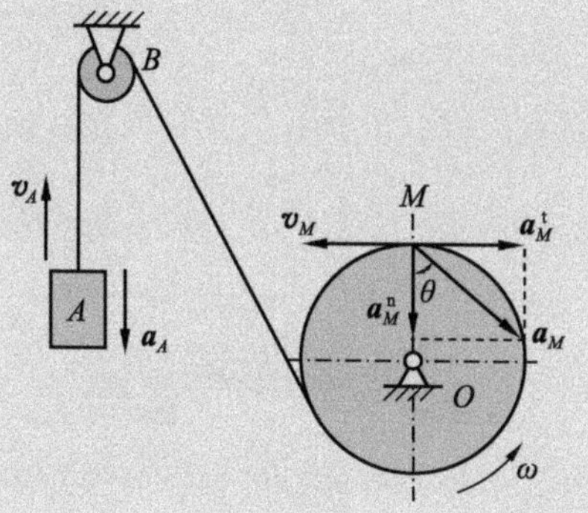

Fig. 6-9

Solution:

The drum wheel is subjected to rotation about a fixed axis passing through point O, we will find the angular velocity and angular acceleration of the drum. Since the equation of angular displacement is known, the angular velocity and angular acceleration of the wheel can be determined

$$\omega=\frac{d\varphi}{dt}=(4-2t)\ \text{rad/s}$$

$$\alpha=\frac{d\omega}{dt}=\frac{d^2\varphi}{dt^2}=-2\ \text{rad/s}^2$$

$$\omega_{|t=1}=4-2\times1=2\ \text{rad/s}$$

$$\alpha_{|t=1}=-2\ \text{rad/s}^2$$

Here α and ω is opposite sign, the drum wheel rotates deceleration. Using this result, the velocity, tangential acceleration and normal acceleration of point M located on its rim can be determined from

$$v_M=r\omega=0.2\times2=0.4\ \text{m/s} \qquad \textit{Ans.}$$

$$a_M^t=r\alpha=0.2\times(-2)=-0.4\ \text{m/s}^2$$

$$a_M^n=r\omega^2=0.2\times2^2=0.8\ \text{m/s}^2$$

The acceleration of point M is

$$a=\sqrt{a_M^t+a_M^n}=\sqrt{(-0.4)^2+0.8^2}=0.894\ \text{m/s}^2 \qquad \textit{Ans.}$$

$$\tan\theta=\frac{|a_M^t|}{a_M^n}=\frac{|-2|}{2^2}=0.5$$

Since the steel rope does not slip on the pulley, the rising distance x_A of weight A is equal to the arch length that point M moves.

$$x_A=s_M=r\varphi$$

Taking the time derivative of distance x_A, we have

$$v_A=r\omega=v_M$$

$$v_A=v_M=0.4\ \text{m/s} \qquad \textit{Ans.}$$

Taking the time derivative of velocity, we have

$$a_A=r\alpha=a_M^t$$

$$a_A=a_M^t=-0.4\ \text{m/s}^2 \qquad \textit{Ans.}$$

The acceleration of weight A is opposite to the sense of its velocity, so the weight is hoisting deceleration.

Exercises

6-1 An angle tray EBF is pin connected with two same crank DE and AB, $DE=AB=0.2$ m, Fig. 6-10. A weight C is put on the tray. If the crank AB rotates with angular velocity $\omega=4$ rad/s and angular acceleration $\alpha=2$ rad/s^2 in this instant, determine the velocity and acceleration of the weight C.

6-2 As shown in Fig. 6-11, the mechanism is consisted with two bars OA and AB that are pin connected, the bar $OA=1.5$ m and $AB=0.8$ m, the workpiece is put on the tray of end B. If the bar OA rotates around the axis O and the velocity of workpiece B is constant $v_B=0.05$ m/s, determine the rotational equation of bar OA and path of workpiece B. When $t=0$, $\varphi=0$.

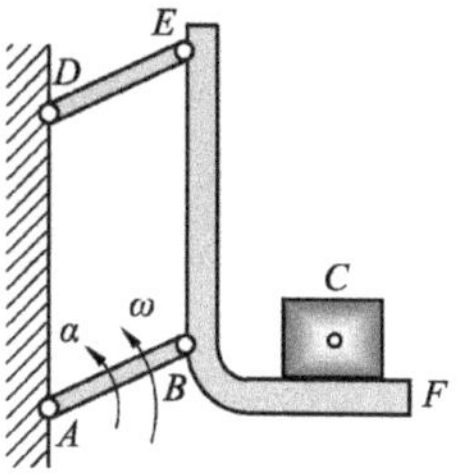

Fig. 6-10

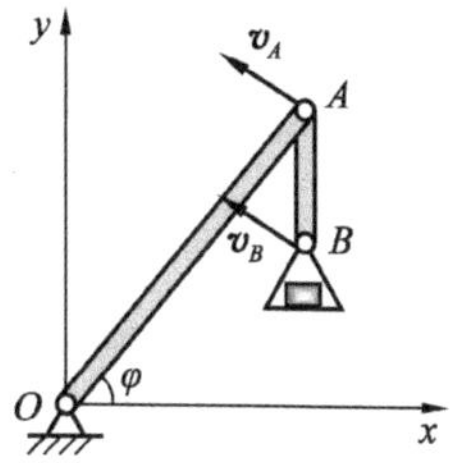

Fig. 6-11

6-3 As shown in Fig. 6-12, a circular tray is supported by three same crank Aa, Bb and Cc with length of 0.15 m. If three cranks rotate with same revolution $n=45$ r/min around the respective axes that are vertical to the same surface. Determine the velocity and acceleration of the center point O of the tray.

6-4 In Fig. 6-13, a rectangular plane plate $ABCD$ with the dimensions $AB=90$ cm and $AD=30$ cm is joined to two fixed pins O_1 and O_2 with two straight rods hinged at their ends. Knowing that the rod O_1A rotates about point O_1 with uniform angular velocity $\omega=2$ rad/s, determine the velocities and accelerations of the points A, B, C, and D. Set $O_1O_2=AB$, $O_1A=O_2B=30$ cm.

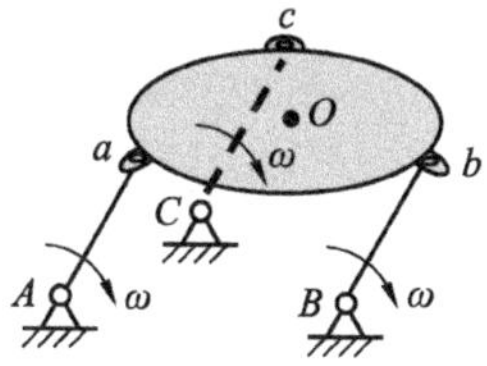

Fig. 6-12

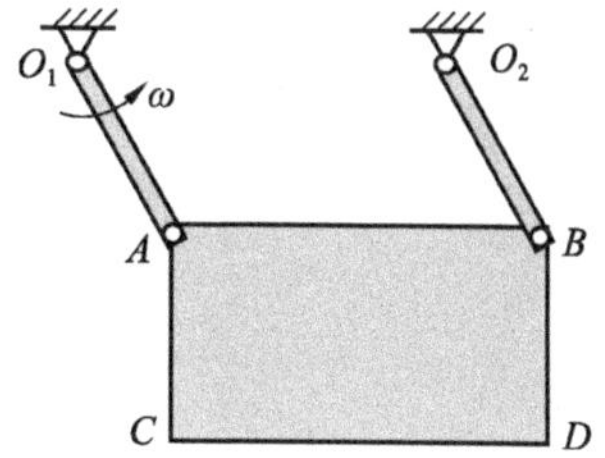

Fig. 6-13

6-5 A cord is wrapped around a wheel, which is initially at rest, Fig. 6-14. If a force is applied to the cord and gives it an acceleration $a=4t$ m/s^2, where t is in seconds, determine, as a function of time, the angular velocity of the wheel.

6-6 In Fig. 6-15, the disk is originally rotating at $\omega_0=8$ rad/s. If it is subjected to a constant angular acceleration of $\alpha=6$ rad/s^2, determine the magnitudes of the velocity and the n and t components of acceleration of point A at the instant $t=0.5$ s.

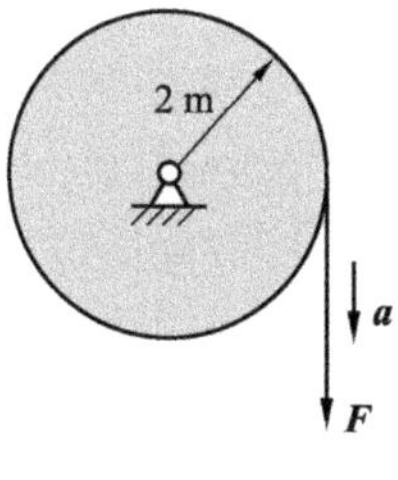

Fig. 6-14

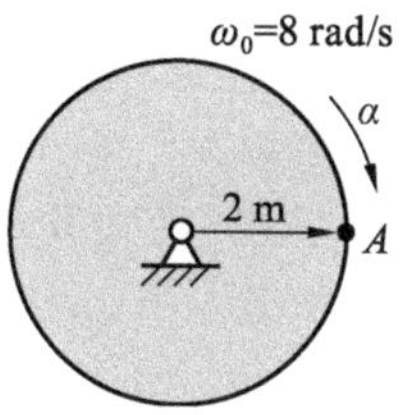

Fig. 6-15

6-7 In Fig. 6-16, the weight A is hoisted by the steel rope that wraps around a drum wheel. The drum has a radius of $R=0.5$ m. If the weight moves with an equation of motion $x=5t^2$ m, where t is in seconds, determine: (1) the angular velocity and angular acceleration of the drum, and (2) the acceleration of point B on the drum.

6-8 In Fig. 6-17, at the instant $\theta=60°$, the rod AB is moving with a velocity $v=10$ m/s and subjected to a deceleration $a=16$ m/s^2. Determine the angular velocity and angular acceleration of link CD at this instant.

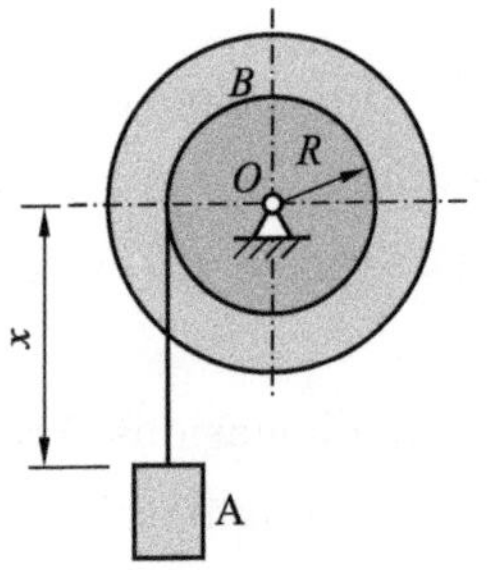

Fig. 6-16

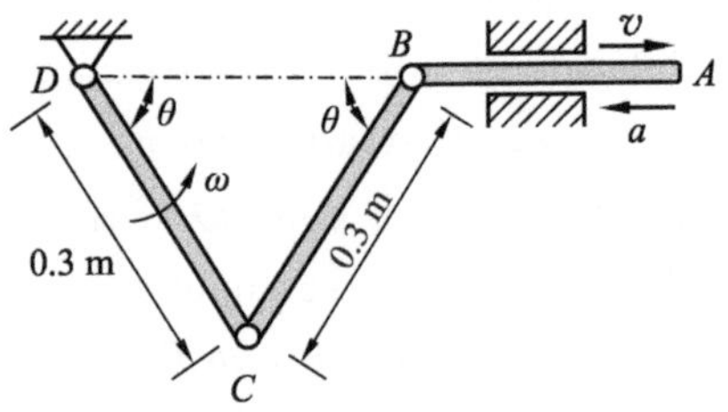

Fig. 6-17

6-9 In Fig. 6-18, the weight is hoisted by the rope that wraps around a drum wheel. If the angular displacement of the wheel is $\theta=(0.5t^3+15t)$ rad, where t is in seconds, determine the velocity and acceleration of the weight when $t=3$ s.

6-10 In Fig. 6-19, the bar DC of length l rotates about the D with a constant angular velocity ω. Determine the velocity and acceleration of the bar AB, which is confined by the guides to move vertically.

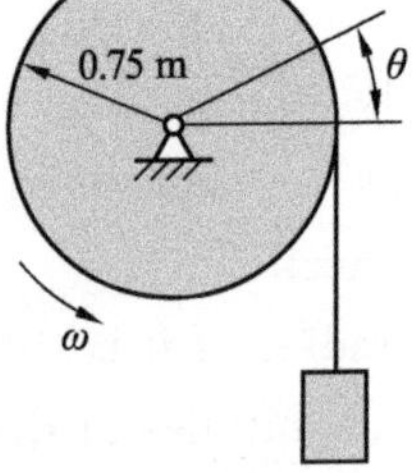

Fig. 6-18

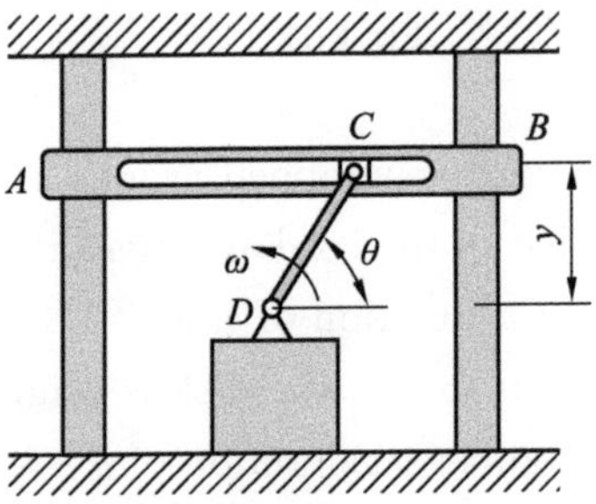

Fig. 6-19

6-11 In Fig. 6-20, the circular plate rotates about the axis pass through the point O with the angular velocity ω and angular acceleration α. Determine the velocity and acceleration of the point A and B on the plate.

6-12 In Fig. 6-21, the circular ring rotates about the axis CD with the angular velocity ω and angular acceleration α. Determine the velocity and acceleration of the point A and B on the ring.

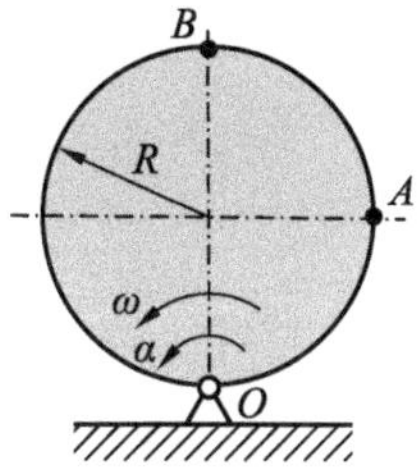

Fig. 6-20

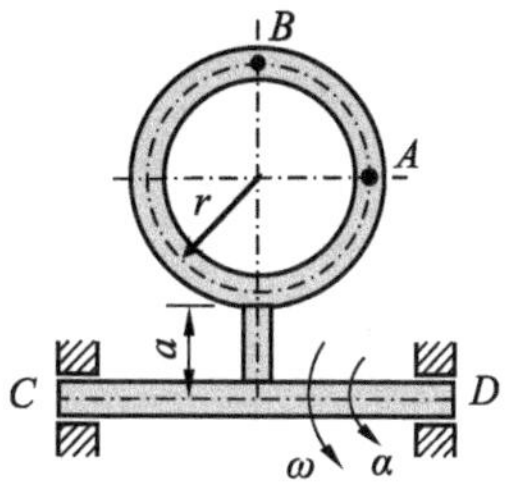

Fig. 6-21

Chapter 7

Composite Motion of Particles

Objectives

In this chapter, we will deal with the relative motion analysis of velocity or acceleration using a translating frame of reference and a rotating frame of reference by using two reference systems, one moving system and a fixed system. After learning, the students can explain and solve the engineering dynamic problems using the knowledge of relative motion of the particle.

7.1 Concepts of Composite Motion

Up to now the absolute motion of a particle has been determined using a single fixed reference frame. But there are many cases, the path of a particle is complicated, so that it may be easier to analyze the motion of a particle by using two or more frames of reference. For example, while the airplane is in flight, the motion of a particle P located at the tip of the propeller, is more easily described if one observes first the motion of the airplane from a fixed reference and then superimposes the circular motion of the particle P measured from a reference attached to the airplane, as shown in Fig. 7-1.

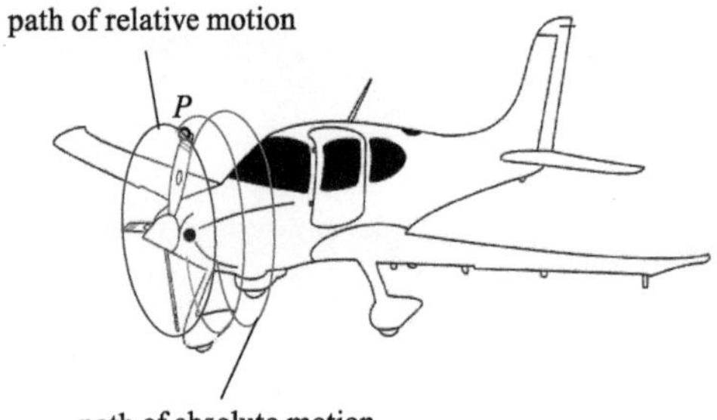

(a)

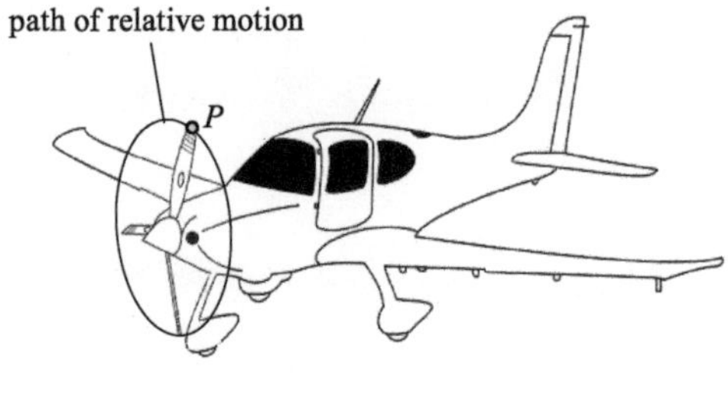

(b)

Fig. 7-1

Consider particles A and B, which move along the arbitrary paths shown in Fig. 7-2. In the following, we will specify the kinematics of particle motion that two reference frames will be considered for the analysis. One is the *fixed reference frame* (FRF), $Oxyz$, usually attached to the earth. The second frame of reference $Ax'y'z'$ is attached to and moves with a moving body, for example the particle A, usually called *moving reference frame* (MRF), shown in Fig. 7-2.

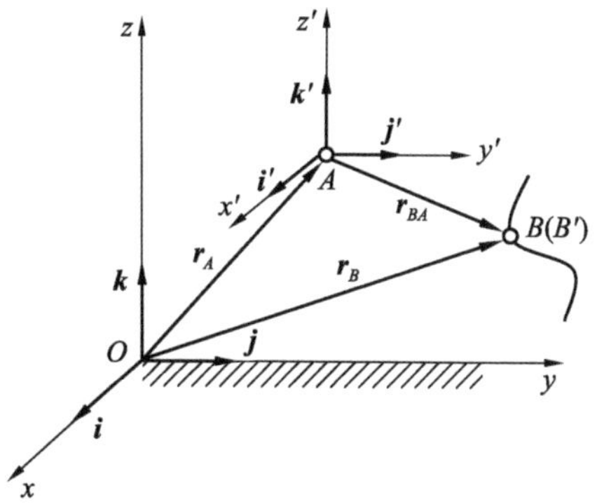

Fig. 7-2

Then we have three kinds of motion to be analyzed.

Absolute motion is the motion of moving particle B observed from the fixed reference frame $Oxyz$. The *absolute position* $\boldsymbol{r}_B$ of moving particle B is measured from the origin O of the fixed reference frame, *absolute velocity* $\boldsymbol{v}_\mathrm{a}$ and *absolute acceleration* $\boldsymbol{a}_\mathrm{a}$ of the moving point B are observed from the fixed frame.

Transport motion is the motion of moving reference frame $Ax'y'z'$ observed from the fixed reference frame $Oxyz$. *Transport point* is the point B' in the moving reference frame *coinciding* with the moving particle B at the instant, Fig. 7-2. The *absolute position* $\boldsymbol{r}_A$ of the origin of the moving reference frame is measured from the origin O of the fixed reference frame. The *transport velocity* $\boldsymbol{v}_\mathrm{e}$ and *transport acceleration* $\boldsymbol{a}_\mathrm{e}$ are the velocity and acceleration of transport point B' in the moving frame, and observed from the fixed reference frame $Oxyz$.

Relative motion is the motion of moving particle B observed from the moving reference frame $Ax'y'z'$. The position of B with respect to the origin A of the moving reference frame is denoted by the *relative-position* vector $\boldsymbol{r}_{BA}$. The *relative velocity* $\boldsymbol{v}_\mathrm{r}$ and *relative acceleration* $\boldsymbol{a}_\mathrm{r}$ of the moving particle B are observed from the moving frame, Fig. 7-2. Thus, if B has relative coordinates (x', y', z'), then

$$\boldsymbol{r}_{BA} = x'\boldsymbol{i}' + y'\boldsymbol{j}' + z'\boldsymbol{k}' \tag{7.1}$$

Using vector addition, the three vectors shown in Fig. 7-2 can be related by the equation

$$\boldsymbol{r}_B = \boldsymbol{r}_A + \boldsymbol{r}_{BA} \tag{7.2}$$

Example 7-1

A rod OA, shown in Fig. 7-3, rotates around axis O with angular velocity $\omega = t^2$, collar B travels outward along the rod with motion $x' = 3t^2$ measured relative to the rod (where ω is in rad/s, x' is in cm and t is in seconds). Determine the relative velocity $\boldsymbol{v}_r$ and relative acceleration $\boldsymbol{a}_r$, the transport velocity $\boldsymbol{v}_e$ and transport acceleration $\boldsymbol{a}_e$ of the collar when $t = 2$ s.

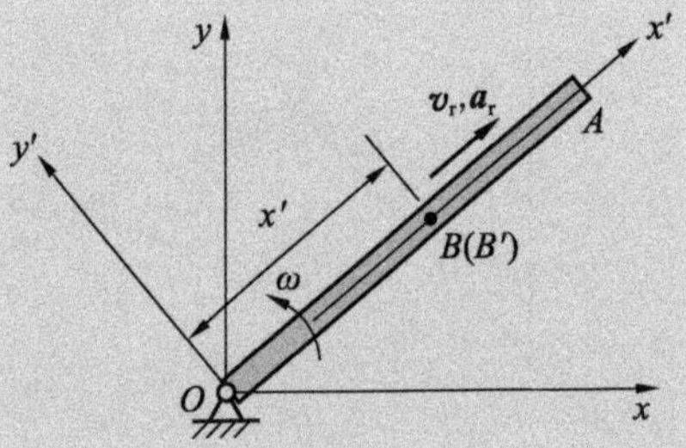

Fig. 7-3

Solution:

The origin of both coordinate systems is located at point O, Fig. 7-3. Since motion of the collar is relative to the rod, the moving frame of reference $Ox'y'z'$ is attached to the rod. The relative motion of the collar with respect to moving reference can be determined.

$$OB' = OB = x' = 3t^2$$

$$\boldsymbol{v}_r = \frac{dx'}{dt} = 6t$$

$$a_r = \frac{d\boldsymbol{v}_r}{dt} = \frac{d^2x'}{dt^2} = 6 \text{ cm/s}^2$$

When $t = 2$ s, we get

$$x' = 3 \times 2^2 = 12 \text{ cm}$$

$$\boldsymbol{v}_r = 6 \times 2 = 12 \text{ cm/s} \qquad \textit{Ans.}$$

$$a_r = 6 \text{ cm/s}^2 \qquad \textit{Ans.}$$

The angular velocity and angular acceleration of the moving frame when $t = 2$ s are obtained

$$\omega|_{t=2} = t^2 = 2^2 = 4 \text{ rad/s}$$

$$\alpha|_{t=2} = \left.\frac{d\omega}{dt}\right|_{t=2} = 2t|_{t=2} = 4 \text{ rad/s}^2$$

Transport motion of the collar when $t = 2$ s are obtained

$$v_e = v_{B'} = OB' \cdot \omega = 12 \times 4 = 48 \text{ cm/s} \qquad \textit{Ans.}$$

$$a_e^t = OB' \cdot \alpha = 12 \times 4 = 48 \text{ cm/s}^2 \qquad \textit{Ans.}$$

$$a_e^n = OB' \cdot \omega^2 = 12 \times 4^2 = 192 \text{ cm/s}^2 \qquad \textit{Ans.}$$

7.2 Composition of the Velocities

Consider the two particles A and B shown in Fig. 7-2. Their location is specified by the position vectors $\boldsymbol{r}_A$ and $\boldsymbol{r}_B$ which are measured with respect to the fixed $Oxyz$ coordinate system. As shown in the figure, the base point A represents the origin of the $O'x'y'z'$ coordinate system, which is assumed to be moving with respect to the $Oxyz$ system. The position of B with respect to A is specified by the *relative-position* vector $\boldsymbol{r}_{BA}$. At the instant considered, transport point B' coinciding with the moving point B, using the Eq. (7.1) and Eq. (7.2), we have

$$\boldsymbol{r}_{B'}=\boldsymbol{r}_B=\boldsymbol{r}_A+x'\boldsymbol{i}'+y'\boldsymbol{j}'+z'\boldsymbol{k}' \tag{7.3}$$

Taking the time derivative of Eq. (7.3), this yields

$$\frac{\mathrm{d}\boldsymbol{r}_B}{\mathrm{d}t}=\frac{\mathrm{d}\boldsymbol{r}_A}{\mathrm{d}t}+\left(x'\frac{\mathrm{d}\boldsymbol{i}'}{\mathrm{d}t}+y'\frac{\mathrm{d}\boldsymbol{j}'}{\mathrm{d}t}+z'\frac{\mathrm{d}\boldsymbol{k}'}{\mathrm{d}t}\right)+\left(\frac{\mathrm{d}x'}{\mathrm{d}t}\boldsymbol{i}'+\frac{\mathrm{d}y'}{\mathrm{d}t}\boldsymbol{j}'+\frac{\mathrm{d}z'}{\mathrm{d}t}\boldsymbol{k}'\right) \tag{7.4}$$

where the first two terms of the equation are the absolute velocities of the two particles B and A,

$$\boldsymbol{v}_a=\boldsymbol{v}_B\frac{\mathrm{d}\boldsymbol{r}_B}{\mathrm{d}t},\quad \boldsymbol{v}_A=\frac{\mathrm{d}\boldsymbol{r}_A}{\mathrm{d}t}$$

The first parenthesis of the Eq. (7.4) contains only the derivatives of the unit vectors of the moving axes, this derivative is obtained as the vector has constant magnitude and direction with respect to the moving reference system. In the right part of the Eq. (7.4), the absolute velocities $\boldsymbol{v}_A$ plus the first parentheses represent the velocity of the point B' as this point moves together with the moving reference system, and it is named *transport velocity* of point B',

$$\boldsymbol{v}_e=\boldsymbol{v}_A+x'\frac{\mathrm{d}\boldsymbol{i}'}{\mathrm{d}t}+y'\frac{\mathrm{d}\boldsymbol{j}'}{\mathrm{d}t}+z'\frac{\mathrm{d}\boldsymbol{k}'}{\mathrm{d}t} \tag{7.5}$$

The second parenthesis of the Eq. (7.4) is the derivative of the vector which considers only the variation in time of the vector without to consider the motion of the moving reference system. This derivative is called *relative derivative* of the vector and it is named *relative velocity* of point B, namely,

$$\boldsymbol{v}_r=\frac{\mathrm{d}x'}{\mathrm{d}t}\boldsymbol{i}'+\frac{\mathrm{d}y'}{\mathrm{d}t}\boldsymbol{j}'+\frac{\mathrm{d}z'}{\mathrm{d}t}\boldsymbol{k}' \tag{7.6}$$

This derivative is obtained as the moving reference system is a fixed one.

Substituting these results into Eq. (7.4), we can write finally:

$$\boldsymbol{v}_a=\boldsymbol{v}_e+\boldsymbol{v}_r \tag{7.7}$$

This equation states that *the absolute velocity of moving point B is the sum of the transport velocity and the relative velocity of point B with respect to A, as measured by the observer fixed in the moving reference frame.*

Eq. (7.7) is a vector equation and can be used to determine two unknowns with the aid of a *velocity diagram*.

◆ **Procedure for Analysis**

① Choose an appropriate location for the origin and proper orientation of the axes for moving reference frames. The moving frame should be selected fixed to the body or device along which the relative motion occurs.

② When specifying the moving point A as the origin of the moving reference frames, this point has a *known* velocity. Draw a *velocity* diagram of the body, if the magnitudes of the velocities are unknown, the sense of direction of these vectors can be assumed.

③ Since vector addition forms a triangle, there can be at most *two unknowns*, represented by the magnitudes or directions of the vector quantities.

④ These unknowns can be solved for either graphically, using trigonometry (law of sines, law of cosines), or by resolving each of the three vectors into rectangular, thereby generating a set of scalar equations.

⑤ If the solution yields a negative answer for an unknown magnitude, it indicates the sense of direction of the vector is opposite to that shown on the velocity diagram.

Example 7-2

The mechanism as shown in Fig. 7-4a, the crank OA of length r rotates around the axis O with an angular velocity of ω, the block A is pin connected to the crank and can slide in the slot of component ABC. At this instant θ, determine the velocity of the component ABC.

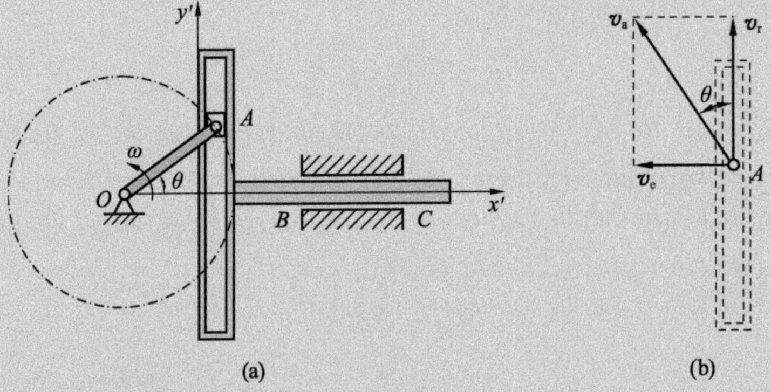

Fig. 7-4

Solution:

The motion of component ABC is translation, so the motion of point A located in component can be used to specify the motion of component ABC. The fixed x, y axes are established at an arbitrary point on the ground and the translating x', y' axes are attached to component ABC, Fig. 7-4a. We chose the block A as the moving point, so that the absolute motion of the block is *circular motion* around the center O, relative motion of the block is *rectilinear motion* along the slot and transport motion is the *translation* of component ABC. Applying the relative-velocity equation gives

$$\boldsymbol{v}_a = \boldsymbol{v}_e + \boldsymbol{v}_r$$

where the direction of *absolute velocity* $\boldsymbol{v}_a$ is perpendicular to the radius OA, the direction of *transport velocity* $\boldsymbol{v}_e$ is horizontal and the direction of relative velocity $\boldsymbol{v}_r$ is along the slot and perpendicular to $\boldsymbol{v}_e$, so the direction of three velocities are known and drawn in Fig. 7-4b. Since the magnitude of absolute velocity $\boldsymbol{v}_a$ is known $v_a = r\omega$, using trigonometry, we have

$$v_e = v_a \sin\theta = r\omega \sin\theta \qquad Ans.$$

At this instant, the magnitude of velocity of the component ABC is the transport velocity $\boldsymbol{v}_e$, the direction is shown in Fig. 7-4b.

Example 7-3

The cam, shown in Fig. 7-5a, with eccentricity $OC = e$ and radium of $r = \sqrt{3}e$, rotates around the axis O with constant angular velocity ω_0. At the instant, $OC \perp CA$ and point O, A and B are collinear, determine the velocity of the rod AB.

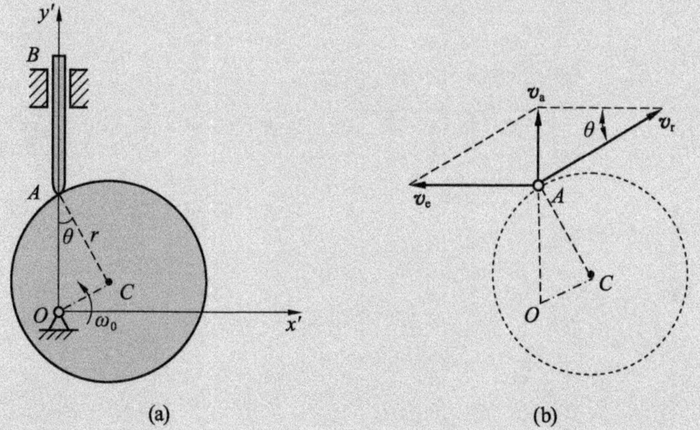

Fig. 7-5

Solution:

The motion of rod AB is translating, so the motion of point A located in rod can be used to specify the motion of rod. The moving reference frame $Ox'y'$ is attached to and rotates with the cam, Fig. 7-5a. We chose the contact point A in the rod AB as the moving point, so that the absolute motion of point A is rectilinear motion, relative motion of point A is the circular motion around the cam center C, and transport motion is the rotation of the cam about the axis O. Applying the relative-velocity equation gives

$$\boldsymbol{v}_a = \boldsymbol{v}_e + \boldsymbol{v}_r$$

where the direction of $\boldsymbol{v}_a$ is along the rod AB, the direction of $\boldsymbol{v}_e$ is perpendicular to OA and the direction of $\boldsymbol{v}_r$ is tangent to the circle, so the direction of three velocities are known and drawn in Fig. 7-5b. The magnitude of transport velocity $\boldsymbol{v}_e$ can be calculated as

$$OA=\sqrt{e^2+(\sqrt{3}e)^2}=2e$$

$$v_e=OA\cdot\omega_0=2e\omega_0$$

Using trigonometry, we have

$$\tan\theta=\frac{OC}{CA}=\frac{e}{\sqrt{3}e}=\frac{\sqrt{3}}{3}$$

$$v_a=v_e\cdot\tan\theta=\frac{2\sqrt{3}}{3}e\omega_0 \qquad \textit{Ans.}$$

At this instant, the magnitude of velocity of the rod AB is the absolute velocity $\boldsymbol{v}_a$, its direction is shown in Fig. 7-5b.

Example 7-4

The triangle A is moving to the right on a horizontal surface with a constant speed of $v=10$ m/s, Fig.7-6a. The block M slides down along the incline with the velocity relative to the triangle $v_r=(2+6t)$ m/s. At the instant $t=1$ s, determine the absolute velocity of the block relative to the ground.

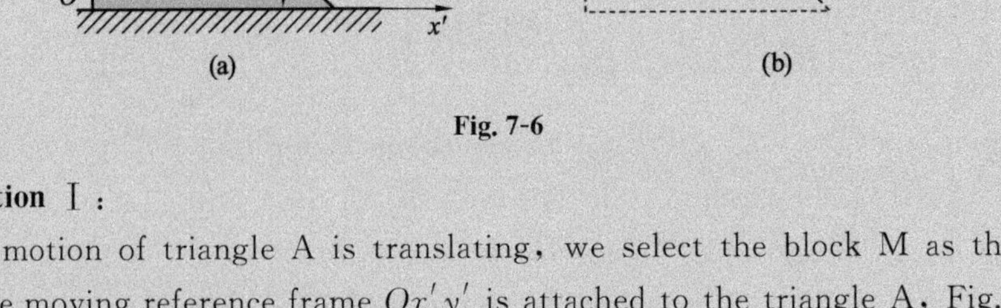

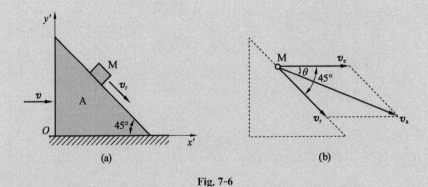

Fig. 7-6

Solution Ⅰ:

The motion of triangle A is translating, we select the block M as the moving point, the moving reference frame $Ox'y'$ is attached to the triangle A, Fig. 7-6a. So that the absolute motion of block M is curvilinear motion, the relative motion of the block is rectilinear motion, and the transport motion is the *translation* of triangle A. Applying the relative-velocity equation gives

$$\boldsymbol{v}_a=\boldsymbol{v}_e+\boldsymbol{v}_r$$

where the direction of $\boldsymbol{v}_e$ is horizontal and its magnitude is 10 m/s, the direction of $\boldsymbol{v}_r$ is along the incline and its magnitude can be calculated when $t=1$ s.

$$v_r=2+6\times1=8 \text{ m/s}$$

Since $\boldsymbol{v}_r$ and $\boldsymbol{v}_e$ are known in both magnitude and direction, the unknowns are the direction and magnitude of $\boldsymbol{v}_a$, in Fig. 7-6b. Using trigonometry, the magnitude of $\boldsymbol{v}_a$ is

$$\begin{aligned} v_a &= \sqrt{v_e^2+v_r^2+2v_e v_r\cos 45^\circ} \\ &= \sqrt{10^2+8^2+2\times10\times8\times\cos 45^\circ} \\ &= 16.6 \text{ m/s} \end{aligned} \qquad \textit{Ans.}$$

The direction of $\boldsymbol{v}_a$ can be obtained by law of sines

$$\frac{v_e}{\sin(45^\circ-\theta)}=\frac{v_r}{\sin\theta}$$

$$\theta=\arctan\frac{v_r\sin 45^\circ}{v_e+v_r\cos 45^\circ}=19.9^\circ \qquad \textit{Ans.}$$

The absolute velocity of the block M is $v_a=16.6$ m/s, its direction is $\theta=19.9^\circ$.

Solution Ⅱ :

The unknown components of absolute velocity $\boldsymbol{v}_a$ can also be determined by applying a scalar analysis.

$$\boldsymbol{v}_a=\boldsymbol{v}_e+\boldsymbol{v}_r$$

Resolving each vector into its x and y components yields

$$v_{ax}=v_{ex}+v_{rx}=10+8\cos 45^\circ=15.656 \text{ m/s}$$

$$v_{ay}=v_{ey}+v_{ry}=0+8\sin 45^\circ=5.656 \text{ m/s}$$

The magnitude and direction of the absolute velocity can be obtained

$$v_a=\sqrt{v_{ax}^2+v_{ay}^2}=\sqrt{15.656^2+5.656^2}=16.6 \text{ m/s} \qquad \textit{Ans.}$$

$$\tan\theta=\frac{v_{ay}}{v_{ax}}=\frac{5.656}{15.656}=0.361$$

$$\theta\approx19.9^\circ \qquad \textit{Ans.}$$

We obtain the previous same results.

7.3 Composition of the Accelerations

7.3.1 Composition of the Accelerations Using Translating Axes

Consider a particles B, which moves along the arbitrary paths. To make the derivation easy to understand, the position vectors $\boldsymbol{r}_A$ and $\boldsymbol{r}_B$ are measured with respect to the fixed x, y coordinate system. The original point of moving reference frame $Ax'y'$ is located in particle A, which is translating with respect to the x, y system, as shown in Fig. 7-7. The position of B measured respect to origin A is denoted by the relative-position vector $\boldsymbol{r}_{BA}$, Eq. (7.1).

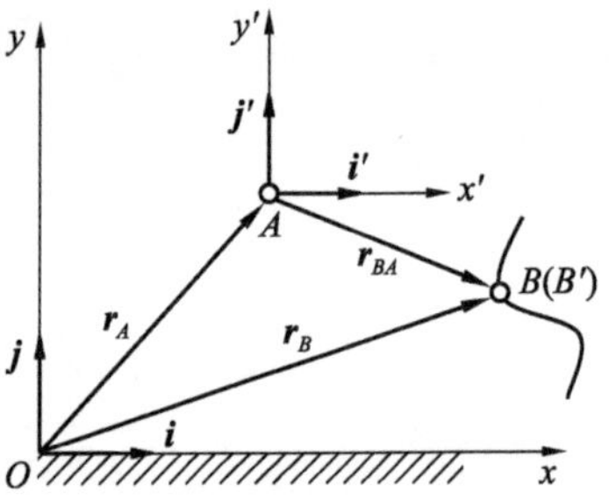

Fig. 7-7

Since the moving reference frame translates, the components of $\boldsymbol{r}_{BA}$ will *not change direction* and therefore the time derivative of $\boldsymbol{r}_{BA}$ will only change the magnitudes. Namely,

$$\frac{d\boldsymbol{i}'}{dt}=\frac{d\boldsymbol{j}'}{dt}=0$$

So the transport velocity Eq. (7.5) is

$$\boldsymbol{v}_e=\boldsymbol{v}_A$$

Using the Eq. (7.4) in Section 7.2, we have

$$\boldsymbol{v}_a=\boldsymbol{v}_A+\left(\frac{dx'}{dt}\boldsymbol{i}'+\frac{dy'}{dt}\boldsymbol{j}'\right) \tag{7.8}$$

The *absolute acceleration* $\boldsymbol{a}_a$ of particle B, observed from the x, y coordinate system, may be expressed by taking the time derivative of Eq. (7.8)

$$\frac{d\boldsymbol{v}_a}{dt}=\frac{d\boldsymbol{v}_A}{dt}+\frac{d}{dt}\left(\frac{dx'}{dt}\boldsymbol{i}'+\frac{dy'}{dt}\boldsymbol{j}'\right)$$

$$\boldsymbol{a}_a=\boldsymbol{a}_A+\left(\frac{d^2x'}{dt^2}\boldsymbol{i}'+\frac{d^2y'}{dt^2}\boldsymbol{j}'\right) \tag{7.9}$$

The first term of the right is the acceleration of the origin A of the x', y' reference, measured from the x, y reference. Since the moving frame translates, all points in moving frame have the *same velocity* and *acceleration*. This acceleration is the *transport acceleration* of particle B'.

$$\boldsymbol{a}_e=\boldsymbol{a}_A=\frac{d\boldsymbol{v}_A}{dt} \tag{7.10}$$

The two terms in the parentheses, represent the acceleration as measured by an observer attached to the moving x', y' coordinate system. It is the *relative acceleration* of particle B

$$\boldsymbol{a}_r=\frac{d\boldsymbol{v}_r}{dt}=\frac{d^2x'}{dt^2}\boldsymbol{i}'+\frac{d^2y'}{dt^2}\boldsymbol{j}' \tag{7.11}$$

Substituting these results into Eq. (7.9), we can obtain a similar vector relation between the absolute and relative accelerations of particles A and B.

$$\boldsymbol{a}_a=\boldsymbol{a}_e+\boldsymbol{a}_r \tag{7.12}$$

Eq. (7.12) states that *the absolute acceleration of particle B is equal to the transport acceleration plus (vectorially) the relative acceleration of B with respect to A.*

Example 7-5

The crank OA has length r and rotates around the axis O with a constant angular velocity of ω, the block is pin-connected with crank and can slide in the slot of arch BC, as shown in Fig. 7-8a. Determine the velocity and acceleration of the component BCD and the acceleration of block A with respect to component BCD at this instant.

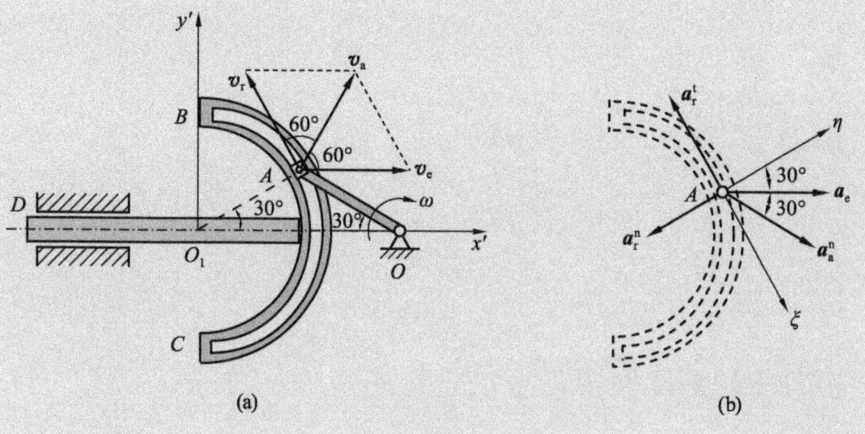

Fig. 7-8

Solution:

① **Velocity.** The component BCD is translating, so the motion of point A located in component can be used to specify the motion of component BCD. The moving reference frame $Ox'y'$ is attached to the rod BCD, Fig. 7-8a. We select the block A as the moving point, so that the absolute motion of block A is circular motion around the center O, relative motion is circular motion around the center O_1 and transport motion is the rectilinear translation of the component BCD. Applying the relative-velocity equation gives

$$\boldsymbol{v}_a = \boldsymbol{v}_e + \boldsymbol{v}_r$$

where the direction of $\boldsymbol{v}_a$ is perpendicular to the radius OA, the direction of $\boldsymbol{v}_e$ is horizontal and the direction of $\boldsymbol{v}_r$ is tangent to the slot and perpendicular to O_1A, so the direction of three velocities are known and drawn in Fig. 7-8a. The magnitude of absolute velocity can be calculated $v_a = r\omega$, using trigonometry, we have

$$v_e = v_r = v_a = r\omega \qquad \textit{Ans.}$$

At this instant, the magnitude of velocity of the component BCD is the transport velocity v_e, the direction is shown in Fig. 7-8a.

② **Acceleration.** Since absolute motion of block A is circular motion with constant angular velocity of ω, the tangential component of absolute acceleration is zero. But block A has both tangential and normal components of relative acceleration. Applying the equation of composite acceleration yields

$$\boldsymbol{a}_a^n = \boldsymbol{a}_e + \boldsymbol{a}_r^t + \boldsymbol{a}_r^n \quad \text{(a)}$$

where the direction of $\boldsymbol{a}_a^n$ is along the rod OA, the direction of $\boldsymbol{a}_e$ is horizontal, the direction of $\boldsymbol{a}_r^t$ is tangent to the slot and perpendicular to the radius O_1A and the direction of $\boldsymbol{a}_r^n$ is along the rod O_1A, so the direction of the four accelerations are known and drawn in Fig. 7-8b. The magnitude of $\boldsymbol{a}_a^n$ and $\boldsymbol{a}_r^n$ are calculated by

$$a_a^n = r\omega^2$$

$$a_r^n = \frac{\boldsymbol{v}_r^2}{r} = r\omega^2$$

The magnitude of $\boldsymbol{a}_e$ and $\boldsymbol{a}_r^t$ can be calculated by resolving each vector of Eq. (a) into its η and ξ components.

$$a_a^n \cos 60° = a_e \cos 30° - a_r^n$$

$$a_a^n \cos 30° = a_e \cos 60° - a_r^t$$

Solving, we obtain

$$a_e = \sqrt{3} r\omega^2 \quad \textit{Ans.}$$

$$a_r^t = 0, \; a_r^n = r\omega^2 \quad \textit{Ans.}$$

At this instant, the acceleration of the component BCD is $a_e = \sqrt{3} r\omega^2$, the acceleration of block A with respect to BCD is $\boldsymbol{a}_r^n$, their directions are shown in Fig. 7-8b.

Example 7-6

The semicircle cam, has radium of $R = 100$ cm, is moving on the horizontal surface, Fig. 7-9a. At the instant $\theta = 60°$, the cam has a translating velocity $\boldsymbol{v} = 60$ cm/s and acceleration $a = 45$ cm/s^2. Determine the velocity and acceleration of the guide rod AB.

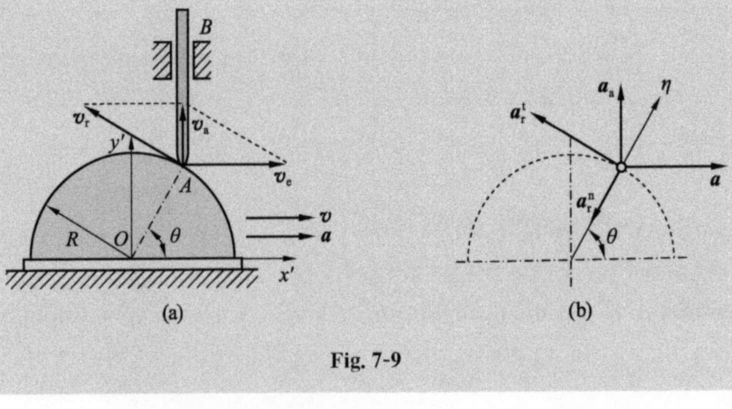

Fig. 7-9

Solution:

① **Velocity.** The guide rod AB is translating, so the motion of point A located in rod AB can be used to specify the motion of guide rod. The moving reference frame $Ox'y'$ is attached to the semicircle cam, Fig. 7-9a. We select the point A on the guide rod AB as the moving point, so that the absolute motion of point A is rectilinear motion up and down, relative motion is the circular motion about the cam center O, and transport motion is the translation of the cam. Applying the relative-velocity equation gives

$$\boldsymbol{v}_a = \boldsymbol{v}_e + \boldsymbol{v}_r$$

where the direction of $\boldsymbol{v}_a$ is along the rod AB, the direction of $\boldsymbol{v}_e$ is same to the $\boldsymbol{v}$ and the direction of $\boldsymbol{v}_r$ is perpendicular to OA and tangent to the circle, so the direction of three velocities are known and drawn in Fig. 7-9a. The magnitude of transport velocity is calculated $\boldsymbol{v}_e = \boldsymbol{v} = 60$ cm/s, using trigonometry, we have

$$v_a = \boldsymbol{v}_e \cot\theta = 60 \cot 60° = 34.6 \text{ cm/s} \qquad \textit{Ans.}$$

$$v_r = \frac{v_e}{\sin\theta} = \frac{v}{\sin\theta} = \frac{60}{\sin 60°} = 69.3 \text{ cm/s}$$

At this instant, the magnitude of velocity of the guide rod AB is the absolute velocity $\boldsymbol{v}_a$, the direction is shown in Fig. 7-9a.

② **Acceleration.** Since relative motion is the circular motion, the point A has both tangential and normal components of relative acceleration. Applying the equation for relative acceleration yields

$$\boldsymbol{a}_a = \boldsymbol{a}_e + \boldsymbol{a}_r^t + \boldsymbol{a}_r^n \qquad (a)$$

where the direction of $\boldsymbol{a}_a$ is along the rod OA, the direction of $\boldsymbol{a}_e$ is horizontal, the direction of $\boldsymbol{a}_r^t$ is tangent to the semicircle cam and the direction of $\boldsymbol{a}_r^n$ is along the radium of the semicircle and perpendicular to $\boldsymbol{a}_r^t$, so the direction of the four accelerations are known and drawn in Fig. 7-9b. The magnitude of $\boldsymbol{a}_e$ and $\boldsymbol{a}_r^n$ are calculated by

$$a_e = a = 45 \text{ cm/s}^2$$

$$a_r^n = \frac{\boldsymbol{v}_r^2}{R} = \frac{69.3^2}{100} = 48 \text{ cm/s}^2$$

The magnitude of $\boldsymbol{a}_a$ can be calculated by resolving each vector of Eq. (a) into its η components.

$$a_a \sin\theta = a_e \cos\theta - a_r^n$$

Solving, we obtain

$$a_a = -30.6 \text{ cm/s}^2 \qquad \textit{Ans.}$$

The acceleration of the guide rod AB is $a_a = -30.6$ cm/s^2 at this instant. The direction is opposite to the direction shown in Fig. 7-9b.

Example 7-7

The crank OA of radius r, rotates with constant angular velocity ω_0, Fig. 7-10a. The collar A which is pin connected with crank OA slides on parallel rod BC. At the instant as shown, determine the angular velocity ω and angular acceleration α of rod BD. Set $BC=DE, BD=CE=l$.

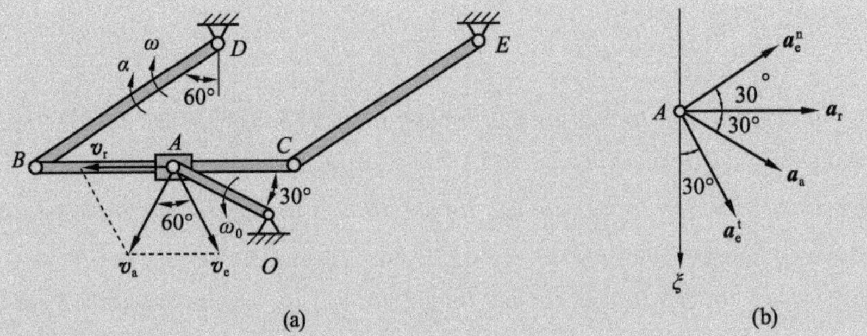

Fig. 7-10

Solution:

① **Velocity.** The rod BC is translating, so the motion of point A located in rod can be used to specify the motion of rod. The fixed axes are established at an arbitrary point on the ground and the moving reference frame is attached to the rod BC. We select the collar A as the moving point, so that the absolute motion of collar A is circular motion around the pin O, relative motion of collar A is the rectilinear motion left and right, and transport motion is the translation of the rod BC. Applying the relative-velocity equation gives

$$\boldsymbol{v}_a = \boldsymbol{v}_e + \boldsymbol{v}_r$$

where the direction of $\boldsymbol{v}_a$ is perpendicular to OA, the direction of $\boldsymbol{v}_r$ is along the rod BC, and the direction of $\boldsymbol{v}_e$ is perpendicular to the rod BD, so the direction of three velocities are known and drawn in Fig. 7-10a. Since the magnitude of absolute velocity can be calculated $v_a = \omega_0 r$, using trigonometry, we have

$$v_e = v_r = v_a = \omega_0 r$$

$$\omega_{BD} = \frac{v_e}{BD} = \frac{r\omega_0}{l} \qquad \textit{Ans.}$$

② **Acceleration.** Since transport motion is the curvilinear translation motion, the point A' has both tangential and normal components of transport acceleration. Applying the equation for relative acceleration yields

$$\boldsymbol{a}_a = \boldsymbol{a}_e^t + \boldsymbol{a}_e^n + \boldsymbol{a}_r \qquad (a)$$

where the direction of $\boldsymbol{a}_a$ is along the rod OA, the direction of $\boldsymbol{a}_r$ is horizontal along the rod BC, the direction of $\boldsymbol{a}_e^t$ is perpendicular to the rod BD and the direction of $\boldsymbol{a}_e^n$ is along the rod BD and perpendicular to $\boldsymbol{a}_e^t$, so the direction of the four accelerations are known and drawn in Fig. 7-10b. The magnitude of $\boldsymbol{a}_a$ and $\boldsymbol{a}_e^n$ are given by

$$a_a = r\omega_0^2$$

$$a_e^n = l\omega_{BD}^2 = \frac{r^2\omega_0^2}{l}$$

The magnitude of $\boldsymbol{a}_e^t$ can be calculated by the projection of equation (a) on ξ axis.

$$a_a \sin 30° = a_e^t \cos 30° - a_e^n \cos 60°$$

Solving, we have

$$a_e^t = \frac{\sqrt{3}\omega_0^2 r(l+r)}{3l}$$

$$\alpha_{BD} = \frac{a_e^t}{BD} = \frac{\sqrt{3}\omega_0^2 r(l+r)}{3l^2} \qquad \textit{Ans.}$$

The angular acceleration of the rod BD is α_{BD} at this instant.

7.3.2 Composition of the Accelerations Using Rotating Axes

When the moving reference frame is the rotating system of coordinates, the components of $\boldsymbol{r}_{BA}$ will *change the direction and magnitude*. The absolute velocity of point B observed from the x, y coordinate system, Fig. 7-7, is determined from the Eq. (7.14).

$$\boldsymbol{v}_a = \boldsymbol{v}_A + \left(x'\frac{d\boldsymbol{i}'}{dt} + y'\frac{d\boldsymbol{j}'}{dt}\right) + \left(\frac{dx'}{dt}\boldsymbol{i}' + \frac{dy'}{dt}\boldsymbol{j}'\right) \tag{7.13}$$

Viewing the axes in two dimensions, and noting that $\boldsymbol{\omega} = \omega\boldsymbol{k}$, then in the first set of parentheses of Eq. (7.13), the instantaneous time rate of change of the unit vectors can express in terms of the cross product as

$$\frac{d\boldsymbol{i}'}{dt} = \boldsymbol{\omega} \times \boldsymbol{i}', \quad \frac{d\boldsymbol{j}'}{dt} = \boldsymbol{\omega} \times \boldsymbol{j}' \tag{7.14}$$

The *absolute acceleration* $\boldsymbol{a}_a$ of point B, observed from the x, y coordinate system, can be expressed by taking the time derivative of Eq. (7.13).

$$\frac{d\boldsymbol{v}_a}{dt} = \frac{d\boldsymbol{v}_A}{dt} + \frac{d}{dt}\left(x'\frac{d\boldsymbol{i}'}{dt} + y'\frac{d\boldsymbol{j}'}{dt}\right) + \frac{d}{dt}\left(\frac{dx'}{dt}\boldsymbol{i}' + \frac{dy'}{dt}\boldsymbol{j}'\right)$$

$$\boldsymbol{a}_a = \boldsymbol{a}_A + \left(x'\frac{d^2\boldsymbol{i}'}{dt^2} + y'\frac{d^2\boldsymbol{j}'}{dt^2}\right) + \left(\frac{d^2x'}{dt^2}\boldsymbol{i}' + \frac{d^2y'}{dt^2}\boldsymbol{j}'\right) + 2\left(\frac{dx'}{dt}\frac{d\boldsymbol{i}'}{dt} + \frac{dy'}{dt}\frac{d\boldsymbol{j}'}{dt}\right) \tag{7.15}$$

The first parenthesis of the Eq. (7.15) contains only the derivatives of the unit vectors of the moving axes, this derivative is obtained as the relative vector has constant magnitude with respect to the moving reference system. The absolute velocities $\boldsymbol{a}_A$ plus the first parentheses represent the acceleration of the point B' in the moving reference frame *coinciding* with the moving particle B at the instant, and it is named *transport acceleration* of particle B:

$$\boldsymbol{a}_e = \boldsymbol{a}_A + \left(x' \frac{d^2 \boldsymbol{i}'}{dt^2} + y' \frac{d^2 \boldsymbol{j}'}{dt^2}\right) \tag{7.16}$$

The second parenthesis of the Eq. (7.15) represents the components of acceleration of point B as measured by an observer attached to the rotating coordinate system x', y', these terms will be named *relative acceleration* of point B,

$$\boldsymbol{a}_r = \frac{d^2 x'}{dt^2}\boldsymbol{i}' + \frac{d^2 y'}{dt^2}\boldsymbol{j}' \tag{7.17}$$

Using the Eq. (7.14), the last parenthesis of the Eq. (7.15) can be rewritten

$$2\left(\frac{dx'}{dt}\frac{d\boldsymbol{i}'}{dt} + \frac{dy'}{dt}\frac{d\boldsymbol{j}'}{dt}\right) = 2\left[\frac{dx'}{dt}(\boldsymbol{\omega}\times\boldsymbol{i}') + \frac{dy'}{dt}(\boldsymbol{\omega}\times\boldsymbol{j}')\right]$$

$$= 2\boldsymbol{\omega}\times\left(\frac{dx'}{dt}\boldsymbol{i}' + \frac{dy'}{dt}\boldsymbol{j}'\right) \tag{7.18}$$

The terms in the parenthesis of the Eq. (7.18) are *relative velocity* $\boldsymbol{v}_r$. Denote

$$\boldsymbol{a}_C = 2\boldsymbol{\omega}\times\boldsymbol{v}_r \tag{7.19}$$

This acceleration $\boldsymbol{a}_C$ called the *Coriolis acceleration*, named after the French engineer Coriolis G. C., who was the first to determine it.

Substituting these results into Eq. (7.15), then we can obtain

$$\boldsymbol{a}_a = \boldsymbol{a}_e + \boldsymbol{a}_r + \boldsymbol{a}_C \tag{7.20}$$

This equation states that the *absolute acceleration* of a moving point is the vector sum of its *transport acceleration*, *relative acceleration* and *Coriolis acceleration*.

◆ **Coriolis Acceleration**

The acceleration $\boldsymbol{a}_C$ represents the combined effect of B moving relative to x, y, z coordinates and rotation of x', y', z' frame, as indicated by the vector cross product, Eq. (7.19). Therefore, Coriolis acceleration will always be perpendicular to both $\boldsymbol{\omega}$ and $\boldsymbol{v}_r$ and the sense is determined by right-hand rule, Fig. 7-11a. It is an important component of the acceleration which must be considered whenever rotating reference frames are used. For example, when studying the accelerations and forces which act on rockets or other bodies whose motions are significantly affected by the rotation of the earth. If the included angle between the $\boldsymbol{\omega}$ and $\boldsymbol{v}_r$ is θ, the magnitude of Coriolis acceleration can be calculated by

$$a_C = 2\omega_e v_r \sin\theta$$

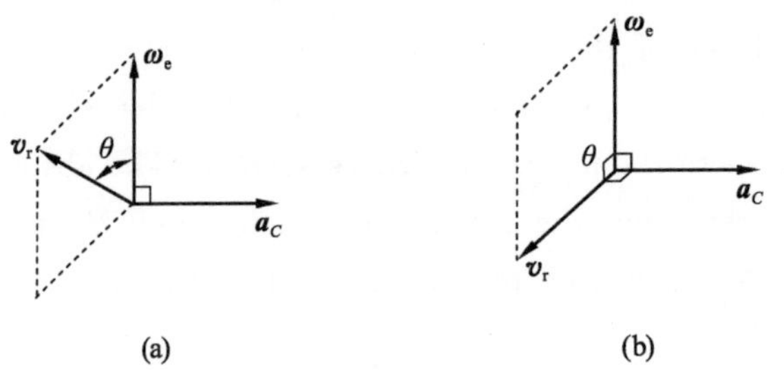

Fig. 7-11

In some case, if $\theta = 90°$ as shown in Fig. 7-11b, the magnitude of Coriolis acceleration becomes

$$a_C = 2\omega_e v_r \tag{7.21}$$

Example 7-8

A rod OA, shown in Fig. 7-12, rotates around axis O, collar B travels outward along the rod. At the instant $\theta = 60°$, the rod has an angular velocity of 3 rad/s and an angular acceleration of 2 rad/s^2. The collar B has velocity of 2 m/s and the acceleration is 3 m/s^2, both measured relative to the rod, when $x = 0.2$ m. Determine the Coriolis acceleration and the velocity and acceleration of the collar at this instant.

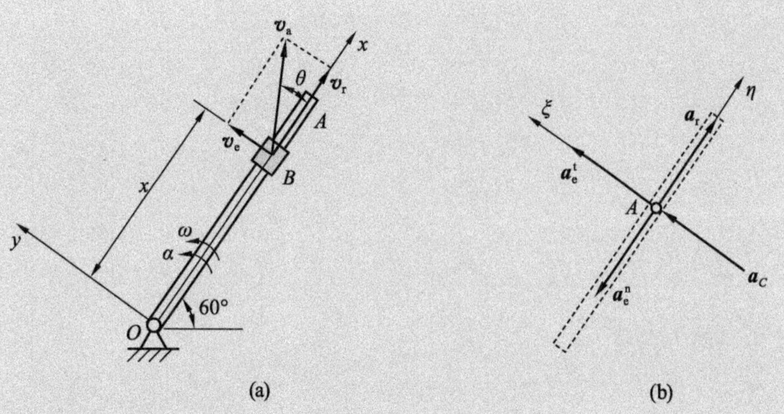

Fig. 7-12

Solution:

① **Velocity.** The moving reference frame $Ox'y'$ is attached to and rotates with the rod OA, and the origin of the moving frames is located at O, Fig. 7-12a. We select the collar A as the moving point, so that the absolute motion of collar A is curvilinear motion, relative motion of collar is rectilinear motion along the rod OA and transport motion is the rotation of the rod OA about the axis O. Applying the relative-velocity equation gives

$$\boldsymbol{v}_a = \boldsymbol{v}_e + \boldsymbol{v}_r$$

where the direction of $\boldsymbol{v}_e$ is perpendicular to rod OA and the direction of $\boldsymbol{v}_r$ is along the rod OA and perpendicular to $\boldsymbol{v}_e$, in Fig. 7-12a. The magnitude of $\boldsymbol{v}_e$ and $\boldsymbol{v}_r$ can be calculated as

$$v_r = 2 \text{ m/s}$$

$$v_e = x \cdot \omega = 0.2 \times 3 = 0.6 \text{ m/s}$$

Since $\boldsymbol{v}_a$ and $\boldsymbol{v}_e$ are known in both magnitude and direction, the unknowns are the direction and magnitude of $\boldsymbol{v}_a$. Using trigonometry, we have

$$v_a = \sqrt{v_e^2 + v_r^2} = \sqrt{0.6^2 + 2^2} = 2.09 \text{ m/s} \qquad Ans.$$

$$\theta=\arctan\frac{v_e}{v_r}=\arctan\frac{0.6}{2}=16.7^\circ \qquad \textit{Ans.}$$

The velocity of the collar at this instant is absolute velocity $v_a=2.09$ m/s and angle $\theta=16.7^\circ$.

② **Acceleration.** Since transport motion is the rotation about the axis O, there are Coriolis acceleration and both tangential and normal components of transport acceleration. Applying the equation for relative acceleration yields

$$\boldsymbol{a}_a=\boldsymbol{a}_e^t+\boldsymbol{a}_e^n+\boldsymbol{a}_r+\boldsymbol{a}_C \qquad \text{(a)}$$

where the direction of $\boldsymbol{a}_e^t$ is perpendicular to OA, the direction of $\boldsymbol{a}_e^n$ is vertical to $\boldsymbol{a}_e^t$ and along the rod OA towards the center O. The direction of $\boldsymbol{a}_r$ is along the rod OA. The direction of $\boldsymbol{a}_C$ is vertical to $\boldsymbol{v}_r$ and sense is same as the angular velocity ω of rod, Fig. 7-12b. The magnitude of some acceleration is calculated as

$$a_r=3 \text{ m/s}^2$$

$$a_C=2\omega v_r=2\times3\times2=12 \text{ m/s}^2 \qquad \textit{Ans.}$$

$$a_e^t=\alpha\cdot x=2\times0.2=0.4 \text{ m/s}^2$$

$$a_e^n=x\cdot\omega^2=0.2\times3^2=1.8 \text{ m/s}^2$$

Since $\boldsymbol{a}_e^t$, $\boldsymbol{a}_e^n$, $\boldsymbol{a}_r$ and $\boldsymbol{a}_C$ are known in both magnitude and direction, the unknowns are the direction and magnitude of $\boldsymbol{a}_a$, in Fig. 7-12b. Resolving each vector of equation (a) into its η and ξ components, yield

$$a_a^\eta=a_r-a_e^n=3-1.8=1.2 \text{ m/s}^2$$

$$a_a^\xi=a_e^t+a_C=0.4+12=12.4 \text{ m/s}^2$$

The magnitude and direction of $\boldsymbol{a}_a$ is

$$a_a=\sqrt{(a_a^\eta)^2+(a_a^\xi)^2}=\sqrt{1.2^2+12.4^2}=12.46 \text{ m/s}^2 \qquad \textit{Ans.}$$

$$\theta=\arctan\frac{a_a^\eta}{a_a^\xi}=\arctan\frac{1.2}{12.4}=5.53^\circ \qquad \textit{Ans.}$$

The Coriolis acceleration is $\boldsymbol{a}_C=12$ m/s^2. The acceleration of the collar at this instant is $\boldsymbol{a}_a=12.46$ m/s^2 and angle $\theta=5.53^\circ$.

Example 7-9

The crank $OA=12$ cm rotates counterclockwise about the axis O with a constant angular velocity $\omega=7$ rad/s, shown in Fig. 7-13a. The collar at A is pin connected to crank OA and slides over rod O_1B. Determine the angular velocity and angular acceleration of rod O_1B at the instant when $\theta=30^\circ$.

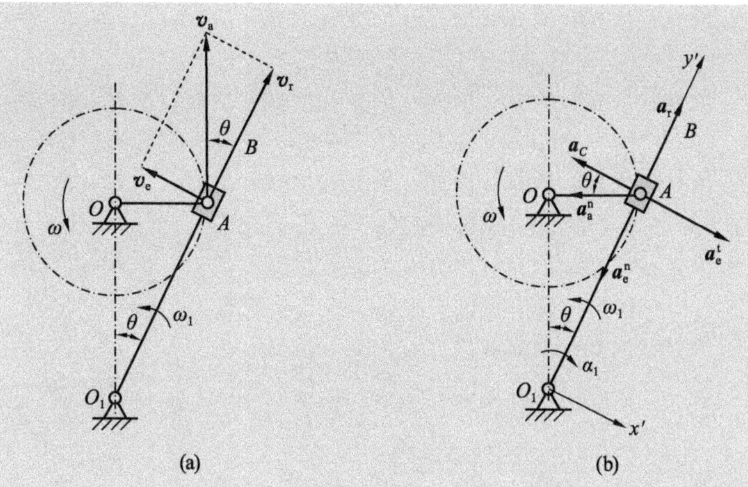

Fig. 7-13

Solution:

① **Velocity.** The moving reference frame $O_1x'y'$ is attached to and rotates with the rod O_1B, and the origin of the moving reference frames is located at O_1, Fig. 7-13a. We select the collar A as the moving point, so that the absolute motion of collar A is circular motion about the center O, relative motion is rectilinear motion along the rod O_1B and transport motion is the rotation of the rod O_1B about the axis O_1. Applying the relative-velocity equation gives

$$\boldsymbol{v}_a = \boldsymbol{v}_e + \boldsymbol{v}_r$$

where the direction of $\boldsymbol{v}_a$ is perpendicular to the crank OA, the direction of $\boldsymbol{v}_e$ is perpendicular to rod O_1B and the direction of $\boldsymbol{v}_r$ is along the rod O_1B and perpendicular to $\boldsymbol{v}_e$. So the direction of three velocities are known and drawn in Fig. 7-13a. The magnitude of absolute velocity $\boldsymbol{v}_a$ can be calculated as

$$v_a = OA \cdot \omega = 12 \times 7 = 84 \text{ cm/s}$$

Using trigonometry, we have

$$v_e = v_a \sin\theta = 84 \times \sin 30° = 42 \text{ cm/s}$$

$$v_r = v_a \cos\theta = 84 \times \cos 30° = 72.7 \text{ cm/s}$$

The angular velocity ω_1 of rod O_1B at this instant can be determined

$$O_1A = \frac{OA}{\sin 30°} = 24 \text{ cm}$$

$$\omega_1 = \frac{v_e}{O_1A} = \frac{42}{24} = 1.75 \text{ rad/s} \qquad \textit{Ans.}$$

② **Acceleration.** Since absolute motion of block A is circular motion with constant angular velocity of ω, the tangential component of absolute acceleration is zero. But transport motion is the rotation about the axis O_1, there are *Coriolis* acceleration and both tangential and normal components of transport acceleration. Applying the equation for relative acceleration yields

$$\boldsymbol{a}_a^n = \boldsymbol{a}_e^t + \boldsymbol{a}_e^n + \boldsymbol{a}_r + \boldsymbol{a}_C \tag{a}$$

where the direction of $\boldsymbol{a}_a^n$ is along the rod OA, the direction of $\boldsymbol{a}_e^t$ is perpendicular to O_1B, the direction of $\boldsymbol{a}_e^n$ is along the rod O_1B towards to O_1, the direction of $\boldsymbol{a}_r$ is along the rod O_1B, the direction of $\boldsymbol{a}_C$ is perpendicular to $\boldsymbol{v}_r$ and sense is same as the angular velocity ω_1 of rod O_1B. So the direction of all accelerations are known and drawn in Fig. 7-13b. The magnitude of $\boldsymbol{a}_a^n$, $\boldsymbol{a}_e^n$ and $\boldsymbol{a}_C$ are given by

$$a_a^n = OA \cdot \omega^2 = 12 \times 7^2 = 588 \text{ cm/s}^2$$

$$a_e^n = O_1A \cdot \omega_1{}^2 = 24 \times 1.75^2 = 73.5 \text{ cm/s}^2$$

$$a_C = 2\omega_1 v_r = 2 \times 1.75 \times 72.7 = 254.5 \text{ cm/s}^2$$

The magnitude of $\boldsymbol{a}_e^t$ can be calculated by resolving each vector of equation (a) into its x' axis, yield

$$-a_a \cos\theta = a_e^t - a_C$$

Solving, we have

$$a_e^t = a_C - a_a \cos\theta = -254.7 \text{ cm/s}^2$$

The angular acceleration α of rod O_1B at this instant can be determined

$$\alpha = \frac{a_e^t}{O_1A} = \frac{-254.7}{24} \text{rad/s}^2 = -10.6 \text{ rad/s}^2 \qquad \textit{Ans.}$$

The negative indicates that the directions of the angular acceleration of rod O_1B is opposite to the direction shown in Fig. 7-13b.

Exercises

7-1 In Fig. 7-14, a rod OA rotates around axis O with angular equation $\varphi = \pi t^2$ (where φ is in rad and t is in seconds), collar B travels along the rod with motion $s = 1 + 2t$ measured relative to the rod (where s is in m). Determine the absolute velocity $\boldsymbol{v}_a$ of the collar when $t = 0.5$ s.

7-2 The collar at A is pin connected to crank OA and slides over angle rod O_1BC, Fig. 7-15. Crank rotates counterclockwise around the axis O with a constant angular velocity ω_0. Determine the angular velocity of angle rod O_1BC at the instant.

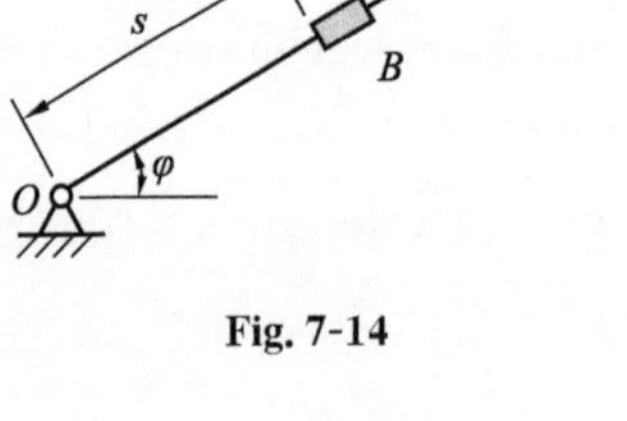

Fig. 7-14

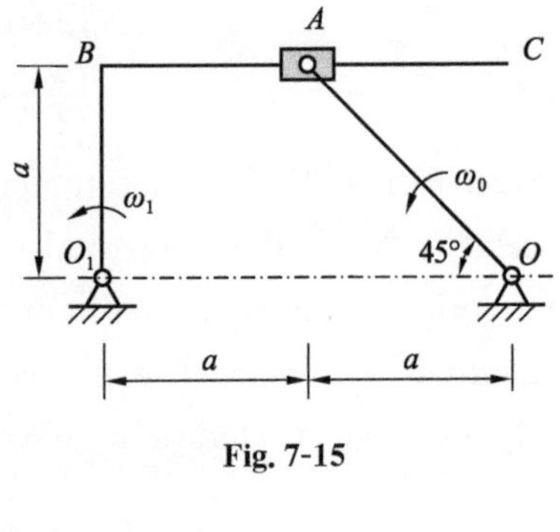

Fig. 7-15

7-3 The collar at A is pin connected to crank O_2A and slides over rod O_1A, Fig. 7-16. Rod O_1A rotates counterclockwise about the axis O_1 with a constant angular velocity $\omega_1 = 3$ rad/s. Determine the angular velocity of rod O_2A at the instant.

7-4 The collar at A is pin connected to crank O_1A and slides over rod O_2A, Fig. 7-17. Rod O_1A rotates counterclockwise about the axis O with a constant angular velocity $\omega_1 = 3$ rad/s. Determine the angular velocity of rod O_2A at the instant.

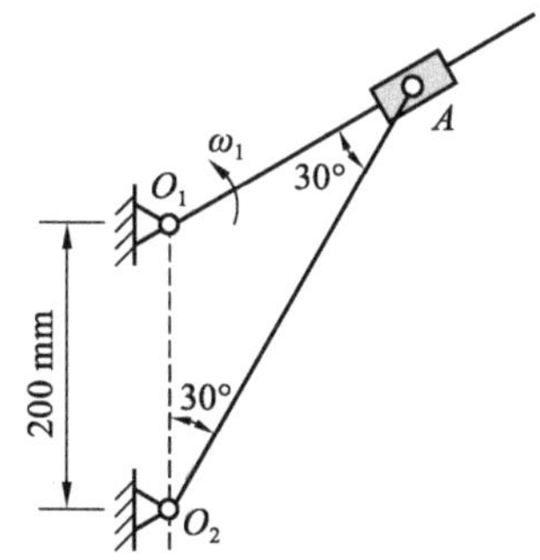

Fig. 7-16

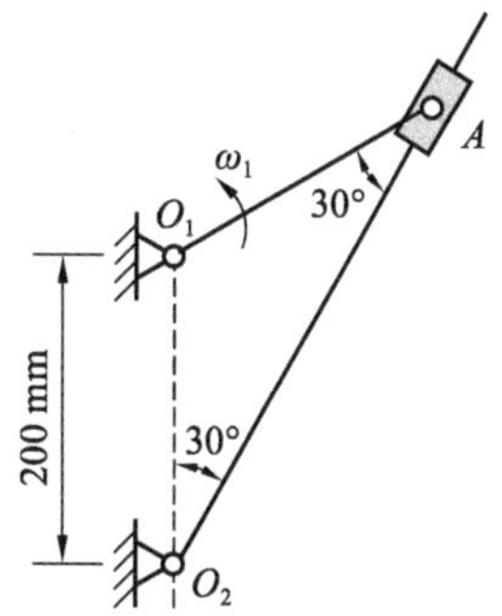

Fig. 7-17

7-5 As shown in Fig. 7-18, angle rod BC is moving to the left with a constant velocity $\boldsymbol{u}$ in the horizontal slot, determine the velocity $\boldsymbol{v}_A$ of the endpoint A of rod OA.

7-6 In Fig. 7-19, the cam, has an eccentricity of $OC = e$ and radium of R, rotates about the axis O with constant angular velocity ω_0. Determine the velocity of the rod AB when $\theta = 0°$.

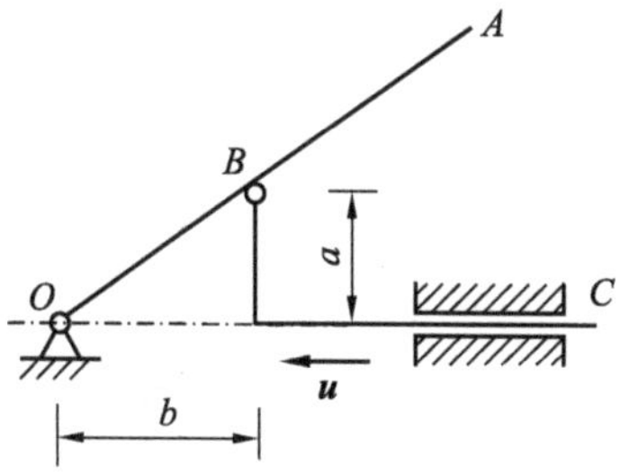

Fig. 7-18

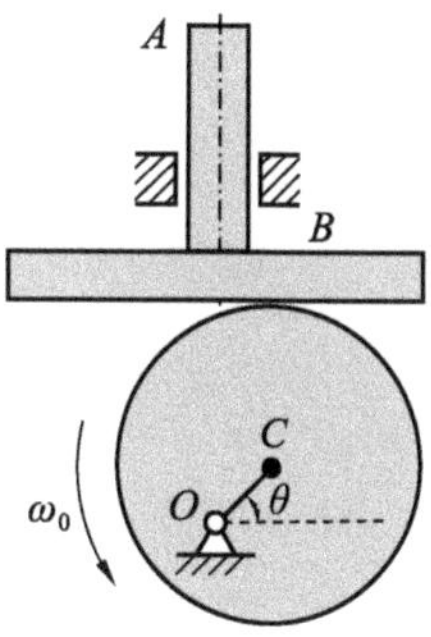

Fig. 7-19

7-7 In Fig. 7-20, car A is traveling with a constant speed of 80 km/h due north, while car B is traveling with a constant speed of 100 km/h due east. Determine the velocity of car B relative to car A.

7-8 In Fig. 7-21, rod AB is moving upward with a constant velocity $\boldsymbol{v}$ in the vertical slot. The collar A slides over rod OC, determine the velocity $\boldsymbol{v}_C$ of the endpoint C of rod OC.

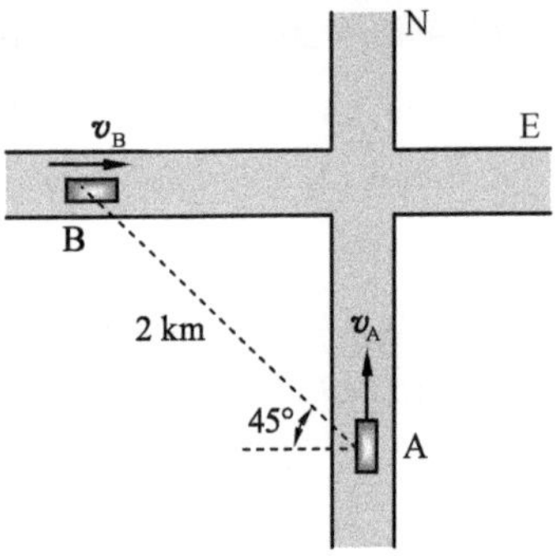

Fig. 7-20

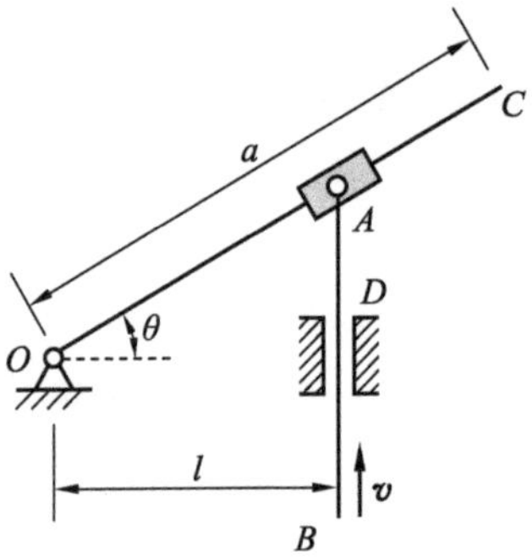

Fig. 7-21

7-9 The angle rod *BEC* is moving horizontal, Fig. 7-22. The collar *A* is pin connected to crank *OA* and slides freely over the rod *ED*. The crank has length $OA=0.1$ m and rotates about the axis *O* with a constant angular velocity of $\omega=20$ rad/s, determine the velocity and acceleration of the angle rod *BEC* at the instant when $\varphi=30°$.

7-10 As shown in Fig. 7-23, at the instant, the triangle *D* is moving to the right on a horizontal surface with velocity of $\boldsymbol{v}$ and acceleration of $\boldsymbol{a}$, determine the velocity and acceleration of rod *AB* as the function of angle θ.

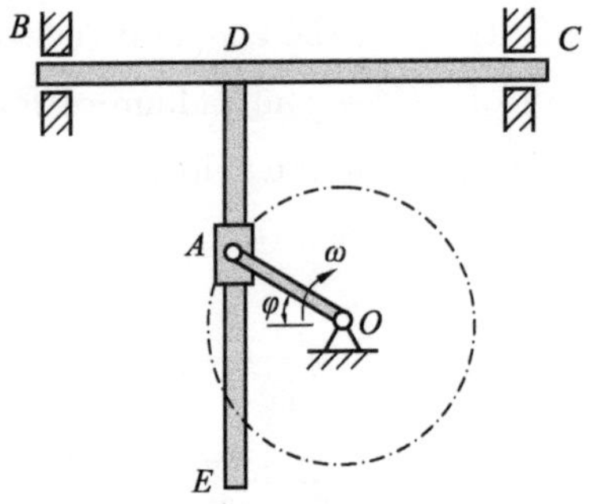

Fig. 7-22

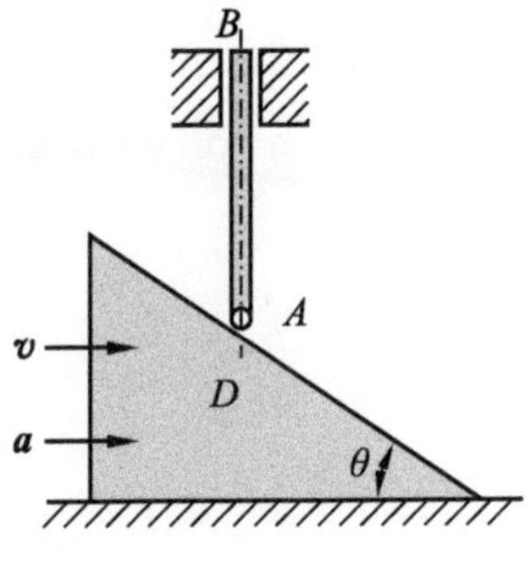

Fig. 7-23

7-11 In Fig. 7-24, the block with an arch slot of radius $r=0.1$ m is moving horizontal, the crank *OA* has length of 0.1 m and rotates around the axis *O*. At the instant $\varphi=30°$, the velocity of block is $v=0.1$ m/s and acceleration is $a=-0.2$ m/s^2 with direction as shown, determine the angular velocity and angular acceleration of the rod *OA*.

7-12 In Fig. 7-25, the crank *OA* of length 0.4 m, rotates counter clockwise about axis *O* with constant angular velocity $\omega=0.5$ rad/s. The slider rod *BC* is at up-down motion. Determine the velocity and acceleration of slider rod *BC* at the instant $\varphi=30°$.

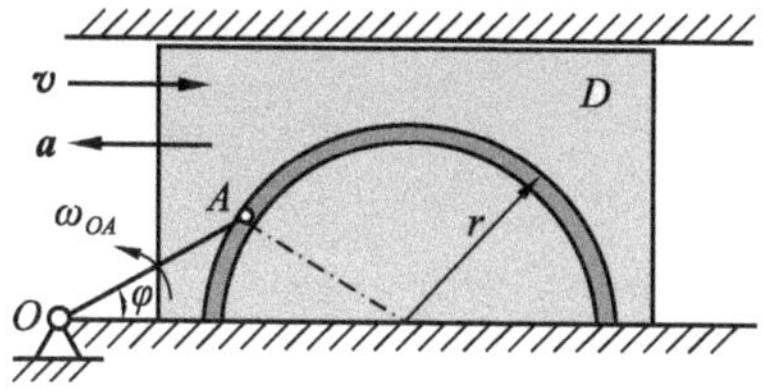

Fig. 7-24

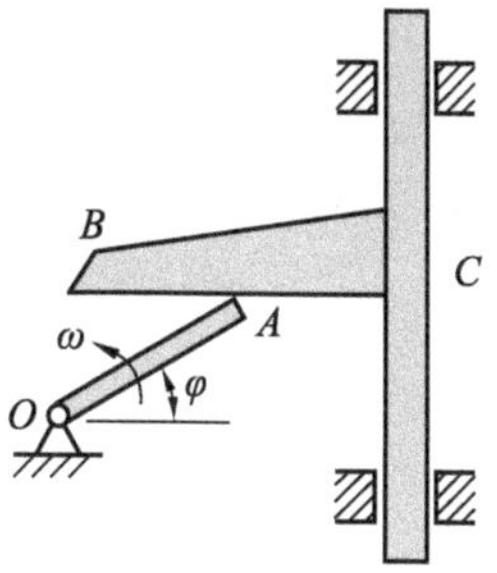

Fig. 7-25

7-13 In Fig. 7-26, the collar C is pin connected to rod CD and slides over rod AB which is supported by two same rod $O_1A = O_2B = 100$ mm, and $O_1O_2 = AB$. Rod O_1A rotates about axis O_1 with a constant angular velocity $\omega = 2$ rad/s. Determine the velocity and acceleration of the rod CD at the instant $\varphi = 60°$.

7-14 In Fig. 7-27, collar D is pin connected to the beam ABD and slides over rod EC. The beam ABD is supported by two same links, $AA' = BB' = 0.25$ m, $AB = A'B'$. The link AA' rotates with uniform angular velocity $\omega = 2$ rad/s about point A'. At the instant $\varphi = 60°$, rod CE is vertical to the surface, determine the angular velocity and angular acceleration of rod EC.

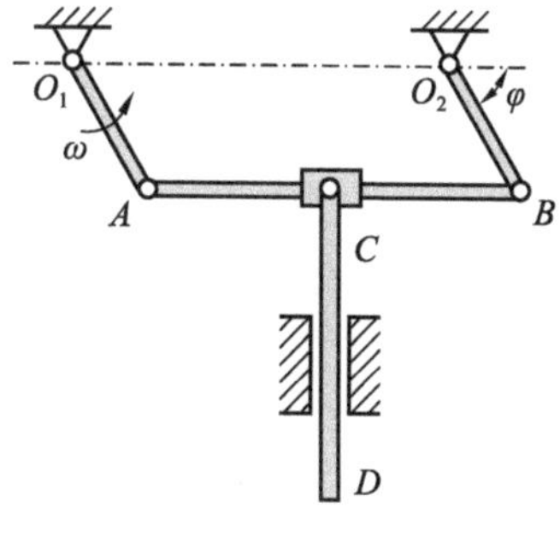

Fig. 7-26

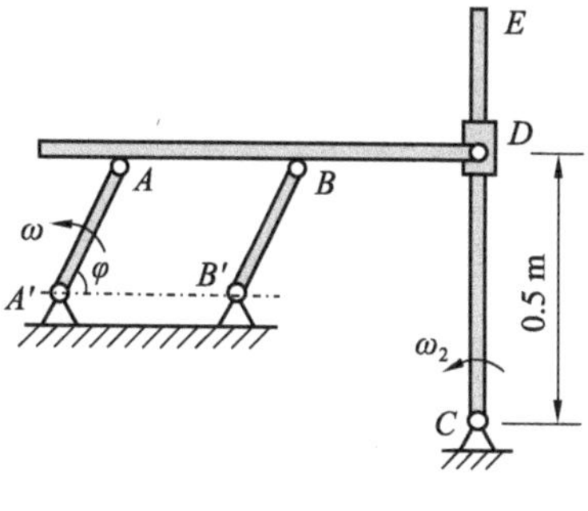

Fig. 7-27

7-15 In Fig. 7-28, the collar at C is pin connected to rod AC and slides over rod DE. Rod AC rotates clockwise and has an angular velocity $\omega_{AC} = 3$ rad/s and angular acceleration $\alpha_{AC} = 4$ rad/s^2 when $\theta = 45°$. Determine the angular velocity and angular acceleration of rod DE at this instant.

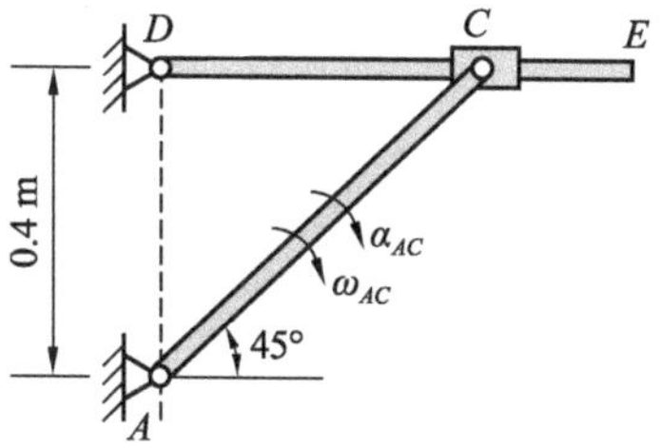

Fig. 7-28

7-16 In Fig. 7-29, the big ring has a radius of $R=0.5$ m and is fixed. The collar M can move in the rod AB and big ring. At the instant $\varphi=30°$, rod AB has an angular velocity $\omega=2$ rad/s and an angular acceleration $\alpha=4$ rad/s^2, determine the velocity and acceleration of collar M.

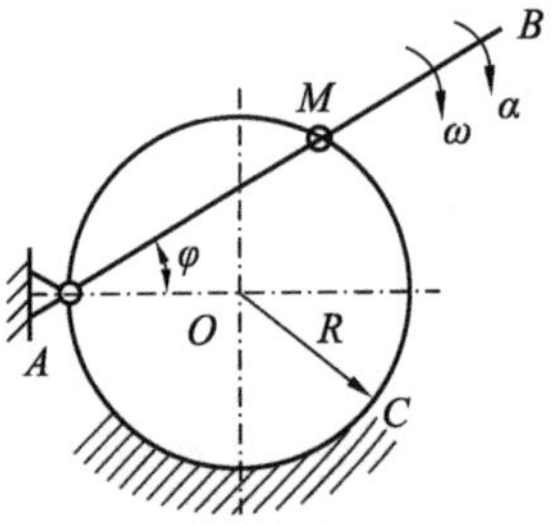

Fig. 7-29

7-17 In Fig. 7-30, the right-angle bar OBC with $OB=0.1$ m rotates about the axis O. The circle ring D slides freely along the fixed straight bar OA. At the instant $\varphi=60°$, the right-angle bar OBC has an angular velocity $\omega=0.5$ rad/s, determine the velocity and acceleration of the ring D.

7-18 In Fig. 7-31, the cam rotates about axis O with a uniformly angular velocity of ω. At the instant of θ, $OA=l$, the radius of curvature at point A is ρ_A, determine the velocity and acceleration of the rod AB.

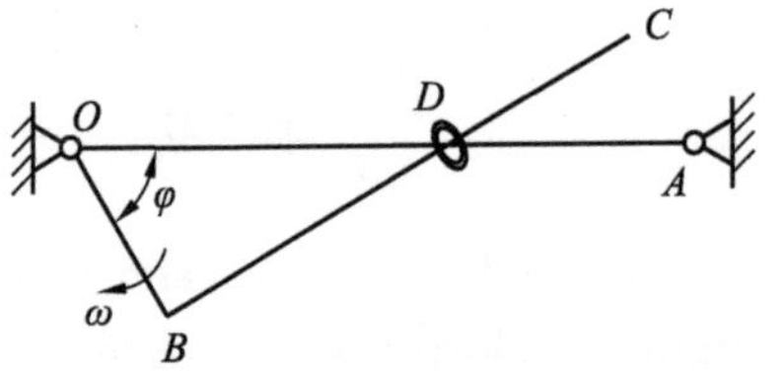

Fig. 7-30

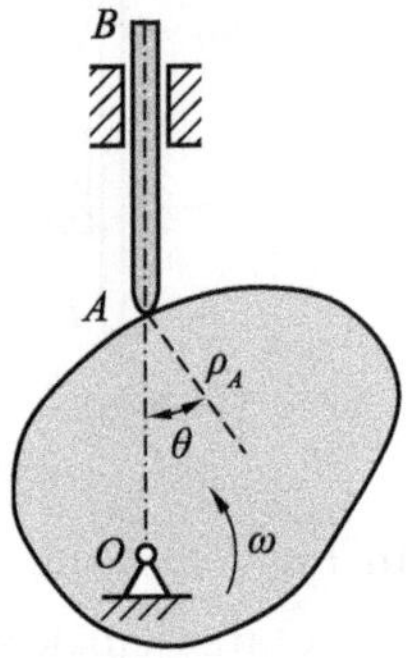

Fig. 7-31

Chapter 8

Plane Motion of Rigid Body

Objectives

In this chapter, the planar motion of a rigid body will be discussed. We will derive the equations which describe the *plane motion* of rigid bodies using a relative motion analysis of velocity and acceleration. Show how to find the instantaneous center of zero velocity and determine the velocity of a point on a body using this method. After studying, the students can use these equations to explain and specific practical problems.

8.1 Planar Rigid Body Motion

The *plane motion* of a body occurs when all the particles of a rigid body move along the paths which are equidistant from a fixed plane. For example, the pure rolls of the wheel in rail, the motion of the gear A in planet gear, these are all plane motion, where all the particles in the bodies move in the x, y-plane or in a plane parallel to it (Fig. 8-1). The study of plane motion is important for many mechanical operations.

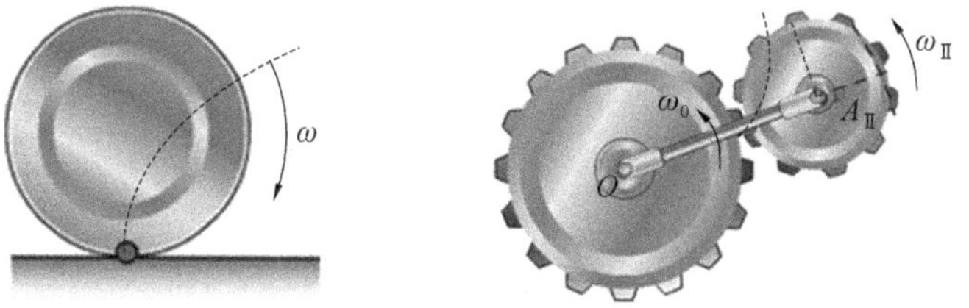

(a) (b)

Fig. 8-1

When studying the plane motion of a rigid body, we do not need to consider its geometrical shape and size, and considering the motion of a *plane figure* is enough to determine the velocity and acceleration of any point in the rigid body. Namely, the motion of rigid body can be represented by the motion of a plane figure which represents the projection of the body on the fixed reference plane, Fig. 8-2a. One way to relate these motions is to use a *rectilinear position* coordinate to locate the point along its path and an angular *position coordinate* to specify the orientation of the line. The plane motion can be completely specified by knowing both the angular rotation of a line fixed in the body $O'C$ and the motion of a point O' on the rigid body, Fig. 8-2b.

$$x_{O'}=f_1(t),\ y_{O'}=f_2(t),\ \varphi=f_3(t) \tag{8.1}$$

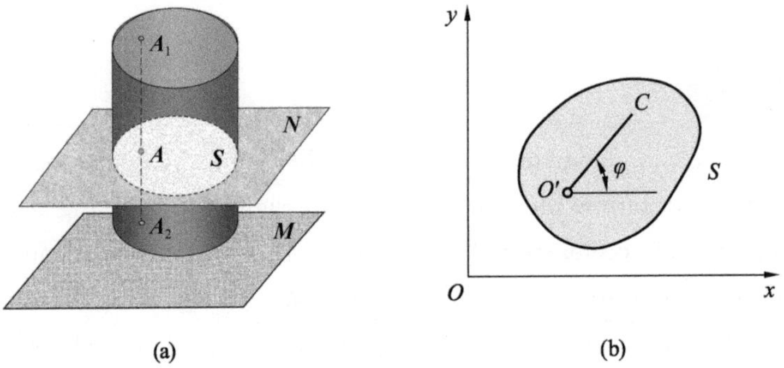

Fig. 8-2

We can also represent the plane motion using the method of composition of motion in Chapter 7. Consider a plate that is subjected to a plane motion, the fixed frame is established at an arbitrary point on the ground and the moving reference frame is attached to the plate and the origin locates in the point O' in the rigid body, named *base point* as shown in Fig. 8-3a. During an instant of time, plate undergoes displacement. If we consider the general plane motion by its component parts, the entire plate first translates so that base point O' moves to its final and point A moves to A', Fig. 8-3b. Then the plate is rotated about O' so that A' undergoes a relative displacement and thus moves to its final position A''. Namely, the plane motion of the plate can be decomposed into the translation with the moving reference frame in the plane of the plate and rotation about the base point O' (an axis pass through the base point and perpendicular to the plane).

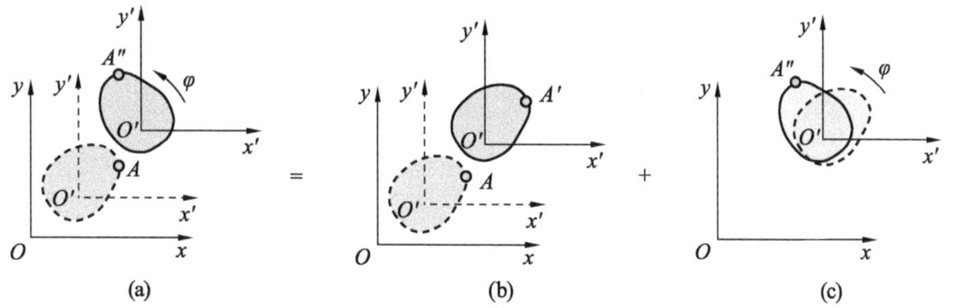

Fig. 8-3

It is obviously that the particular values of these two motions will generate the particular motions of the body. If the angular velocity of the plate is zero, that is, no rotation of the plate, then the plate is in *translation* motion. If the velocity of the plate is zero, then, no translation of the plate. In this case the motion of the plate is called *rotation* motion about a fixed axis. Examples of bodies undergoing these motions are shown in Fig. 8-4, there are three types of rigid body planar motion, crank OA is *rotation* about the axis pass through point O, the link AB is general *plane motion*, and the piston is in translation, but these motions are all planar motion.

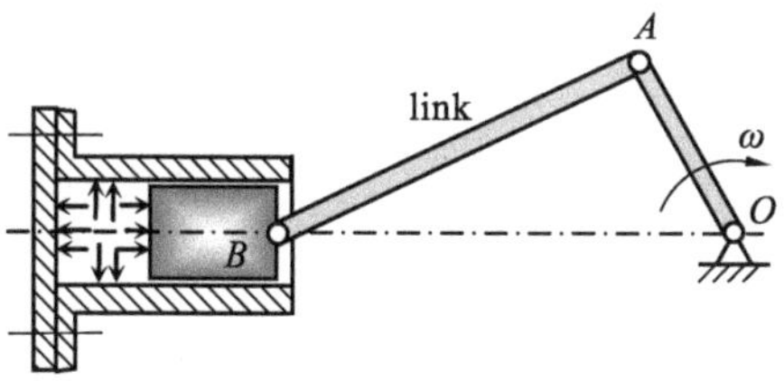

Fig. 8-4

8. 2 Relative Velocity Analysis

8.2.1 Method of Base Point

The *general plane motion* of a rigid body can be described as a combination of *translation* and *rotation*. We will use a relative-motion analysis involving two sets of coordinate axes to institute these component motions separately. The x, y coordinate system is fixed and measures the absolute position of two points A and B on the body, Fig. 8-5a. The origin of the moving x', y' coordinate system will be attached to the selected "base point" A, which generally has a *known* motion or it can be obtained *easy*. The axes of moving coordinate system *translate* with respect to the fixed frame but do not rotate with the body. The difference with the composition of motion that the moving point and moving reference frame must locate in two *different* rigid bodies, here the point B and base point A are all located in *same* rigid body.

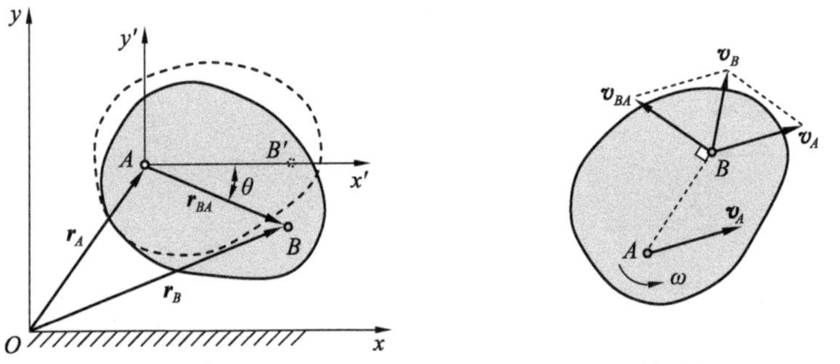

(a) (b)

Fig. 8-5

The position vector $\boldsymbol{r}_A$ in Fig. 8-5a specifies the location of the "base point" A, and the relative-position vector $\boldsymbol{r}_{BA}$ locates point B with respect to point A. By vector addition, the position vector of B is then

$$\boldsymbol{r}_B=\boldsymbol{r}_A+\boldsymbol{r}_{BA} \tag{8.2}$$

Take the time derivative of the position equation Eq. (8.2), this yields

$$\frac{\mathrm{d}\boldsymbol{r}_B}{\mathrm{d}t}=\frac{\mathrm{d}\boldsymbol{r}_A}{\mathrm{d}t}+\frac{\mathrm{d}\boldsymbol{r}_{BA}}{\mathrm{d}t} \tag{8.3}$$

The first two terms are measured with respect to the fixed x, y axes and represent the absolute velocities $\boldsymbol{v}_A$ and $\boldsymbol{v}_B$ of points A and B, respectively.

$$\frac{\mathrm{d}\boldsymbol{r}_B}{\mathrm{d}t}=\boldsymbol{v}_B,\quad \frac{\mathrm{d}\boldsymbol{r}_A}{\mathrm{d}t}=\boldsymbol{v}_A$$

Since the moving reference frame is translation, the relative displacement is caused only by a *rotation* about A, then $\mathrm{d}\boldsymbol{r}_{BA}=r_{BA}\,\mathrm{d}\theta$, the magnitude of the third term in Eq. (8.3) is

$$\frac{\mathrm{d}\boldsymbol{r}_{BA}}{\mathrm{d}t}=\boldsymbol{r}_{BA}\cdot\frac{\mathrm{d}\theta}{\mathrm{d}t}=\boldsymbol{r}_{BA}\boldsymbol{\omega} \tag{8.4}$$

where ω is the angular velocity of the body at the instant. Since the third term represents the velocity of B with respect to A as measured by an observer fixed to the translating x', y' axes, this term is denoted as the relative velocity $\boldsymbol{v}_{BA}$. We therefore have

$$\boldsymbol{v}_B=\boldsymbol{v}_A+\boldsymbol{v}_{BA} \tag{8.5}$$

Where the magnitude of $\boldsymbol{v}_{BA}$ is $\boldsymbol{v}_{BA}=AB\cdot\omega$, its direction is always perpendicular to AB, sense is in accordance with the rotation ω of the body, Fig. 8-5b. This equation states that *the velocity of any point B in the figure is the vector sum of the velocity of the base point and the relative velocity with respect to the base point*. The Eq. (8.5) can be used to determine the velocity of point B graphically with the aid of a *velocity diagram*.

◆ **Procedure for Analysis**

① Choosing the appropriate base point, which generally has a known motion.

② Draw a *velocity diagram* of the body. Indicate on it the velocities $\boldsymbol{v}_A$ and $\boldsymbol{v}_B$ of points A and B, the angular velocity ω, and the relative velocity.

③ If the magnitudes of $\boldsymbol{v}_A$ and $\boldsymbol{v}_B$ or ω are unknown, the sense of direction of these vectors can be assumed.

④ If the solution yields a negative answer for an unknown magnitude, it indicates the sense of direction of the vector is opposite to that shown on the *velocity diagram*.

Example 8-1

The link of $AB=l$ shown in Fig. 8-6a is guided by two blocks at A and B, which slide in the vertical fixed slots. If the velocity of block A is v_A to the left, determine the velocity of block B and angular velocity of link AB at the instant.

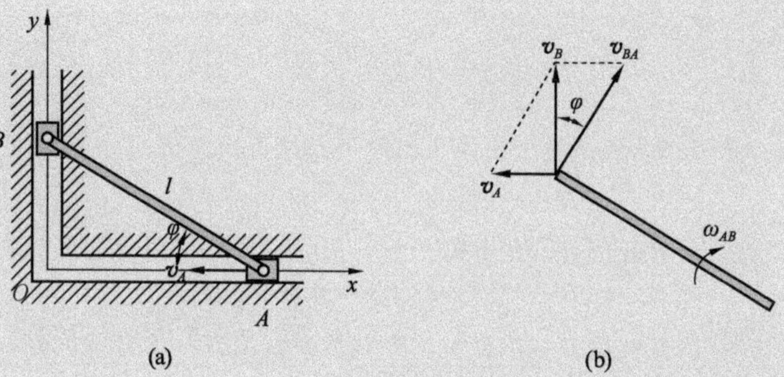

Fig. 8-6

Solution:

Since blocks A and B are restricted to move along the fixed slots and $\boldsymbol{v}_A$ is directed horizontally to the left, the velocity $\boldsymbol{v}_B$ must be directed upward, Fig. 8-6b. Knowing the magnitude and direction of $\boldsymbol{v}_A$ and the lines of action of $\boldsymbol{v}_B$, so we can take the point A as the base point and apply the velocity equation

$$\boldsymbol{v}_B=\boldsymbol{v}_A+\boldsymbol{v}_{BA}$$

where the direction of $\boldsymbol{v}_B$ is along the vertical slot, the direction of $\boldsymbol{v}_A$ is along the horizontal slot and perpendicular to $\boldsymbol{v}_B$, the direction of $\boldsymbol{v}_{BA}$ is vertical to link BA, so the direction of three velocities are known and drawn in the *velocity diagram*, Fig. 8-6b. The magnitude of $\boldsymbol{v}_A$ is known as v_A, using trigonometry, we have

$$v_B=\frac{v_A}{\tan\varphi} \qquad \textit{Ans.}$$

$$v_{BA}=\frac{v_A}{\sin\varphi}$$

Thus,

$$\omega_{BA}=\frac{v_{BA}}{l}=\frac{v_A}{l\sin\varphi} \qquad \textit{Ans.}$$

Example 8-2

The crank-rod mechanism in Fig. 8-7, link AB has length of l. Crank OA has length of r and rotates with an angular velocity of ω. At this instant $\varphi=60°$, rod OA is vertical to link AB determine the velocity of B and the angular velocity of link AB.

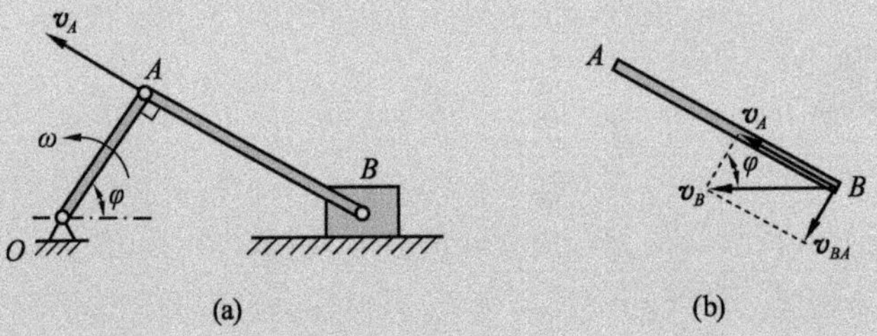

Fig. 8-7

Solution:

The rod AB is in *plane motion* and the magnitude and direction of $\boldsymbol{v}_A$ can be determined by the rotation of crank OA about fixed axis O,

$$v_A = OA \cdot \omega = \omega r$$

The direction of $\boldsymbol{v}_A$ is perpendicular to the crank OA, so we can take the point A as the base point and apply the velocity equation

$$\boldsymbol{v}_B = \boldsymbol{v}_A + \boldsymbol{v}_{BA}$$

where the direction of $\boldsymbol{v}_B$ is in horizontal, and the direction of $\boldsymbol{v}_{BA}$ is perpendicular to the link AB, so the direction of three velocities are known and drawn in Fig. 8-7. Using trigonometry, we have

$$v_B = \frac{v_A}{\sin 60°} = \frac{2\sqrt{3}}{3}\omega r \qquad \textit{Ans.}$$

$$v_{BA} = v_A \tan 30° = \omega r \tan 30° = \frac{\sqrt{3}}{3}\omega r$$

The angular velocity of rod AB at this instant can be determined

$$\omega_{AB} = \frac{v_{BA}}{AB} = \frac{\sqrt{3}\omega r}{3l} \qquad \textit{Ans.}$$

Example 8-3

The stone-crushing mechanism as shown in Fig. 8-8a, crank OA has length of $R=0.5$ m, link AB has length of 1 m, and the plate BC has a length of 1.15 m. When $\theta=60°$, the link AB is vertical to plate BC, and the driving wheel is rotating with an angular velocity of $\omega=4$ rad/s, determine the velocity of B and the angular velocities of member BC and AB at this instant.

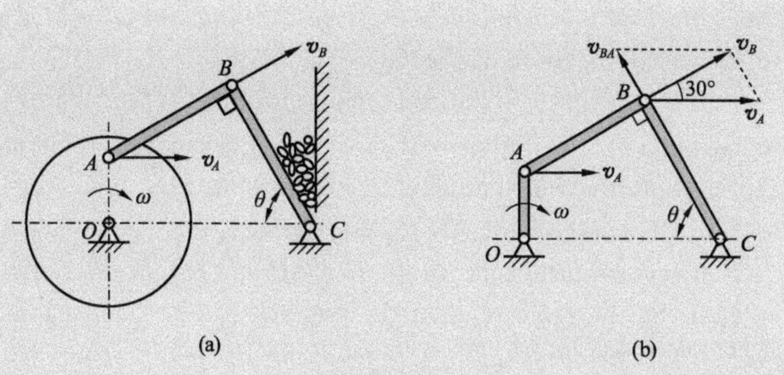

Fig. 8-8

Solution:

The stone-crushing mechanism can be simplified to a four-rod linkage in Fig. 8-8b. The velocities of points A and B are defined by the rotation of crank OA and the plate BC about their fixed axes O and C, respectively. The rod AB is in *plane motion* and the magnitude and direction of $\boldsymbol{v}_A$ can be determined firstly, so we can take the point A as the base point and apply the velocity equation

$$\boldsymbol{v}_B = \boldsymbol{v}_A + \boldsymbol{v}_{BA}$$

where the direction of $\boldsymbol{v}_B$ is perpendicular to the plate BC, the direction of $\boldsymbol{v}_A$ is perpendicular to the crank OA and the direction of $\boldsymbol{v}_{BA}$ is perpendicular to the link AB, so the direction of three velocities are known and drawn in Fig. 8-8b. The magnitude of $\boldsymbol{v}_A$ can be calculated as

$$v_A = OA \cdot \omega = 0.5 \times 4 = 2 \text{ m/s}$$

Using trigonometry, we have

$$v_B = v_A \cos 30° = 2\cos 30° = 2 \times \frac{\sqrt{3}}{2} = 1.732 \text{ m/s} \qquad \textit{Ans.}$$

$$v_{BA} = v_A \sin 30° = R\omega \sin 30° = 0.5 \times 4 \times \frac{1}{2} = 1 \text{ m/s}$$

The angular velocity of link AB at this instant can be determined

$$\omega_{AB} = \frac{v_{BA}}{AB} = \frac{1}{1} = 1 \text{ rad/s} \qquad \textit{Ans.}$$

The angular velocity of plate BC at this instant can be determined

$$\omega_{BC} = \frac{v_B}{BC} = \frac{1.73}{1.15} = 1.5 \text{ rad/s} \qquad \textit{Ans.}$$

Since both results are positive, the sense of two angular velocities is indeed correct as shown in Fig. 8-8b.

8.2.2 Method of Velocities' Projection

We will show propriety of the distribution of velocities in the general plane motion of the rigid body, that is, the *theorem of velocities' projection*: *the projections of the velocities of the points on a straight line passing through the points, are equal.* The propriety results from the relation of the velocity Eq. (8.5).

$$\boldsymbol{v}_B = \boldsymbol{v}_A + \boldsymbol{v}_{BA}$$

The demonstration of this propriety is made when calculating the scalar product of the relation representing the distribution of velocities, Fig. 8-9. We make the projection of the Eq. (8.5) on the straight-line AB passing through the two points A and B. The third term $\boldsymbol{v}_{BA}$ is always perpendicular to the line AB and consequently its projection on AB is equal to *zero*. So the projections of the two velocities $\boldsymbol{v}_A$ and $\boldsymbol{v}_B$ on the straight-line AB is *equal*.

$$\boldsymbol{v}_B \cos\theta_B = \boldsymbol{v}_A \cos\theta_A \tag{8.6}$$

This propriety can be used in the plane motion of the body to determine the unknown directly.

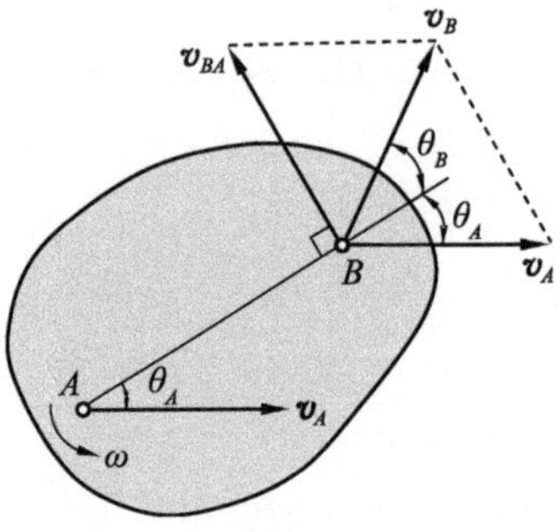

Fig. 8-9

Example 8-4

The rod of $AB = l$ shown in Fig. 8-10 performs a motion in vertical plane so that the two ends A and B move on the two perpendicular straight lines, one horizontal and one vertical. If the point A moves to the right on the horizontal line with the constant velocity v_A, determine the velocity of block B at the instant.

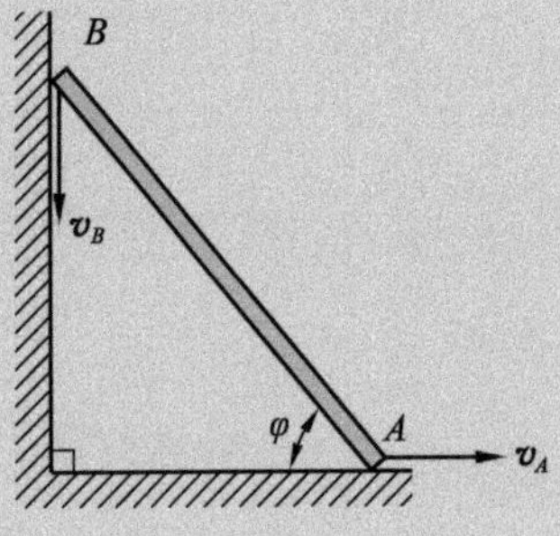

Fig. 8-10

Solution:

The motion of the rod is a *plane motion*. Since the direction of $\boldsymbol{v}_A$ and $\boldsymbol{v}_B$ are restricted to move along the horizontal and one vertical surface, the include angle between $\boldsymbol{v}_A$ and rod AB is φ, the include angle between the velocity $\boldsymbol{v}_B$ and rod AB is $(90°-\varphi)$, Fig. 8-10. Knowing the magnitude and direction of v_A and the lines of action of v_B, so we can apply the method of velocity projection, that the projections of the velocities of the points A and B on the AB are equals:

$$\boldsymbol{v}_B \cos(90°-\varphi) = v_A \cos\varphi$$

$$\boldsymbol{v}_B = \frac{\boldsymbol{v}_A \cos\varphi}{\sin\varphi} = \frac{\boldsymbol{v}_A}{\tan\varphi} \qquad \textit{Ans.}$$

It is obviously that we can obtain the previous result easily. Note that we *cannot* obtain directly the relative velocity $\boldsymbol{v}_{AB}$ of point B respect to point A and *cannot* obtain the angular velocity ω of the general plane motion by using the method of velocities projection.

8.3 Instantaneous Center of Zero Velocity

8.3.1 Instantaneous Center of Velocity

According to Section 8.2.1, general plane motion of a rigid body is composed of a *translation* and a *rotation* and the base point is selected arbitrary. And we can determine the unknowns using the Eq. (8.5). Obviously, if we choose the base point A to be a point that has *zero* velocity at the instant, the velocity of any point B located on a rigid body can be obtained in a very direct way, that is, *a plane motion may be considered at each instant to be a pure rotation about a certain point, this point is referred to as the instantaneous center of zero velocity* (IC). This point locates in the body or outside the body.

We will verify that there always exists a point (= *instantaneous center of zero velocity*) which has a zero velocity. Consider a plane motion shown in Fig. 8-11 with an angular velocity ω, the velocity $\boldsymbol{v}_A$ of point A is known, taking the point A as the base point, the velocity of an arbitrary point M can be solved using the velocity equation.

$$\boldsymbol{v}_M = \boldsymbol{v}_A + \boldsymbol{v}_{MA}$$

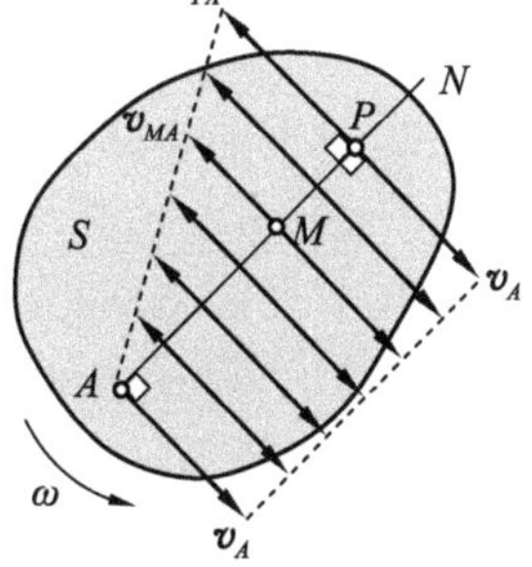

Fig. 8-11

Drawing a straight line AN at point A which is perpendicular to the action line of $\boldsymbol{v}_A$ (the direction from the $\boldsymbol{v}_A$ to line AN is coincide to the sense of the ω). If the point M locates in the line AN, then the $\boldsymbol{v}_{MA}$ and $\boldsymbol{v}_A$ are located in a same line and opposite to each other, so the magnitude of $\boldsymbol{v}_M$

$$v_M = v_A - AM \cdot \omega$$

From the above the equation, the magnitude of $\boldsymbol{v}_{MA}$ is changed with the location of the point M. So there must be point P that has *zero* velocity at the instant. Let the $AP = v_A/\omega$, we have

$$v_P = v_A - AP \cdot \omega = 0$$

This uniquely determines the location of the *instantaneous center of zero velocity*, which has an important propriety used to find the distribution of velocities in plane motion, that is: the instantaneous motion of a rigid body may indeed be considered as being a *pure rotation* about *instantaneous center*, as shown in Fig. 8-12. In this way, if we know the position of the instantaneous center of velocity, all the proprieties of the distribution of velocities in rotation motion are used. For example, in Fig. 8-12, the body rotates about the instantaneous axis, the velocities of arbitrary points A, B and C can be calculated by

$$v_A = \omega \cdot AP, v_B = \omega \cdot BP, v_C = \omega \cdot CP \tag{8.7}$$

where ω is the angular velocity of the body. Due to the circular motion, the direction of each velocity must always be perpendicular to each joint line. Note that the instantaneous center of zero velocity may lie outside the body and it is not a fixed point, namely the IC changes its position.

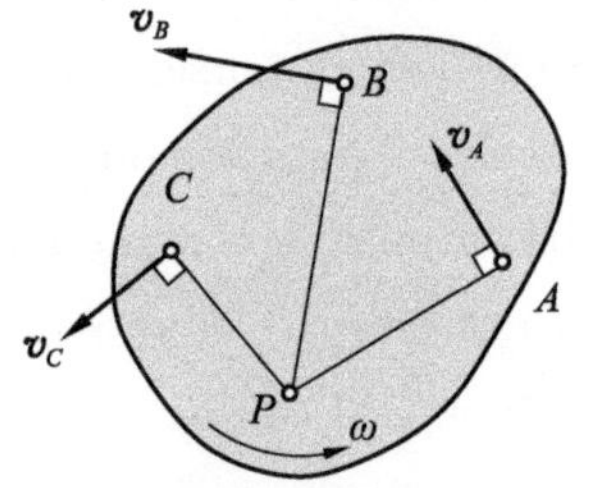

Fig. 8-12

8.3.2 Location of the IC

The way to find the position of IC, is based on the proprieties of the distribution of velocities in instantaneous rotation motion. Following represents a few cases of determination of the instantaneous center of zero velocity (IC).

① In the case of the wheel in Fig. 8-13, which rolls without slipping, point P on the wheel contacting the ground has *zero* velocity since the ground does not move, it is the instantaneous center of velocity.

② *If the directions of the velocities of two points of the rigid body are known, and the lines of action of* $\boldsymbol{v}_A$ *and* $\boldsymbol{v}_B$ *are nonparallel*, Fig. 8-14. Construction at point A and B line segments that are perpendicular to $\boldsymbol{v}_A$ and $\boldsymbol{v}_B$. Extending the two straight lines to their intersection point P as shown locates the IC at the instant.

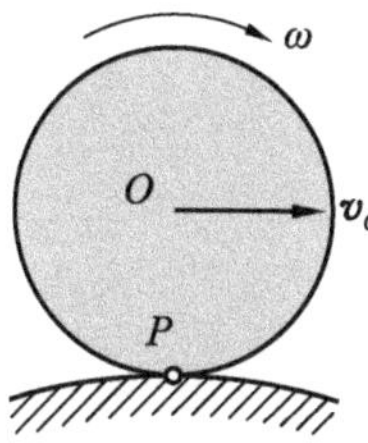

Fig. 8-13

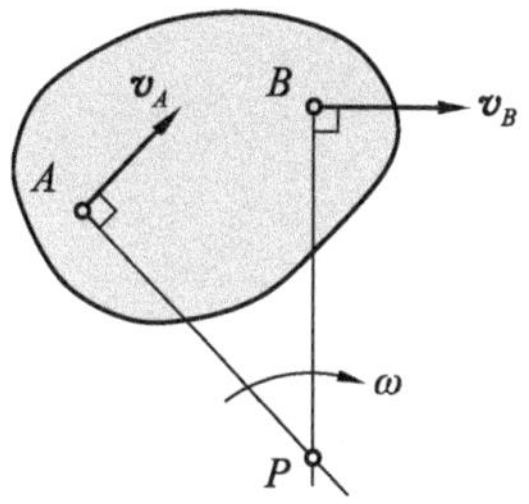

Fig. 8-14

③ *If the magnitude and direction of two parallel velocities* $\boldsymbol{v}_A$ *and* $\boldsymbol{v}_B$ $(\boldsymbol{v}_A \neq \boldsymbol{v}_B)$ *are known and both are perpendicular to the line AB*. The location of IC is determined at the intersection between the common perpendicular line AB and the straight line that joins the tops of the two velocities $\boldsymbol{v}_A$ and $\boldsymbol{v}_B$. Examples are shown in Fig. 8-15.

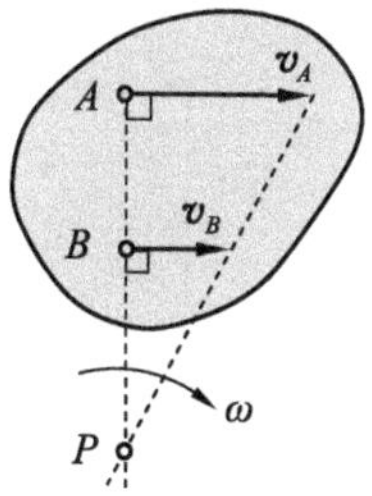

(a)

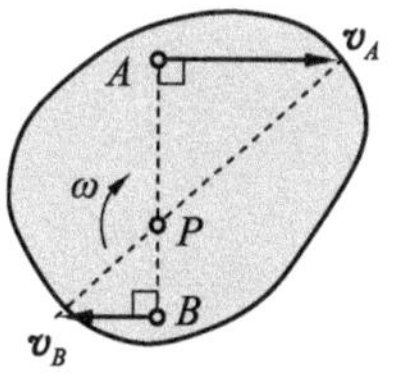

(b)

Fig. 8-15

④ *If the magnitude and direction of two parallel velocities* $\boldsymbol{v}_A$ *and* $\boldsymbol{v}_B$ *are same* $(\boldsymbol{v}_A = \boldsymbol{v}_B)$ *and both are perpendicular to the joining line AB*, Fig. 8-16a, *or the perpendicular lines to the velocity vectors are parallel*, Fig. 8-16b, which means that the instantaneous center is an infinite distance from points A and B. This leads to the conclusion that angular velocity of rigid body is zero, that is, the body is in *instantaneous translation*. At this instant, the velocities of every point in rigid body are *same*, but the accelerations are *different*.

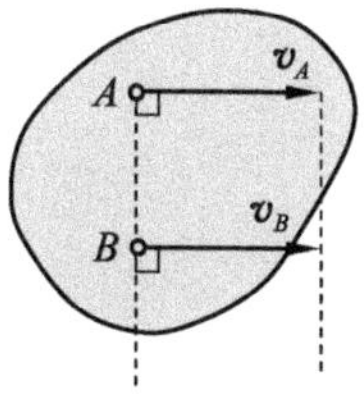

(a)

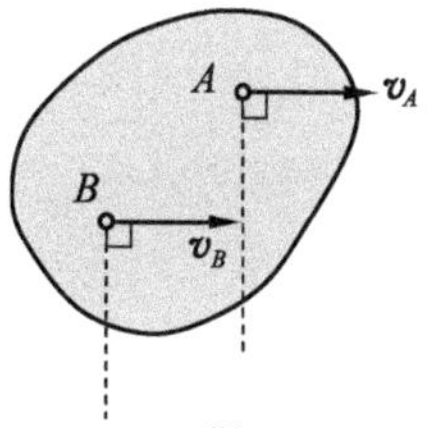

(b)

Fig. 8-16

The point chosen as the instantaneous center of zero velocity for the rigid body can only be used at the instant since the body changes its position from one instant to the next. Although the IC may be conveniently used to determine the velocity of any point in a body, it generally does not have *zero acceleration* and therefore it should not be used to find the accelerations of points in a rigid body.

Example 8-5

Solve Example 8-1 using the method of instantaneous center of velocity and determine the velocity of middle point D of link AB, Fig. 8-17a.

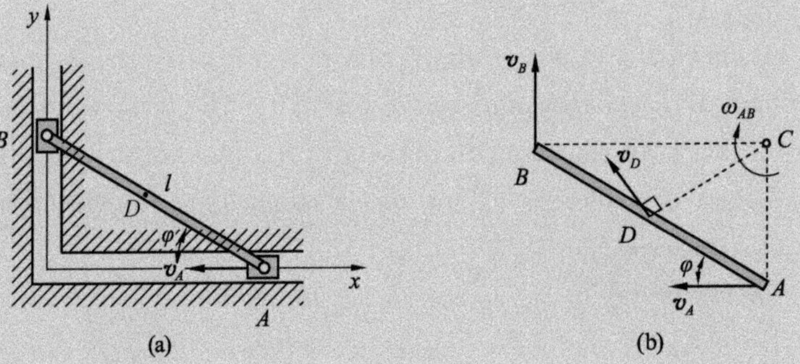

Fig. 8-17

Solution:

The link AB is in plane motion, the block moves horizontal with a velocity $\boldsymbol{v}_A$, the unknown velocity $\boldsymbol{v}_B$ is vertical. We construct two line segments AC and BC perpendicular to $\boldsymbol{v}_A$ and $\boldsymbol{v}_B$, respectively, the instantaneous center of zero velocity for link AB is located at the intersection C of two straight lines, Fig. 8-17b. From the geometry, we have

$$AC = l\sin\varphi, BC = l\cos\varphi, CD = \frac{l}{2}$$

Since the magnitude of $\boldsymbol{v}_A$ is known, the angular velocity of link AB is

$$\omega_{AB} = \frac{v_A}{AC} = \frac{v_A}{l\sin\varphi}$$

The velocity of B is therefore

$$v_B = BC \cdot \omega_{AB} = l\cos\varphi \cdot \frac{v_A}{l\sin\varphi} = \frac{v_A}{\tan\varphi} \qquad Ans.$$

The velocity of middle point D is

$$v_D = CD \cdot \omega_{AB} = \frac{v_A}{2\sin\varphi} \qquad Ans.$$

Example 8-6

A wheel shown in Fig. 8-18, has radium of R, is rolling without slipping on a horizontal plane. If its center O has the velocity $\boldsymbol{v}$, determine the velocity of the point A, B, and C on the wheel.

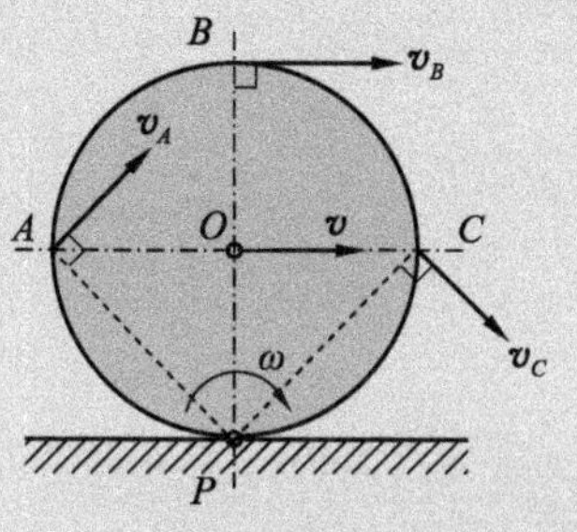

Fig. 8-18

Solution:

Since the wheel rolls without slipping and in plane motion, the contact point P with the ground momentarily has zero velocity (no slipping), it is the instantaneous center of zero velocity (Fig. 8-18). Since the magnitude of v is known, the angular velocity of wheel is

$$\omega=\frac{v_O}{OP}=\frac{v}{R}$$

According to Eq. (8.7), the velocity of the point A, B, and C on the wheel is calculated by

$$v_A=\omega\cdot AP=\frac{v}{R}\cdot\sqrt{2}R=\sqrt{2}v \qquad Ans.$$

$$v_B=\omega\cdot BP=\frac{v}{R}\cdot 2R=2v \qquad Ans.$$

$$v_C=\omega\cdot CP=\frac{v}{R}\cdot\sqrt{2}R=\sqrt{2}v \qquad Ans.$$

The directions of the velocities are shown in Fig. 8-18.

Example 8-7

The cylinder of radius $r=0.125$ m rolls without slipping between the two moving plates E and D which have velocity $\boldsymbol{v}_E=0.25$ m/s and $\boldsymbol{v}_D=0.4$ m/s, respectively, shown in Fig. 8-19a. Determine the angular velocity of the cylinder and the velocity of its center O.

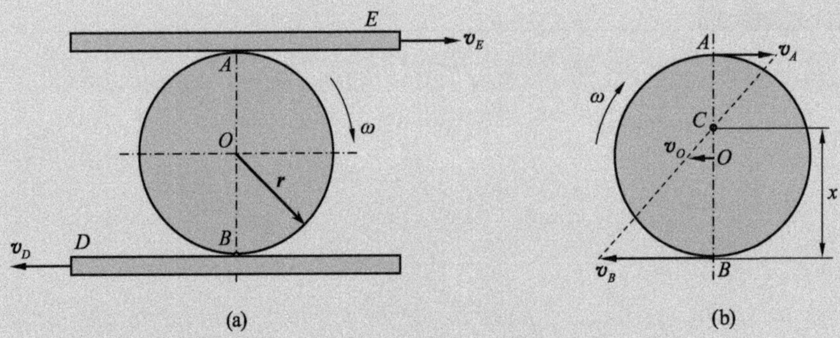

Fig. 8-19

Solution:

Since no slipping occurs, the contact points A and B on the cylinder have the same velocities as the plates E and D, respectively. Furthermore, the velocities $\boldsymbol{v}_A$ and $\boldsymbol{v}_B$ are parallel, so the location of IC is located at the intersection point C between the common perpendicular line AB and the straight line that joins the tops of the two velocities $\boldsymbol{v}_A$ and $\boldsymbol{v}_B$, Fig. 8-19b. Assuming this point C to be a distance x from B, we have

$$\boldsymbol{v}_B = \boldsymbol{v}_D = \omega \cdot x$$

$$\boldsymbol{v}_A = \boldsymbol{v}_E = \omega \cdot (2r - x)$$

Substituting the $\boldsymbol{v}_E$, $\boldsymbol{v}_D$ and r into the equations, solving, we obtain

$$x = 0.153\ 8\ \text{m}$$

$$\omega = 2.6\ \text{rad/s} \qquad \textit{Ans.}$$

The velocity of center point O is therefore

$$v_O = \omega \cdot OC = 2.6 \times (0.153\ 8 - 0.125)$$

$$= 0.075\ \text{m/s} \qquad \textit{Ans.}$$

Example 8-8

Solve Example 8-3 using the method of instantaneous center of zero velocity.

Solution:

The link AB is in *plane motion*, as crank OA rotates about point O, it causes rod BC to rotate about point C. Hence, $\boldsymbol{v}_A$ and $\boldsymbol{v}_B$ is directed perpendicular to OA and BC, respectively, then the instantaneous center of zero velocity for link AB is located at the intersection P of two line segments drawn perpendicular to $\boldsymbol{v}_A$ and $\boldsymbol{v}_B$, Fig. 8-20. The magnitudes of AP and BP can be obtained from the geometry of the triangle,

$$AP = \frac{AB}{\cos 60^\circ} = 2\ \text{m}, BP = \frac{AB}{\tan 60^\circ} = \sqrt{3}\ \text{m}$$

The crank OA rotates about a fixed axis, so the velocity of point A is

$$v_A = OA \cdot \omega = 0.5 \times 4 = 2\ \text{m/s}$$

The sense of ω_{AB} must be the same as the rotation caused by $\boldsymbol{v}_A$ about the IC, the angular velocity of link AB is

$$\omega_{AB}=\frac{v_A}{AP}=\frac{2}{2}=1\ \text{rad/s}$$

The velocity of point B is therefore

$$v_B=\omega_{AB}\cdot BP=1\times\sqrt{3}=1.732\ \text{m/s}$$

Using this result, the angular velocity of rod BC is

$$\omega_{BC}=\frac{v_B}{BC}=\frac{1.73}{1.15}=1.5\ \text{rad/s} \qquad \textit{Ans.}$$

We obtained the previous same result.

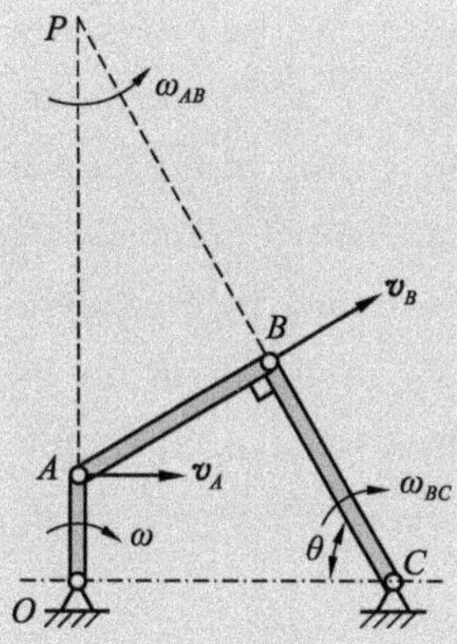

Fig. 8-20

8.4 Relative Acceleration Analysis

The relative acceleration analysis also involves two sets of coordinate axes, Fig. 8-21a. The x, y coordinate system is fixed and the origin of the *translating* x', y' coordinate system will be attached to the selected "*base point*" A, which generally has a *known* acceleration or it can be obtained *easy*. An equation that relates the accelerations of two points A and B on a rigid body subjected to plane motion may be determined by differentiating Eq. (8.5) with respect to time. This yield

$$\frac{\mathrm{d}\boldsymbol{v}_B}{\mathrm{d}t}=\frac{\mathrm{d}\boldsymbol{v}_A}{\mathrm{d}t}+\frac{\mathrm{d}\boldsymbol{v}_{BA}}{\mathrm{d}t} \tag{8.8}$$

The first two terms are measured with respect to the fixed x, y axes and represent the absolute accelerations of points B and A, respectively.

$$\frac{\mathrm{d}\boldsymbol{v}_B}{\mathrm{d}t}=\boldsymbol{a}_B,\quad \frac{\mathrm{d}\boldsymbol{v}_A}{\mathrm{d}t}=\boldsymbol{a}_A$$

The last term represents the acceleration of B with respect to A as measured with respect to translating x', y' axes which have their origin at the base point A. In Fig.

8-21a, since the relative displacement is caused only by a *rotation* about A, so $d\boldsymbol{v}_{BA}/dt$ is the acceleration $\boldsymbol{a}_{BA}$ of B with respect to A, and $\boldsymbol{a}_{BA}$ can be expressed in terms of its tangential and normal components; i. e. ,

$$\boldsymbol{a}_{BA}=\boldsymbol{a}_{BA}^{t}+\boldsymbol{a}_{BA}^{n}$$

Hence, the relative acceleration Eq. (8. 8) can be written in the form

$$\boldsymbol{a}_{B}=\boldsymbol{a}_{A}+\boldsymbol{a}_{BA}^{t}+\boldsymbol{a}_{BA}^{n} \tag{8. 9}$$

where, $\boldsymbol{a}_{BA}^{t}$—tangential acceleration component of B with respect to A. The magnitude is a_{BA}^{t}, the direction is perpendicular to AB and with the same sense of the angular acceleration α;

$\boldsymbol{a}_{BA}^{n}$—normal acceleration component of B with respect to A. The magnitude is $\boldsymbol{a}_{BA}^{n}$, and the direction is always from B towards *base point* A, Fig. 8-21b.

This equation states that *the acceleration of any point in a plane figure equals to the vector sum of the acceleration of the base point A, tangential acceleration and normal acceleration of this point rotating about the base point.*

The terms in Eq. (8. 9) are represented graphically in Fig. 8-21b. At a given instant, the acceleration of point B is determined by considering the body to translate with an acceleration $\boldsymbol{a}_{A}$, and simultaneously rotate about the base point A with an instantaneous angular velocity ω and angular acceleration α. It should be noted that since points A and B move along curved paths, the accelerations of these points will have both *tangential* and *normal components*.

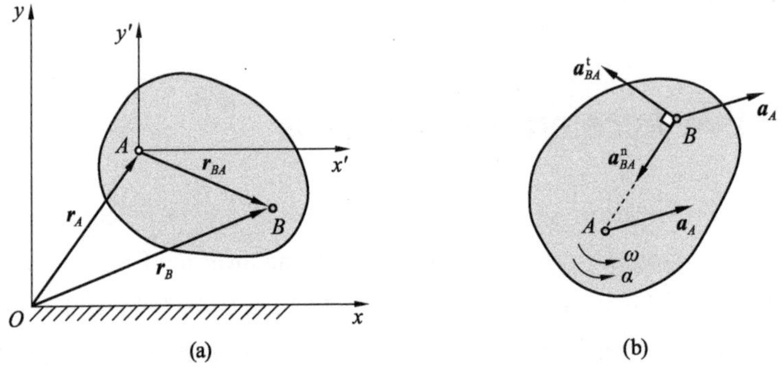

Fig. 8-21

Example 8-9

In Example 8-4, the acceleration of point A is $\boldsymbol{a}_{A}$, Fig. 8-22a, determine the acceleration of B and angular acceleration of link AB at the instant.

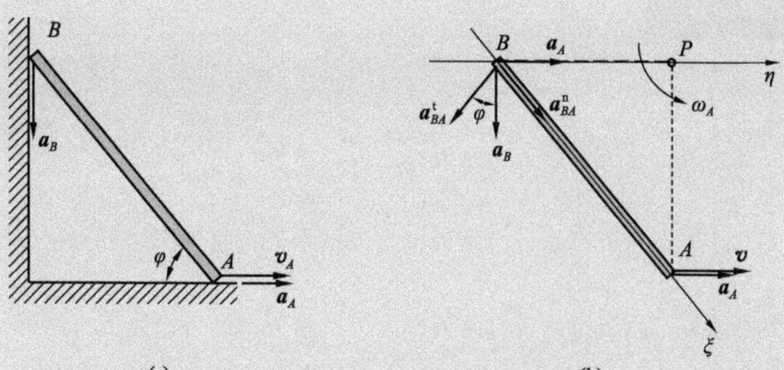

(a) (b)

Fig. 8-22

Solution:

The velocity of B and angular velocity of link AB at the instant are determined in Example 8-4.

$$\omega_{AB}=\frac{v_A}{l\sin\varphi},\quad v_B=\frac{v_A}{\tan\varphi}$$

Knowing the magnitude and direction of $\boldsymbol{a}_A$, so we can take the point A as the *base point*. Since points A and B both move along straight-line paths, they have no components of acceleration normal to the paths. Apply the acceleration equation

$$\boldsymbol{a}_B=\boldsymbol{a}_A+\boldsymbol{a}_{BA}^{t}+\boldsymbol{a}_{BA}^{n} \tag{a}$$

where the direction of $\boldsymbol{a}_B$ is directed downward, the direction of $\boldsymbol{a}_{BA}^{t}$ is perpendicular to the link AB and the direction of $\boldsymbol{a}_{BA}^{n}$ is along the link AB and from B to A, so the directions of all acceleration are known and the acceleration diagram is drawn in Fig. 8-22b. The magnitude of $\boldsymbol{a}_{BA}^{n}$ can be calculated by

$$\boldsymbol{a}_{BA}^{n}=AB\cdot\omega_{AB}^{2}=l\times\left(\frac{v_A}{l\sin\varphi}\right)^2=\frac{v_A^2}{l\sin^2\varphi}$$

There are two unknowns of $\boldsymbol{a}_B$ and $\boldsymbol{a}_{AB}^{t}$, their magnitude can be calculated by the projection of equation (a) on η and ξ axes, respectively.

$$a_B\sin\varphi=a_{BA}^{n}+a_A\cos\varphi$$

$$0=a_A+a_{BA}^{n}\cos\varphi-a_{BA}^{t}\sin\varphi$$

Solving, we have

$$a_B=\frac{v_A^2}{l\sin^3\varphi}+\frac{a_A}{\tan\varphi} \qquad \textit{Ans.}$$

$$a_{BA}^{t}=\frac{a_A}{\sin\varphi}+\frac{v_A^2\cos\varphi}{l\sin^3\varphi}$$

Thus,

$$\alpha_{BA}=\frac{a_A}{l\sin\varphi}+\frac{v_A^2\cos\varphi}{l^2\sin^3\varphi} \qquad \textit{Ans.}$$

Example 8-10

In Example 8-3, the length of BC is 1.15 m. In Fig. 8-23a, at the instant $\theta=60°$, the link AB is vertical to rod BC, determine the angular acceleration of rod BC.

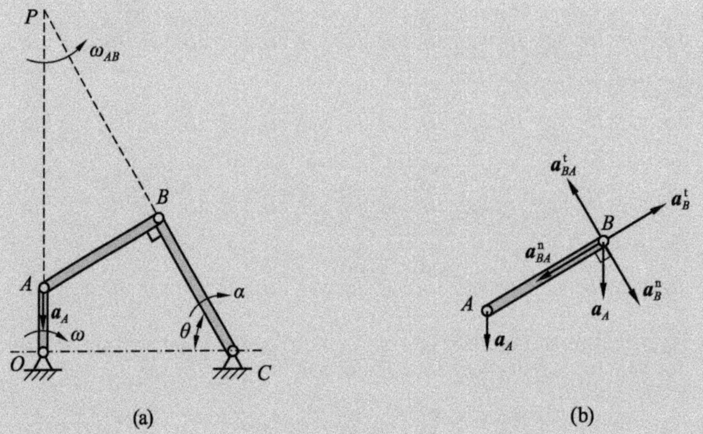

Fig. 8-23

Solution:

The velocity of B and angular velocity of link AB and rod BC at the instant are determined in Example 8-3.

$$\omega_{AB}=1 \text{ rad/s},\ v_B=1.732 \text{ m/s},\ \omega_{BC}=1.5 \text{ rad/s}$$

Since the crank OA rotates with uniform angular velocity, so the tangential acceleration component of point A is zero, the direction of normal acceleration component is along the crank OA and from A to O, the magnitude of $\boldsymbol{a}_A^n$ can be calculated by

$$a_A^n=R\cdot\omega^2=0.5\times4^2=8 \text{ m/s}^2$$

So we can take the point A as the *base point*. Point B is in circular motion, there are *tangential and normal acceleration components*. Apply the acceleration equation

$$\boldsymbol{a}_B^t+\boldsymbol{a}_B^n=\boldsymbol{a}_A^n+\boldsymbol{a}_{BA}^t+\boldsymbol{a}_{BA}^n \qquad \text{(a)}$$

where the direction of $\boldsymbol{a}_{BA}^t$ is perpendicular to the rod BC, the direction of $\boldsymbol{a}_{BA}^n$ is along the rod BC and from B to C, the direction of $\boldsymbol{a}_{BA}^t$ is perpendicular to the link AB and the direction of $\boldsymbol{a}_{BA}^n$ is along the link AB and from B to A. So the direction of all accelerations is known and the acceleration diagram is drawn in Fig. 8-23b. The magnitude of $\boldsymbol{a}_B^t$ and $\boldsymbol{a}_B^n$ are calculated by

$$a_B^n=BC\cdot\omega_{BC}^2=1.15\times1.5^2=2.59 \text{ m/s}^2$$

$$a_{BA}^n=AB\cdot\omega_{AB}^2=1\times1^2=1 \text{ m/s}^2$$

The magnitude of $\boldsymbol{a}_B^t$ can be calculated by the projection of equation (a) on $\boldsymbol{a}_B^t$ axis.

$$a_B^t = -a_A \sin 30° - a_{BA}^n \tag{b}$$

Solving, we have

$$a_B^t = -5 \text{ m/s}^2$$

The angular acceleration α of rod BC at this instant can be determined

$$\alpha_{BC} = \frac{a_B^t}{BC} = -\frac{5}{1.15} \text{ rad/s}^2 = -4.33 \text{ rad/s}^2 \qquad Ans.$$

The negative indicates that the real sense of angular acceleration of rod BC is opposite to the sense shown in Fig. 8-23b.

Example 8-11

A wheel of radius R rolls on a horizontal surface without slipping, shown in Fig. 8-24a. At a given instant, the center of the wheel has a velocity $\boldsymbol{v}_O$ and acceleration $\boldsymbol{a}_O$, determine the angular velocity and angular acceleration of the wheel and acceleration of the contact point at P.

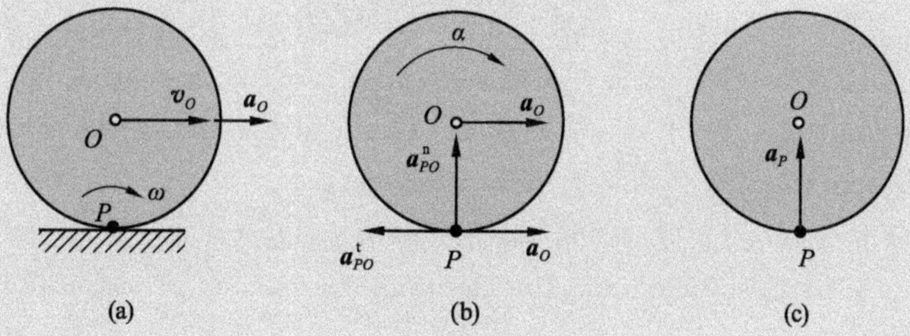

Fig. 8-24

Solution:

Since the wheel rolls without slipping and in plane motion, at the instant the contact point P is the instantaneous center of zero velocity, $v_P = 0$. Since the magnitude of v_O is known, the angular velocity of wheel is

$$\omega = \frac{v_O}{R}$$

The angular acceleration can be determined by taking the time derivative of its angular velocity, which gives

$$\alpha = \frac{d\omega}{dt} = \frac{d}{dt}\left(\frac{v_O}{R}\right) = \frac{1}{R}\frac{dv_O}{dt}$$

Since the motion of center O is always along a straight line, then dv_O/dt is the acceleration a_O of the center, which gives

$$\alpha = \frac{a_O}{R}$$

Knowing the magnitude and direction of $\boldsymbol{a}_O$, so we can take the point O as the *base point* and apply the acceleration equation

$$\boldsymbol{a}_P=\boldsymbol{a}_O+\boldsymbol{a}_{PO}^{t}+\boldsymbol{a}_{PO}^{n}$$

where the direction of $\boldsymbol{a}_{PO}^{t}$ is perpendicular to the radius and the direction of $\boldsymbol{a}_{PO}^{n}$ is along the radius and from P to O, the acceleration diagram is drawn in Fig. 8-24b. The magnitude of $\boldsymbol{a}_{PO}^{n}$ and $\boldsymbol{a}_{PO}^{t}$ are calculated by

$$a_{PO}^{t}=PO\cdot\alpha=R\cdot\frac{a_O}{R}=a_O$$

$$a_{PO}^{n}=PO\cdot\omega^2=R\cdot\omega^2=\frac{v_O^{\ 2}}{R}$$

From the figure, the direction of $\boldsymbol{a}_{PO}^{t}$ is opposite to $\boldsymbol{a}_O$ and cancel out each other, which yields

$$a_P=a_{PO}^{n}=\frac{v_O^{\ 2}}{R} \qquad \textit{Ans.}$$

Exercises

8-1 In Fig. 8-25, the crank OA of radium 0.2 m, rotates with uniform angular velocity of $\omega=10$ rad/s about axis O, linkage $AB=1$ m, determine the velocity of piston B and the angular velocity of linkage AB at the instant.

8-2 In Fig. 8-26, the crank AB of radium 0.3 m, rotates with uniform angular velocity of $\omega_{AB}=4$ rad/s about axis A, determine the velocity of piston C at the instant shown.

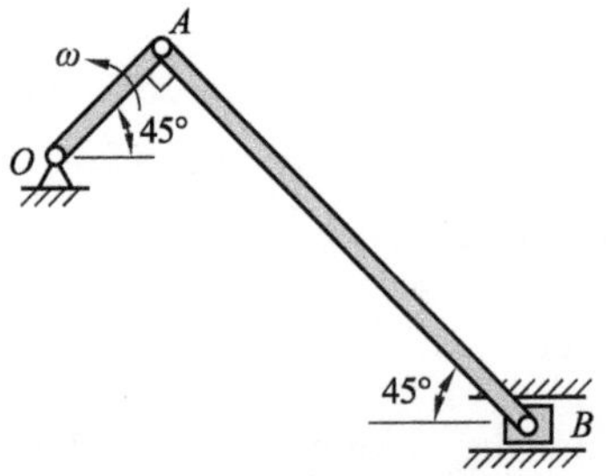

Fig. 8-25

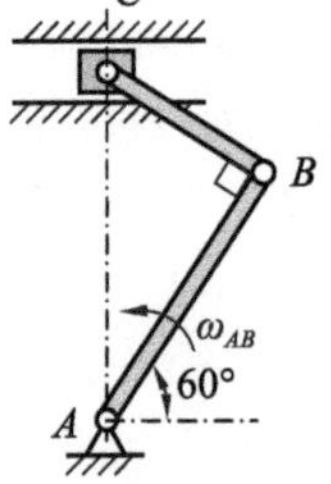

Fig. 8-26

8-3 The collar C is moving downward with a velocity of 2 m/s, Fig. 8-27. Determine the angular velocity of link CB at this instant.

8-4 In Fig. 8-28, the crank AB has a clockwise angular velocity of 30 rad/s when $\theta=60°$. Determine the angular velocities of link BC and the wheel D at this instant. Given: $AB=BC=0.2$ m, $r=0.1$ m.

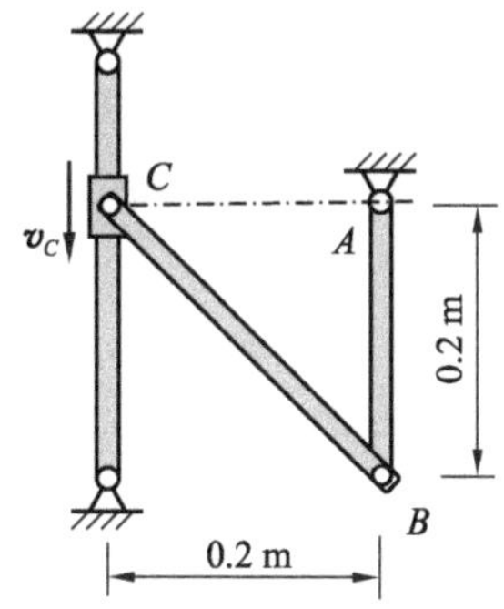

Fig. 8-27

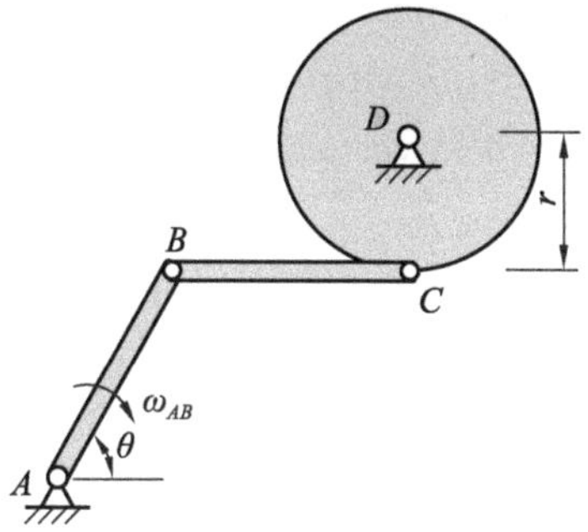

Fig. 8-28

8-5 In Fig. 8-29, if crank OA rotates with an angular velocity of $\omega = 12$ rad/s, determine the velocity of piston B and the angular velocity of link AB at the instant shown.

8-6 In Fig. 8-30, the plate BC is supported by two same links, $O_1C = O_2B$ and crank O_1C is parallet to crank O_2B. At this instant shown, crank OA of length 0.30 m, has rotating revolution $n_{OA} = 40$ r/min, and the link AB is vertical to crank OA, plate BC is collinear with point O, determine the velocity of plate BC.

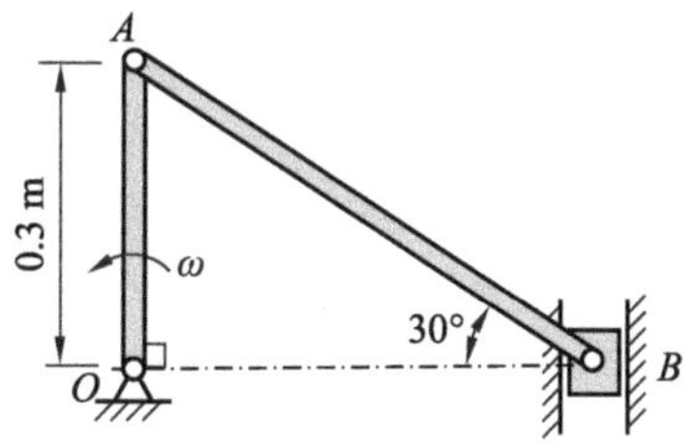

Fig. 8-29

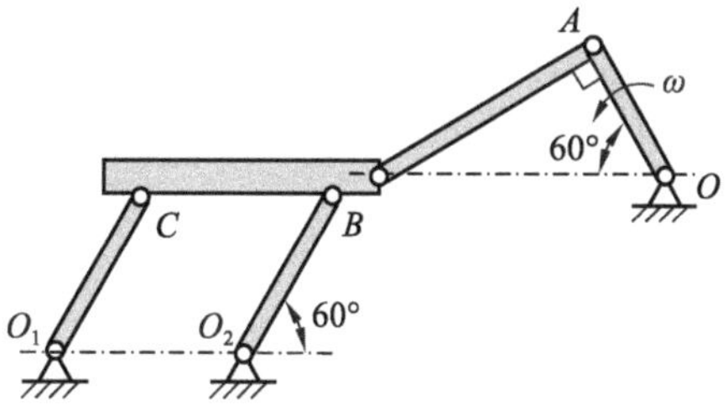

Fig. 8-30

8-7 A four-bar linkage, crank AB has a clockwise angular velocity $\omega = 1$ rad/s at the instant, Fig. 8-31. Determine the velocity of point C at this instant.

8-8 Block C moves with a speed of $v = 3$ m/s at horizontal surface, Fig. 8-32. Determine the angular velocities of links BC and crank AB, at the instant shown.

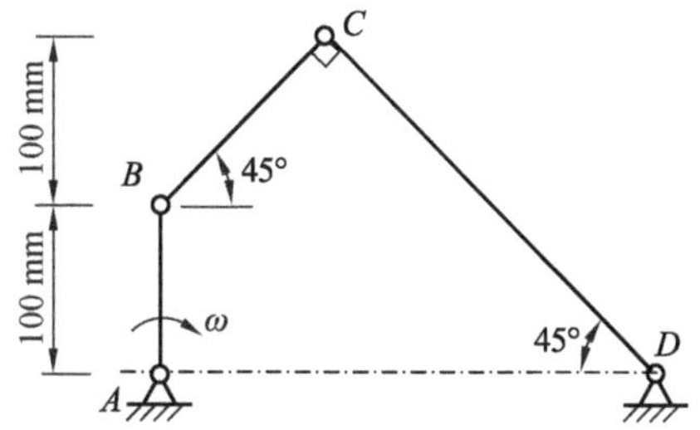

Fig. 8-31

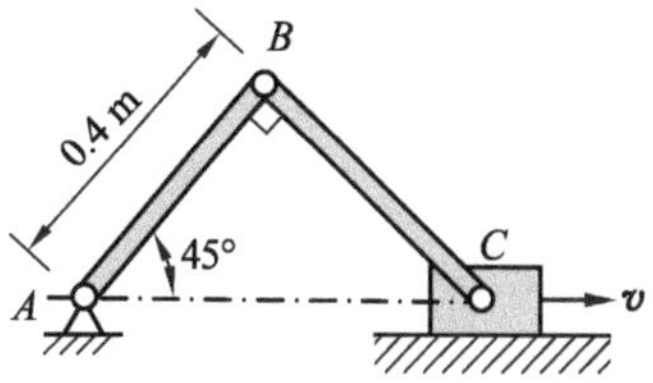

Fig. 8-32

8-9 In Fig. 8-33, the wheel has a radius of $R=15$ cm. The crank OA of radius 15 cm, rotates with uniform angular velocity of $\omega=2$ rad/s about axis O. When $\alpha=60°$, the link AB is vertical to crank OA, determine angular velocity ω_B of the wheel at this instant.

8-10 In Fig. 8-34, crank OA has a clockwise angular velocity $\omega=4$ rad/s when crank OA is parallel to OO_1, link AB is vertical to crank OA, and crank O_1B is vertical to link BC, determine the velocity of B and C and the angular velocities of member BC and AB at this instant. Given: $OA=0.15$ m, $AB=0.20$ m, $BC=0.30$ m.

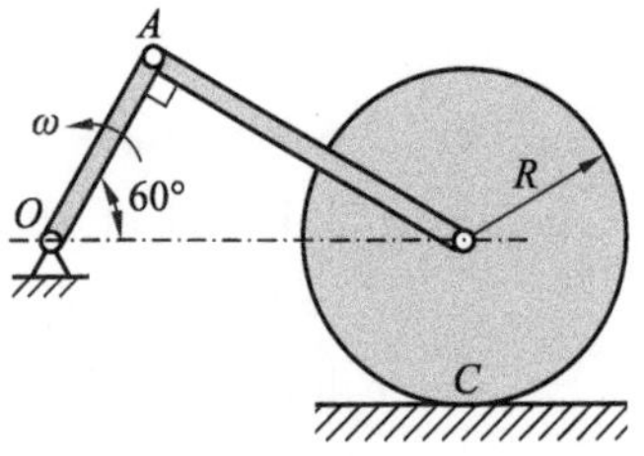

Fig. 8-33

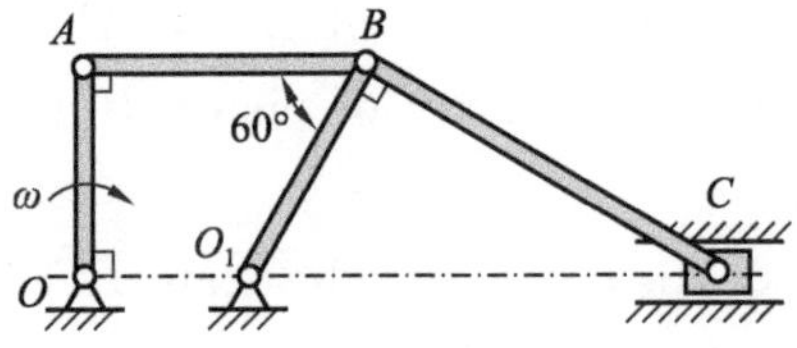

Fig. 8-34

8-11 In Fig. 8-35, if crank CD has an angular velocity of $\omega_{CD}=6$ rad/s, determine the velocity of point E on link BC and the angular velocity of rod AB at the instant shown.

8-12 In Fig. 8-36, gear Ⅰ is fixed, while arm OA rotates clockwise with an angular velocity of ω_{OA} = 6 rad/s, determine the angular velocity of gear Ⅱ at the instant shown.

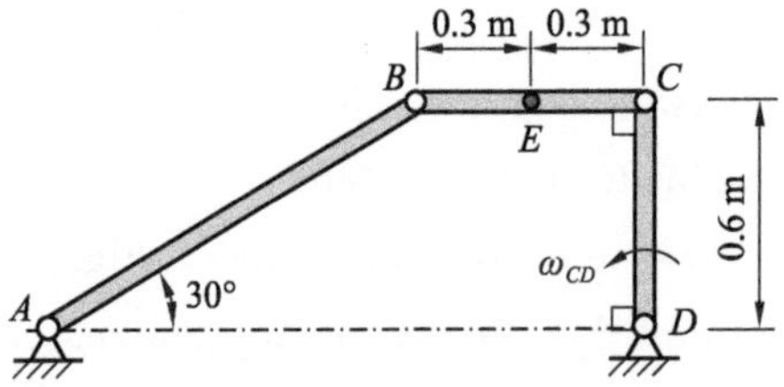

Fig. 8-35

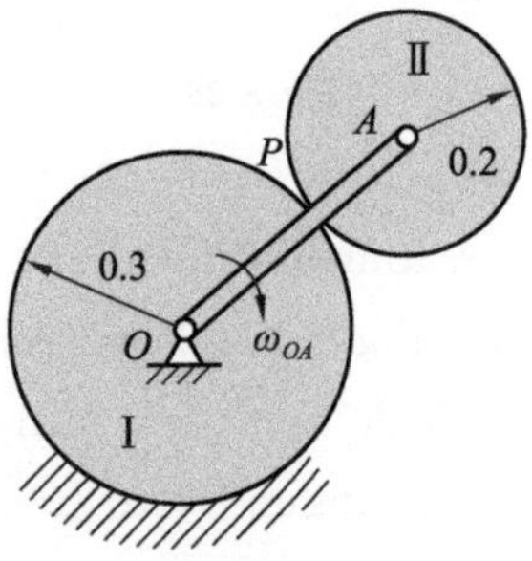

Fig. 8-36

8-13 In Fig. 8-37, the disk of radius r is confined to roll without slipping at plates A and B. If the plates have the velocities shown, determine the angular velocity of the disk.

8-14 In Fig. 8-38, crank OA of the mechanism rotates with angular velocity ω. Determine the velocities of points A and B and the angular velocities of links AB and rod BC at the instant shown.

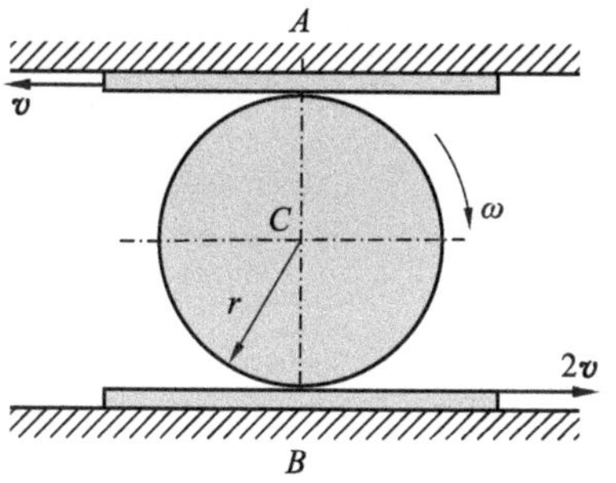

Fig. 8-37

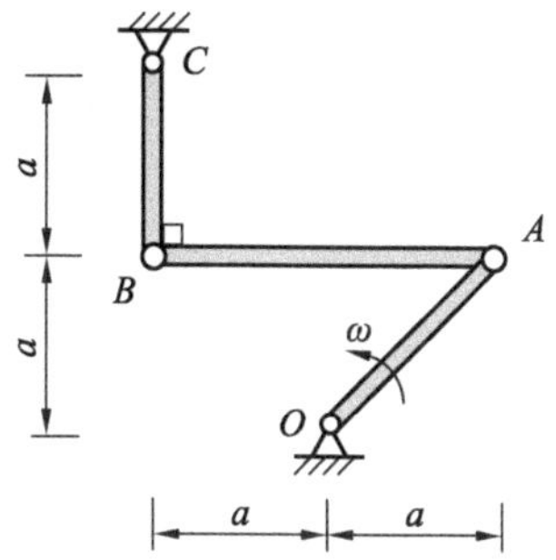

Fig. 8-38

8-15 In Fig. 8-39, crank OA has an angular velocity $\omega=3$ rad/s and crank O_1B is collinear with OO_1, when $\varphi=90°$. Determine the angular velocities of member O_1B and AB and the angular acceleration of O_1B at this instant. Set $OA=O_1B=0.5AB$.

8-16 Crank OA rotates with constant angular velocity ω, Fig. 8-40. At the instant shown, the crank O_1B is vertical to link AB, determine the angular velocities and the angular acceleration of member O_1B. Set $OA=O_1B=r$.

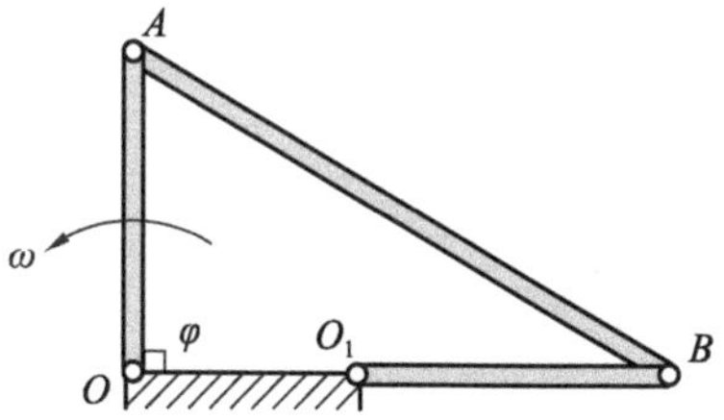

Fig. 8-39

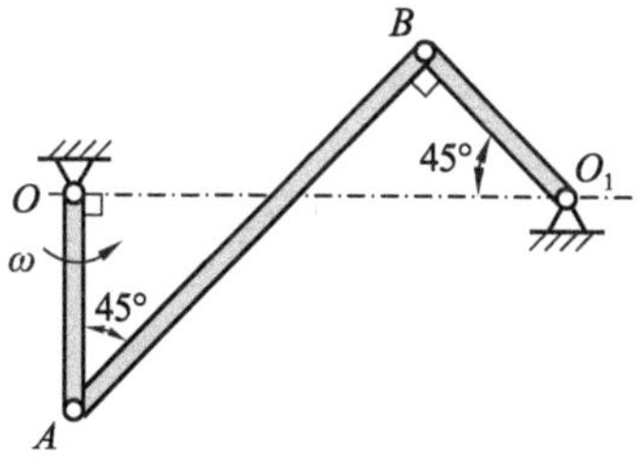

Fig. 8-40

8-17 In Fig. 8-41, the crank OA of radium 0.20 m, rotates with constant angular velocity of $\omega=10$ rad/s, linkage $AB=1$ m. At the instant $\varphi=45°$, the crank OA is vertical to link AB, determine the velocity and acceleration of piston B and the angular velocity of linkage AB.

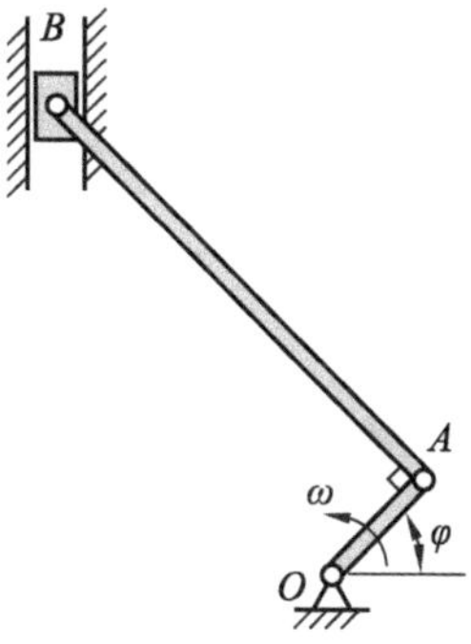

Fig. 8-41

8-18 In Fig. 8-42, the crank OA of radium r, rotates with constant angular velocity ω, link $AB=6r$, determine the velocity and acceleration of block B at the instant shown.

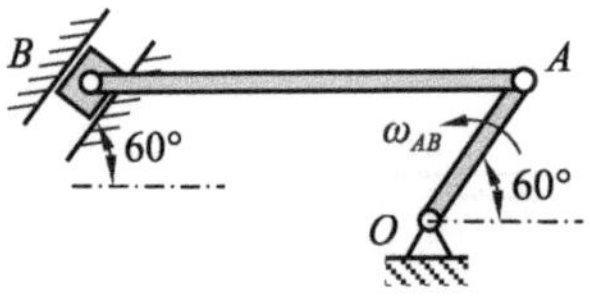

Fig. 8-42

8-19 In Fig. 8-43, at a given instant, the cylinder of radius r, has an angular velocity ω and angular acceleration α. Determine the velocity and acceleration of its center O and the acceleration of the contact point at P if it rolls without slipping.

8-20 In Fig. 8-44, the spool shown unravels from the cord, such that at the instant shown it has an angular velocity of $\omega=3$ rad/s and an angular acceleration of $\alpha=4$ rad/s^2. Determine the acceleration of point B. Given: $R=0.75$ m, $r=0.5$ m.

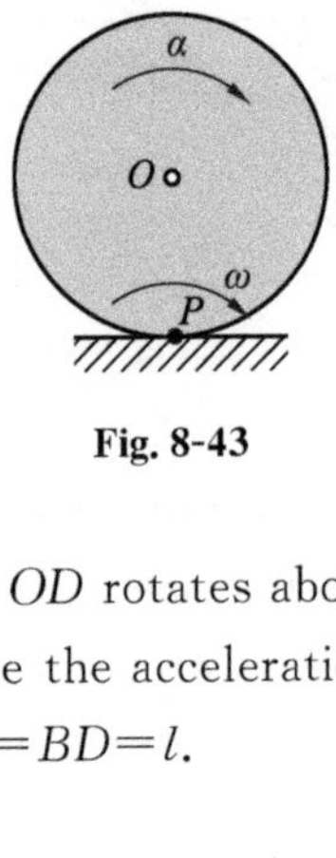

Fig. 8-43

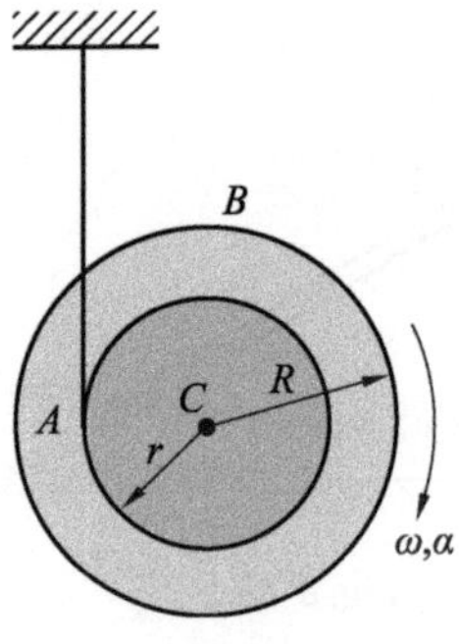

Fig. 8-44

8-21 Crank OD rotates about point O with a uniform ω, Fig. 8-45. At the instant $\varphi=60°$, determine the acceleration of slider A and the angular acceleration of rod AB. Given: $OD=AD=BD=l$.

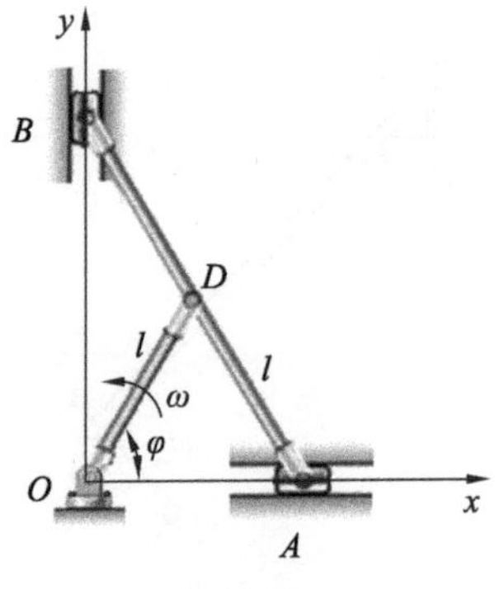

Fig. 8-45

Part 3 Kinetics

Up to now, we have already studied the concept of forces in detail in statics and the kinematics of the particle and rigid body. Statics is concerned with the equilibrium of a body without considering the motion. Kinematics only utilized kinematic quantities (position, velocity, acceleration) to describe motion without involving the forces. However, the motion involves forces in general, it is necessary to couple the concept of force to the kinematic quantities. *Kinetics* is a branch of dynamics that deals with the relationship between the change in motion of a body and the forces that cause this change. That is, *Kinetics* studies the mechanical motion of the bodies considering the *masses* of them and the forces which act about them.

In the study of *Kinetics*, we will use many concepts that we have already introduced in the statics and kinematics, e. g. , mass, force, moment, displace, velocity and acceleration. Fundamental concepts from statics and kinematics such as the action-reaction law, the force parallelogram law and coordinate systems will also be employed. In the Kinetics, the free-body diagrams will play a critical role similarly, just as they did in the study of statics and kinematics. And we will further introduce a new fundamental variable and define new dynamical concepts and dynamical laws.

***A particle*:** A body possessing mass but having such small dimensions that it can be treated as a point in a geometric sense. For example, the size of the satellite is insignificant compared to the size of its orbit, and therefore the satellite can be modeled as a particle when studying its orbital motion. When a body is idealized as a particle, the principles of mechanics reduce to a rather simplified form since the geometry of the body *will not be involved* in the analysis of the problem. Therefore, the body can be represented as a point with a fixed mass m.

***A system of particles*:** A collection of particles; such as solar system, a pile of sand. The rigid body can be considered as a special case of a *system of particles*, in which the distances between different particles of the body remain *constant* under the action of the forces.

We will begin our study of kinetics by discussing the kinetics of a particle, we shall define the specific basic notions of the dynamics. After we shall study three theorems, *linear momentum*, *angular momentum* and *work-energy theorem*, and use them to study the motions of the particle, rigid body and systems of particles. Besides of these we shall study two kinds of notions, *D'Alembert's* principle and principle of *virtual work* which are used to analyze some of the previous problems.

Chapter 9

Kinetics of a Particle

Objectives

In this chapter, we shall study the motion of the particle considering its mass and the action of the forces. We will state Newton's three laws and to define mass and weight. Newton's second law is formulated in rectangular, normal and tangential coordinates. Forward and inverse dynamics problems are defined and analyzed. The motion of a particle subjected to the action of forces is studied using the equation of motion with different coordinate systems.

9.1 Newton's Laws of Motion

The foundations of kinetics are established in Newton's three laws that were formulated by Issac Newton. They are drawn from a summary of all experimental experience and are compatible with common experience.

9.1.1 Newton's 1st Law

A body at rest not acted upon by an external force (the resultant force acting on a particle is zero) will remain at rest, and a body in motion moving at a constant speed along a straight line will remain in motion unless acted upon by an external force.

This law says that a body has the property of maintaining its original state of static or uniform rectilinear motion as long as it experiences no external force. This law can also be named as *Law of Inertia*.

The *statics* is the special case that is contained in *Newton's 1st Law* by considering the case where $\boldsymbol{v}=\mathbf{0}$(i. e., the body remains rest for all times).

9.1.2 Newton's 2nd Law

> *The acceleration of a particle is proportional to the external force acting on the particle; the direction and the sense of acceleration are identical to the forces.*

This law states that when an *unbalanced* force acts on a particle, the particle will accelerate in the direction of the force with a magnitude that is proportional to the force. Newton's second law of motion can be formulated in the following form:

$$\frac{\mathrm{d}(m\boldsymbol{v})}{\mathrm{d}t}=\boldsymbol{F} \tag{9.1}$$

If it is assumed that the mass is constant, Eq. (9.1) can also be written as

$$\boldsymbol{F}=m\boldsymbol{a} \tag{9.2}$$

where m is the *mass* of the particle, measure of the resistance of the particle to a change in its velocity, that is, the larger the mass of a particle, the less likely it is to change its motion, so the mass is its *inertia*. The acceleration $\boldsymbol{a}$ has the same direction as the force $\boldsymbol{F}$. The above equation of motion is one of the most important formulations in mechanics.

If a particle of mass m located at or near the surface of the earth, the only gravitational force between the earth and the particle, is termed the *weight* $\boldsymbol{G}$. In this case, the particle moves under the influence of the *earth's gravitational acceleration* $\boldsymbol{g}$ in the direction of the center of the earth (here the value $g=9.81\ \mathrm{m/s^2}$ is used for calculations). Substituting $\boldsymbol{g}$ into Eq. (9.2), then we have

$$\boldsymbol{G}=m\boldsymbol{g} \tag{9.3}$$

In the SI system, the base units are the units of length(meter, m), mass(kilogram, kg), and time (second, s). Therefore, according to Eq. (9.3), the force is a derived quantity. The common unit for force is the *Newton* ($1\ \mathrm{N}=1\ \mathrm{kg\cdot m/s^2}$). As a result, a body of mass 1 kg has a weight of 9.81 N.

9.1.3 Newton's 3rd Law

> *The mutual forces of action and reaction between two bodies are equal, opposite, and collinear.*

Newton's third law is also known as the *law of action and reaction*. It is valid both for bodies in contact and for bodies interacting at a distance. This law gives the relation of the interaction among the particles in the system of particles, so that the dynamics of particle can be extended and applied to the particles' system or body.

Newton's Laws of Motion are valid only in *inertial frames of reference*. When applying the equation of motion, the reference frame that the acceleration of the particle be measured with respect to must be either fixed or translates with a constant velocity. In this way, the observer will not accelerate and measurements of the particle's acceleration will be the same from any reference. For example, when studying the motions of rockets and satellites, we will consider the inertial reference frame as fixed to the stars, whereas for the majority of applications, the earth can be considered as an inertial system.

One restriction of Newton's 2nd Law is that in the case where the velocity of body approaches the speed of light ($c \approx 300\ 000$ km/s), one needs to consider the special theory of relativity due to *Einstein*. But this is seldom the case in engineering. For the mechanical motion problems in general engineering, the results obtained by applying *Newton's Laws* are enough precision.

9.2 The Equation of Motion

9.2.1 Differential Equation of Motion

At the instant considered, the arbitrary particle, having a mass m, is subjected to a force $\boldsymbol{F}$ and moves along a path. The position of a particle in space is described by its *position vector* $\boldsymbol{r}$, as shown in Fig. 9-1a. When more than one forces act on a particle, the resultant force is determined by a vector summation of all the forces; i. e. , $\boldsymbol{F}_R = \sum \boldsymbol{F}_i$, Fig. 9-1b. For this general case, the equation of motion may be written as

$$m\frac{\mathrm{d}^2\boldsymbol{r}}{\mathrm{d}t^2} = \sum \boldsymbol{F}_i \tag{9.4}$$

This represents a differential vector equation by second order in $\boldsymbol{r}$, which emphasizes that the force depends on the position of a particle defined by a radius vector on time. Eq. (9.4) is a second-order non-linear differential equation.

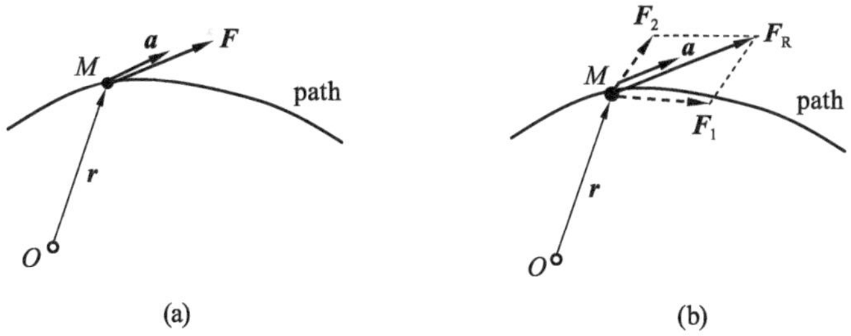

Fig. 9-1

9.2.2 Equations of Motion: Rectangular Coordinates

We know that a free particle in space has three degrees of freedom that means the motion of the particle is defined by three scalars. If a particle moves relative to an inertial rectangular coordinate, the forces acting on the particle, as well as its position vector, can be expressed in terms of their x, y, z components, Fig. 9-2. Projecting the vector equation of motion Eq. (9.4) onto the axes of the system, we may write the following three scalar equations:

$$m\frac{d^2x}{dt^2}=\sum F_x$$
$$m\frac{d^2y}{dt^2}=\sum F_y \tag{9.5}$$
$$m\frac{d^2z}{dt^2}=\sum F_z$$

This means that we can determine *the motion of the particle* using three scalar equations. In particular, if the particle is moving only in the x-y plane, then the first two of these equations are used to specify the motion.

$$m\frac{d^2x}{dt^2}=\sum F_x,\ m\frac{d^2y}{dt^2}=\sum F_y \tag{9.6}$$

In the equation each vector is replaced with three scalars in space and with two scalars in plane.

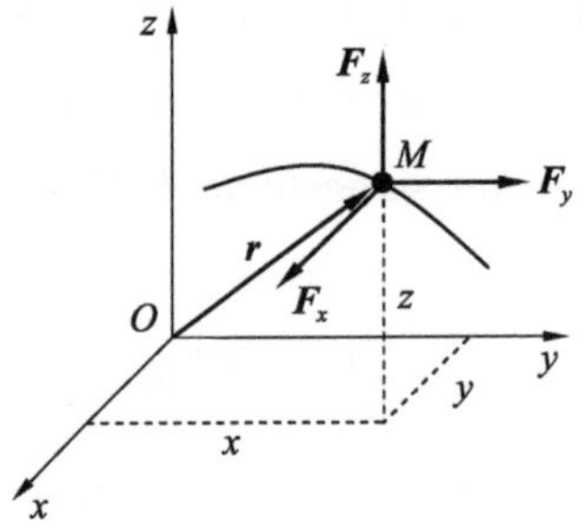

Fig. 9-2

9.2.3 Equations of Motion: Normal and Tangential Coordinates

According to previous discussion in Chapter 5, Section 5.3, in the *natural coordinates* we can resolve the acceleration and force into *normal* and *tangential components*. Now consider a particle moves along a curved path which is known, we can construct *normal* and *tangential* coordinates, Fig. 9-3, and then the equation of motion for the particle Eq.(9.4) may be projected in the tangential, normal, and binormal directions. Note that there is no motion of the particle in the binormal direction, since the particle is constrained to move along the path. We have

$$ma_t = m\frac{\mathrm{d}v}{\mathrm{d}t} = \sum F_t$$

$$ma_n = m\frac{v^2}{\rho} = \sum F_n \tag{9.7}$$

$$0 = \sum F_b$$

In the equation, a_t represents the time rate of change in the magnitude of velocity. So if $\sum F_t$ acts in the direction of motion, the particle's speed will increase, whereas if it acts in the opposite direction, the particle will slow down. Likewise, a_n represents the time rate of change in the velocity's direction. And $\sum F_n$ always acts toward the path's center of curvature. So it is often referred to as the *centripetal force*.

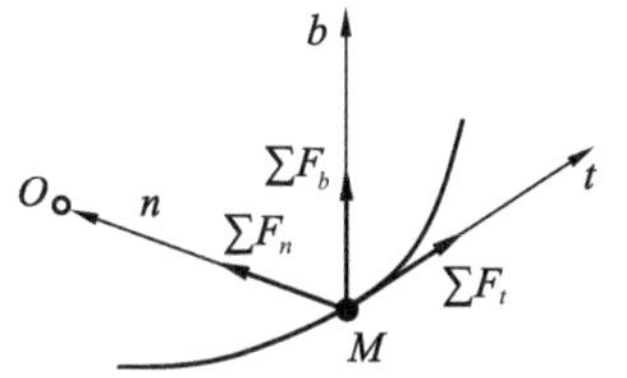

Fig. 9-3

The equation of motion of the particle can be projected on an arbitrary reference system. According to the motion characteristics of the particle, Newton's second law can be formulated in cylindrical, spherical, and polar coordinates to simplify the analysis.

9.2.4 Classification of Dynamics Problems

The differential equation of a particle's motion deals with two classes of problems, *forward problems* and *inversed problems*. In the first case the motion of a particle is known, and the force that causes the motion is determined. In the second case, the force is known, and the motion of a particle is to be determined. There are also mixed problems in which we have to solve the problem in the both senses.

The integration of the differential equation of the motion is generally difficult to make, so the *inverse dynamics problem* is more complicated. To determine the motion of a particle knowing the force, one should integrate the differential equation twice. Then we obtain a general solution of the system of equations. In order to uniquely determine the motion of the particle, one should determine two integration constants from the initial conditions referring to the initial position and initial velocity of the particle.

For the *forward* problem of the dynamics of the particle the solution is find easier because it will be obtained through derivation of the given law of motion and replacing in the differential equation results the resultant force acting on the particle.

Example 9-1

The block A with a weight of $\boldsymbol{G}$ is put in crate. If the crate is lifting upward with constant acceleration of $\boldsymbol{a}$ by a crane, Fig. 9-4a, determine the reaction force of block.

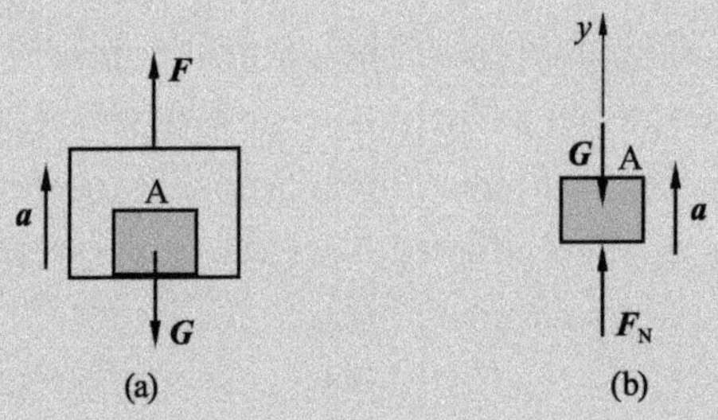

Fig. 9-4

Solution:

The free-body diagram of the block is shown in Fig. 9-4b. The weight of the block is $\boldsymbol{G}$, the block is subjected to a constant upward acceleration $\boldsymbol{a}$ in the positive y direction. Only one unknown $\boldsymbol{F}_N$ to be determined, using the equation of motion in the vertical direction, we have

$$\sum F_y = ma_y, \quad \frac{G}{g}a = F_N - G$$

Solving, we obtain

$$F_N = \frac{G}{g}a + G = G\left(1 + \frac{a}{g}\right) \qquad Ans.$$

From the result, it can be seen that reaction force of block equals to the weight of block when the acceleration $a = 0$. The reaction force of block is bigger than the weight of block when the acceleration $a > 0$.

Example 9-2

A particle D of mass m moves along a horizontal path under the action of the force $F = F_0 \cos \omega t$ (F_0, ω are all constant), as shown in Fig. 9-5, If it is initially located at the origin O, $v_0 = 0$, determine the displacement equation of the particle.

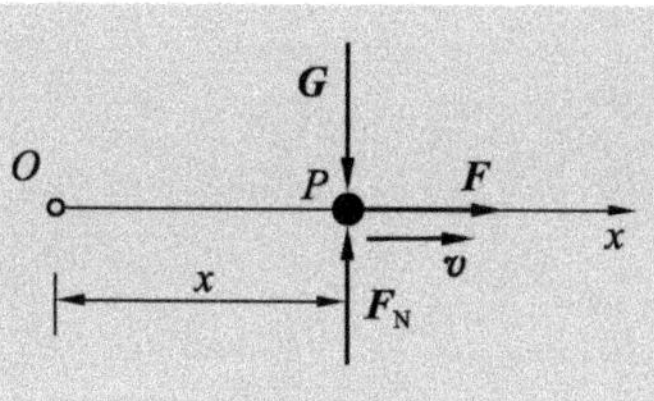

Fig. 9-5

Solution:

The free-body diagram of the block is shown in Fig. 9-5. The weight of the particle $\boldsymbol{G}$ and constraint force $\boldsymbol{F}_N$ are pair of balance forces, the direction of velocity of particle is assumed to horizontally, in the positive x direction. Using the differential equation of motion in the x direction, we have

$$\sum F_x = ma_x, \quad m\frac{dv}{dt} = F_0 \cos \omega t$$

Separating the variables and integrating, using that initially $x_O = 0$, $v_0 = 0$, we have

$$\int_0^v dv = \frac{F_0}{m}\int_0^t \cos \omega t \, dt$$

$$v = \frac{F_0}{m\omega}\sin \omega t \tag{a}$$

Using $v = \frac{dx}{dt}$, the Eq. (a) can be written as

$$\frac{dx}{dt} = \frac{F_0}{m\omega}\sin \omega t$$

Separating the variables and integrating again, we have

$$\int_0^x dx = \frac{F_0}{m\omega}\int_0^t \sin \omega t \, dt$$

$$x = \frac{F_0}{m\omega}(1 - \cos \omega t) \qquad \textit{Ans.}$$

Example 9-3

The grinder with internal diameter of $D = 3.2$ m, rotates uniformly around center O. The iron ball rises by the frictional force of grinder, Fig. 9-6a. To make iron ball getting enough energy to smash the ore, the iron ball should fall down when $\varphi = 54°40'$. Determine the angular velocity of the grinder (in r/min) at this instant.

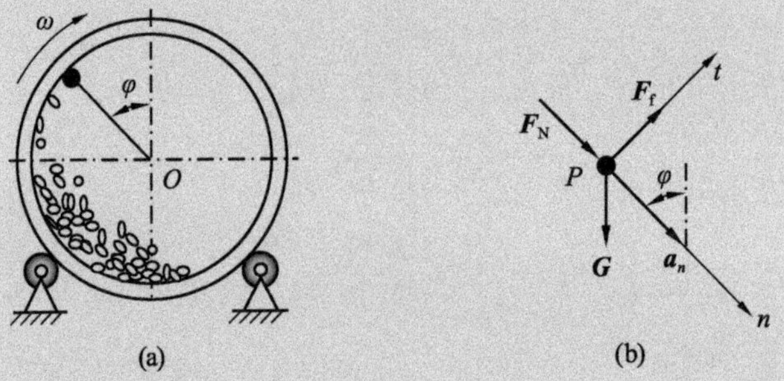

Fig. 9-6

Solution:

The free-body diagram of the iron ball when it is at an arbitrary position φ is shown in Fig. 9-6b. The iron ball is subjected to the weight $\boldsymbol{G}$, normal force $\boldsymbol{F}_N$ and frictional force $\boldsymbol{F}_f$. Since normal component of acceleration a_n can be related to angular velocity, it is calculated

$$a_n = \frac{D}{2}\omega^2$$

And

$$\omega = \frac{2n\pi}{60} = \frac{n\pi}{30} \tag{a}$$

Using the differential equation of motion in the n direction, we have

$$\sum F_n = ma_n, \quad \frac{G}{g}\frac{D}{2}\omega^2 = F_N + G\cos\varphi$$

As the iron ball leaves the cylinder, $F_N = 0$, we obtain

$$\frac{D}{2}\omega^2 = g\cos\varphi$$

$$\omega = \sqrt{\frac{2g\cos\varphi}{D}} \tag{b}$$

Substituting Eq. (b) into Eq. (a), we have

$$n = \frac{30}{\pi}\sqrt{\frac{2g\cos\varphi}{D}} = \frac{30}{\pi}\sqrt{\frac{2\times 9.8\times\cos 54°40'}{3.2}}$$

$$= 18 \text{ r/min}$$ *Ans.*

Example 9-4

A small ball has the mass of $m=0.1$ kg and is attached to the cord of length $l=0.3$ m. The cord is tied at the top, in Fig. 9-7. If the ball travels around the circular path and makes the cord has an angle $\theta=60°$ with the vertical line OC, determine the velocity of the small ball and the tensile force of the cord. Neglect air resistance and the size of the ball.

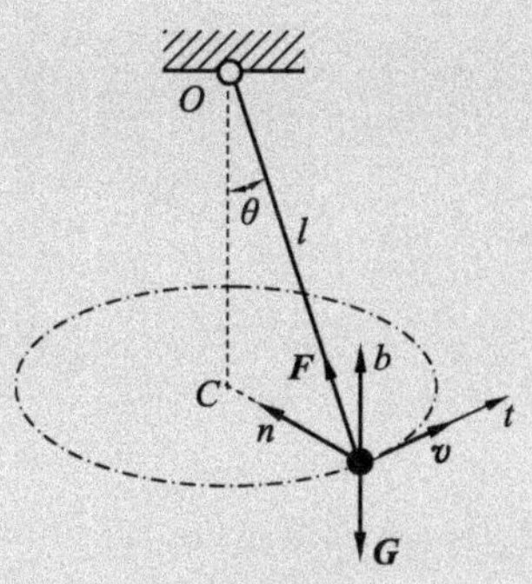

Fig. 9-7

Solution:

The free-body diagram of the small ball is shown in Fig. 9-7. The ball is subjected to the weight $\boldsymbol{G}=mg$, and tensile force $\boldsymbol{F}$. *The unknowns are* $\boldsymbol{v}$ and $\boldsymbol{F}$. Using the differential equation of motion in the $\boldsymbol{n}$, $\boldsymbol{b}$ direction, we have

$$\sum F_n = ma_n,\ m\frac{v^2}{l\sin\theta} = F\sin\theta \tag{a}$$

$$\sum F_b = 0,\ F\cos\theta - mg = 0 \tag{b}$$

Eq.(b) can be solved for the tensile force $\boldsymbol{F}$ of the cord. Solving all the equations, we obtain

$$F=\frac{mg}{\cos\theta}=1.96\ \mathrm{N} \qquad \textit{Ans.}$$

$$v=\sqrt{\frac{Fl\sin^2\theta}{m}}=2.1\ \mathrm{m/s} \qquad \textit{Ans.}$$

The example belongs to the mixed problems in which we solved the problem both force $\boldsymbol{F}$ and velocity $\boldsymbol{v}$.

Exercises

9-1 As shown in Fig. 9-8, the motor winds in the cable with a constant acceleration, such that the 1 500 kg block moves a distance $s = 1.8$ m in 3 s, starting from rest. Determine the tension developed in the cable. (Neglect the weight of cable)

9-2 In Fig. 9-9, the block M with mass m is put on an inclined plane. The motor winds in the cable with a constant acceleration α, starting from rest. Determine the tension developed in the cable. The coefficient of kinetic friction between the block and the plane is μ_k.

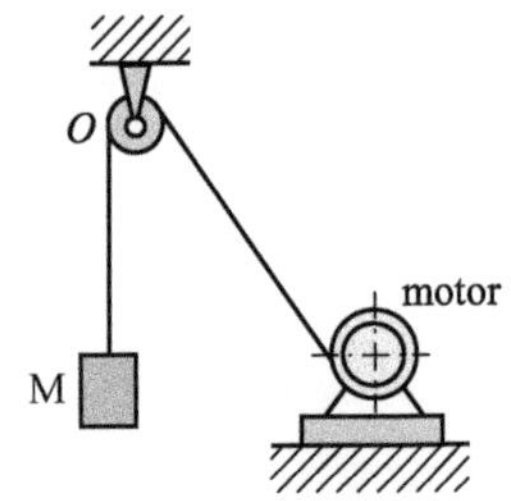

Fig. 9-8

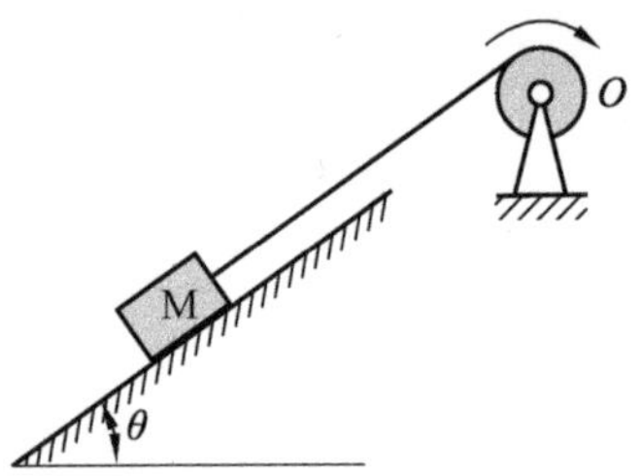

Fig. 9-9

9-3 As shown in Fig. 9-10, the 50 kg crate rests on a horizontal surface for which the coefficient of kinetic friction is $\mu_k=0.3$. If the crate is subjected to a 400 N towing force, determine the velocity of the crate in 3 s starting from rest.

9-4 In Fig. 9-11, the motor winds in the cable with a constant acceleration, such that the 20 kg crate moves a distance $s=6$ m in 3 s, starting from rest. Determine the tension developed in the cable. The coefficient of kinetic friction between the crate and the plane is $\mu_k=0.3$.

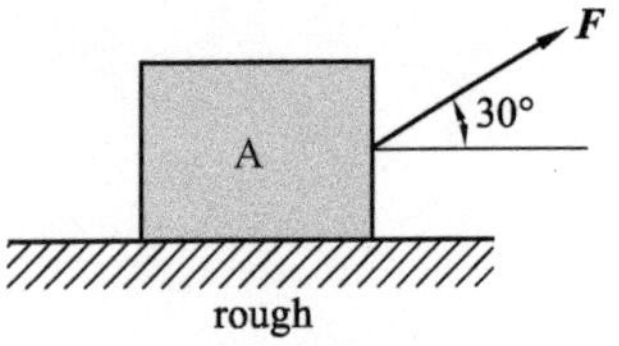

Fig. 9-10

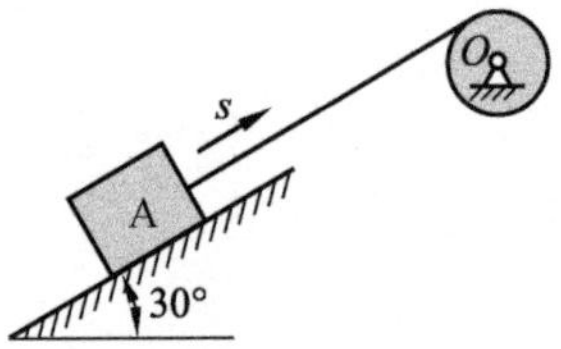

Fig. 9-11

9-5 In Fig. 9-12, blocks A and B have a mass m_A and m_B, respectively, where $m_A>m_B$. The distance between two blocks is h at the rest. Determine the time it takes for two blocks to reach a same height after releasing from rest. Neglect the mass of the pulley.

9-6 In Fig. 9-13, the blocks A and B of same mass m, are placed on the smooth vertical and horizontal plane respectively. If blocks release from the rest with $\theta=60°$, determine the force developed in the link AB. Neglect the mass of the link.

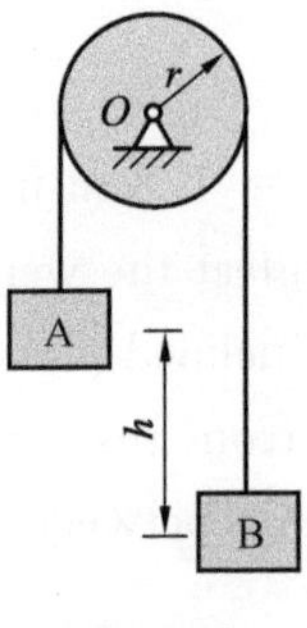

Fig. 9-12

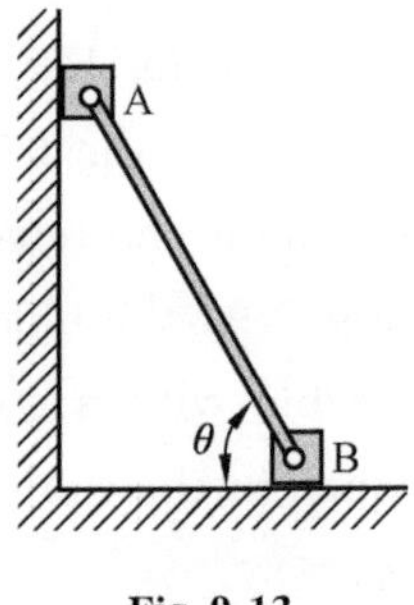

Fig. 9-13

9-7 In Fig. 9-14, the block A with mass m rests at a distance of r from the center of the platform that rotates about axis. If the coefficient of static friction between the block and the platform is μ_k, determine the maximum angular velocity ω which the block can attain before it begins to slip. Assume the angular motion of the disk is slowly increasing.

9-8 In Fig. 9-15, the 10 kg ball M moves along a vertical circular path with radius $r=2$ m. If the ball has a velocity of $v=3$ m/s when it is at the position shown, determine the tension in the cord at this position.

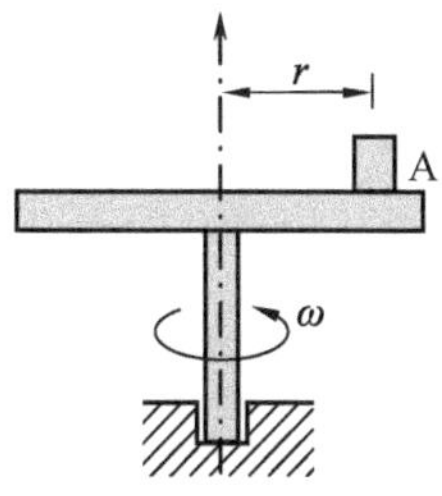

Fig. 9-14

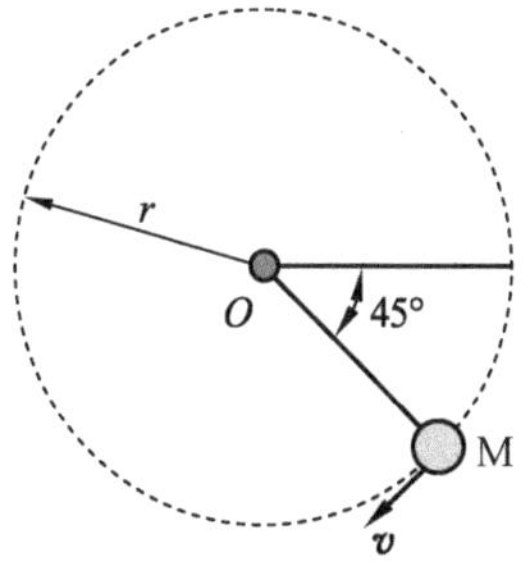

Fig. 9-15

9-9 As shown in Fig. 9-16, determine the maximum speed at which the car A with mass m can pass over the top point of the hill and still keep contact with the road.

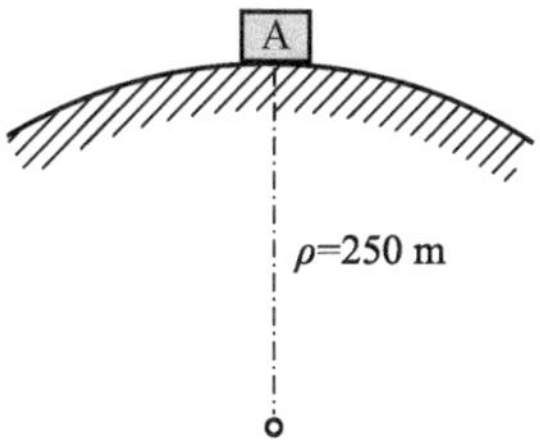

Fig. 9-16

9-10 Determine the tension in wire BC just after wire AB is cut, as shown in Fig. 9-17. The small ball has a mass m.

9-11 In Fig. 9-18, at the instant shown, the 50 kg projectile travels in the vertical plane with a speed of $v=40$ m/s. Determine the tangential component of its acceleration and the radius of curvature ρ of its trajectory at this instant.

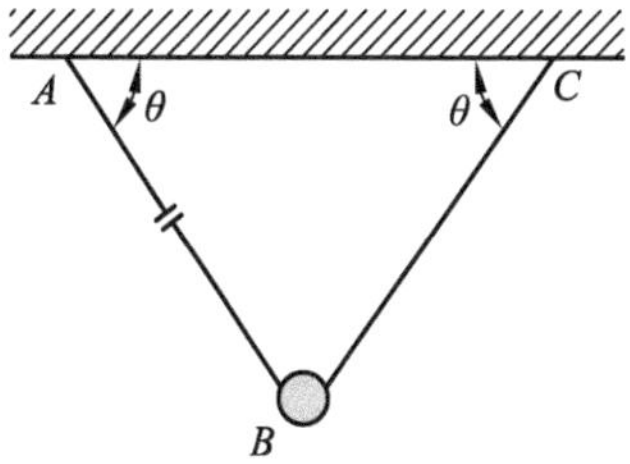

Fig. 9-17

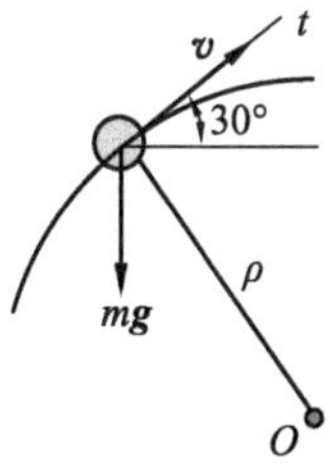

Fig. 9-18

Chapter 10

Principle of Linear Momentum

Objectives

In this chapter, we will first formalize the methods for obtaining a body's linear momentum, then develop the principle of linear momentum and apply it to solve problems that involve force, velocity, and time. Further, we will study the conservation of linear momentum and its application. Finally, we will study the law of motion of the mass center.

10.1 Linear Momentum and Impulse

Dynamics problem for one particle can use the differential equations of motion to solve them. But dynamics problem for the particles system, we can write $3n$ differential equations for a system of n particles to solve them. Practical problems are:

① Combining and solving $3n$ differential equations (performing the integral operation) is very difficult.

② In a great number of problems, we only need to investigate the motion of the particles system as a whole without the necessity to know the motion of every particle in the system.

Following, we introduce the general theorems of dynamics (including the theorem of momentum, the theorem of kinetic energy, the theorem of angular momentum and some others theorems) for solving dynamical problems.

10.1.1 Linear Momentum

We consider a particle of mass m that has the instantaneous velocity $\boldsymbol{v}$ at any instant of the motion. The *linear momentum* $\boldsymbol{p}$ of a particle is the vector quantity and equal to the product of the mass m and the instantaneous velocity $\boldsymbol{v}$ of the particle,

$$\boldsymbol{p}=m\boldsymbol{v} \tag{10.1}$$

Since mass m is a positive scalar, the linear momentum vector has the same direction as the velocity $\boldsymbol{v}$ of the particle, and its magnitude has units of kg · m/s. Linear momentum is a physical quantity measuring the intensity of the mechanical motion of a body.

For a system of particles, the linear momentum is equal to the vector sum of the linear momentum of all the particles in the system.

$$\boldsymbol{p} = \sum m_i \boldsymbol{v}_i \tag{10.2}$$

To simplify the calculation of the linear momentum of a particles system, we introduce the position vector $\boldsymbol{r}_C$ for the system's *center of mass* or center of gravity C.

$$\boldsymbol{r}_C = \frac{\sum m_i \boldsymbol{r}_i}{\sum m_i} = \frac{\sum m_i \boldsymbol{r}_i}{m}$$
$$m\boldsymbol{r}_C = \sum m_i \boldsymbol{r}_i \tag{10.3}$$

Here, $m = \sum m_i$ is the total mass of the particles system. Taking time derivatives of Eq. (10.3) and using definitions: $\boldsymbol{v} = \frac{\mathrm{d}\boldsymbol{r}}{\mathrm{d}t}$, yields

$$m\boldsymbol{v}_C = \sum m_i \boldsymbol{v}_i \tag{10.4}$$

Substituting into Eq. (10.2), the linear momentum $\boldsymbol{p}$ of the particles system can be expressed as

$$\boldsymbol{p} = \sum m_i \boldsymbol{v}_i = m\boldsymbol{v}_C \tag{10.5}$$

where $\boldsymbol{v}_C$ is velocity of the center of mass. Thus, the total linear momentum of the system of particles can be determined from the product of the total mass m and the velocity $\boldsymbol{v}_C$ of mass center. If for a rigid body, the linear momentum can also be determined using the Eq. (10.5). Namely, the *rigid body*'s linear momentum is a vector quantity having a *magnitude* $m\boldsymbol{v}_C$ and a *direction* defined by the velocity $\boldsymbol{v}_C$ of the body's mass center.

Example 10-1

Determine the linear momentum of the rigid body as shown in Fig. 10-1. The mass of each rigid body is m.

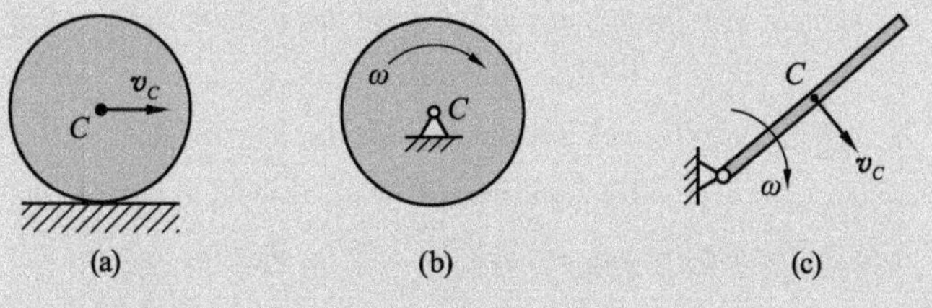

Fig. 10-1

Solution:

① The wheel is rolling without slipping along the horizontal surface, the velocity of the mass center is $\boldsymbol{v}_C$. According to the Eq.(10.5), the linear momentum of the wheel is

$$p = mv_C$$

② The wheel rotates around the axis passing through mass center C with constant angular velocity ω. Since the velocity of the mass center is zero, so the linear momentum of the wheel is zero.

$$p = 0$$

③ The rod with length l rotates about the fixed end axis with constant angular ω. The linear momentum of the rod is

$$p = mv_C = \frac{1}{2}ml\omega$$

Example 10-2

The system consists of the crank OA of mass m_1, link BD of mass $2m_1$, and block B, D of mass m_2 as shown in Fig. 10-2. The crank OA rotates about point O with a uniform angular velocity ω. At the instant of angle φ, determine the linear momentum of system. Set $OA = AB = AD = l$.

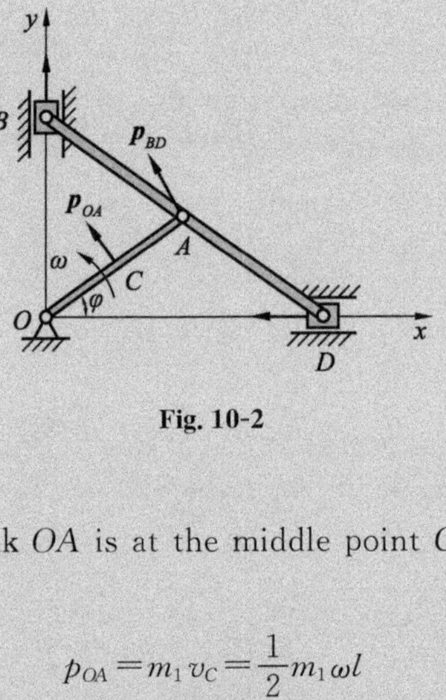

Fig. 10-2

Solution:

The mass center of crank OA is at the middle point C, and linear momentum of it is

$$p_{OA} = m_1 v_C = \frac{1}{2}m_1 \omega l$$

The link BD and block B, D are considered as a single system, and its mass center located in point A, the linear momentum of this system of particles is

$$p' = p_B + p_D + p_{BD} = 2(m_1 + m_2)v_A = 2(m_1 + m_2)\omega l$$

By inspection, the direction of the linear momentum of crank OA is same as the direction of the single system (consists of the link BD and block B, D), so the total linear momentum of the system is

$$p = p_{OA} + p' = \left(\frac{5}{2}m_1 + 2m_2\right)\omega l \qquad Ans.$$

10.1.2 Linear Impulse

The *linear impulse* is a *vector quantity which measures the effect of a force during the time the force acts*. Since time is a positive scalar, the impulse acts in the same direction as the force, and its magnitude has units of force-time, N · s.

If the force is expressed as a function of time, the impulse can be determined by direct evaluation of the integral.

$$I = \int_{t_1}^{t_2} \boldsymbol{F} \mathrm{d}t \tag{10.6}$$

In particular, if the force is constant in both magnitude and direction, the resulting impulse becomes

$$I = \int_{t_1}^{t_2} \boldsymbol{F} \mathrm{d}t = \boldsymbol{F}(t_2 - t_1) \tag{10.7}$$

Linear impulse is used to characterize the accumulated effect on a body of force acting during a certain time interval.

For the system of particles, by Newton's third law the internal forces occur in equal but opposite collinear pairs and therefore cancel out. So for the system of particles,

$$\sum \boldsymbol{F}^{(i)} = 0, \ \sum \boldsymbol{M}(\boldsymbol{F}^{(i)}) = 0$$

That means the impulse is only caused by all the *external forces* acting on the system from t_1 to t_2.

10.2 Principle of Linear Momentum

In this section we will integrate the equation of motion with respect to time and thereby obtain the principle of linear momentum and impulse.

10.2.1 Principle of Linear Momentum for a Particle

According to the Newton's second law, the equation of motion for a particle of mass m can be written as

$$\frac{\mathrm{d}}{\mathrm{d}t}(m\boldsymbol{v}) = \boldsymbol{F} \tag{10.8a}$$

Eq. (10.8a) states that *the time rate of change of the total linear momentum is equal to the total resultant of the external forces*. The preceding equation can be written in the form

$$\mathrm{d}(m\boldsymbol{v}) = \boldsymbol{F}\mathrm{d}t \tag{10.8b}$$

where $\boldsymbol{F}\mathrm{d}t$ is *the elementary impulse of a force*. The preceding formula has the following interpretation: *An infinitesimal increment of linear momentum of a particle is equal to the elementary impulse of forces acting on the particle*.

Integrating both sides of Eq. (10.8b) from t_1 to t_2, we obtain:

$$m\boldsymbol{v}_2 - m\boldsymbol{v}_1 = \int_{t_1}^{t_2} \boldsymbol{F}\mathrm{d}t = \boldsymbol{I} \tag{10.9}$$

where m = const was assumed. This equation is referred to as the *principle of linear momentum and impulse*. It states that *the change in the linear momentum between time t_0 and an arbitrary time t is equal to the impulses created by the external force system which acts on the system during this time interval*. It provides a direct means of obtaining the particle's final velocity $\boldsymbol{v}_2$ after a specified time period when the particle's initial velocity is known and the forces acting on the particle are either constant or can be expressed as functions of time.

10.2.2 Principle of Linear Momentum for a System of Particles

For a system of particles moving relative to an inertial reference in Fig. 10-3, each particle m_i is subject to both internal as well as external forces. From the equations of motion Eq. (10.8), the *principle of linear momentum* is obtained for the individual particle:

$$\mathrm{d}(m_i\boldsymbol{v}_i) = (\boldsymbol{F}_i^{(i)} + \boldsymbol{F}_i^{(e)})\mathrm{d}t = \boldsymbol{F}_i^{(i)}\mathrm{d}t + \boldsymbol{F}_i^{(e)}\mathrm{d}t \tag{10.10}$$

If we sum the equations of motion over all n particles, then we have

$$\sum \mathrm{d}(m_i\boldsymbol{v}_i) = \sum \boldsymbol{F}_i^{(e)}\mathrm{d}t + \sum \boldsymbol{F}_i^{(i)}\mathrm{d}t \tag{10.11}$$

By Newton's third law, the internal forces occur in equal but opposite collinear pairs and therefore cancel out, so $\sum \boldsymbol{F}_i^{(i)}\mathrm{d}t = 0$. Supposing $\sum \mathrm{d}(m_i\boldsymbol{v}_i) = \mathrm{d}\boldsymbol{p}$, yield

$$\mathrm{d}\boldsymbol{p} = \sum \boldsymbol{F}_i^{(e)}\mathrm{d}t = \sum \mathrm{d}\boldsymbol{I}_i^{(e)} \tag{10.12a}$$

or

$$\frac{\mathrm{d}}{\mathrm{d}t}\boldsymbol{p} = \sum \boldsymbol{F}_i^{(e)} = \boldsymbol{F} \tag{10.12b}$$

where $\boldsymbol{F}$ is the resultant force of all the external forces acting on the system. The equation says: *the time rate of change of the total linear momentum of the system of particles is equal to the total resultant of the external forces.*

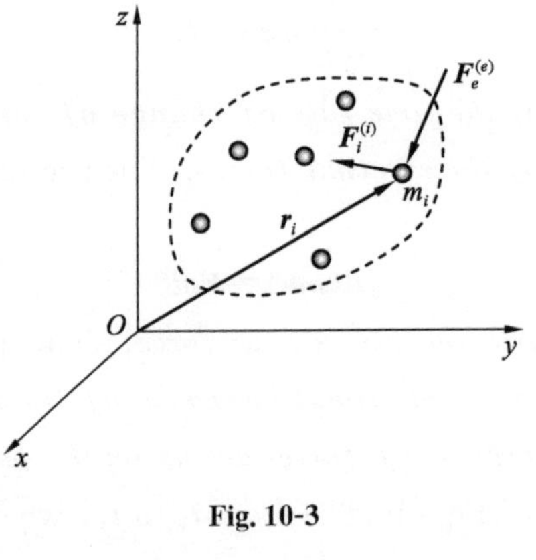

Fig. 10-3

We can integrate Eq. (10. 12) with respect to time to give

$$\int_{p_1}^{p_2} \mathrm{d}\boldsymbol{p} = \sum \int_{t_1}^{t_2} \boldsymbol{F}_i^{(e)} \, \mathrm{d}t \tag{10. 13a}$$

or

$$\boldsymbol{p}_2 - \boldsymbol{p}_1 = \sum \boldsymbol{I}_i^{(e)} \tag{10. 13b}$$

This equation states that *the change of the linear momentum of the system is equal to the impulses of all the external forces acting on the system from t_1 to t_2.*

If each of the vectors in Eq. (10. 12) and Eq. (10. 13) is resolved into its x, y, z components, we can write the following three scalar equations of linear momentum and impulse.

$$\frac{\mathrm{d}p_x}{\mathrm{d}t} = \sum F_x^{(e)}, \quad \frac{\mathrm{d}p_y}{\mathrm{d}t} = \sum F_y^{(e)}, \quad \frac{\mathrm{d}p_z}{\mathrm{d}t} = \sum F_z^{(e)} \tag{10. 14}$$

and

$$p_{2x} - p_{1x} = \sum I_x^{(e)}, \quad p_{2y} - p_{1y} = \sum I_y^{(e)}, \quad p_{2z} - p_{1z} = \sum I_z^{(e)} \tag{10. 15}$$

Example 10-3

The 100 kg block shown in Fig. 10-4a is original at rest on the horizontal smooth surface. If a towing force of 200 N, acting at an angle of 45°, is applied to the block for 10 s, determine the final velocity and the normal force which the surface exerts on the block during the time interval.

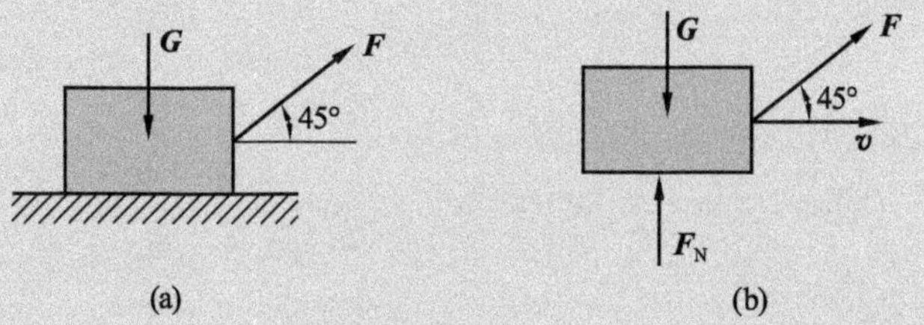

Fig. 10-4

Solution:

Free-body diagram of block is shown as Fig. 10-4b, since all the forces are constant, the impulses are simply the product of the force $\boldsymbol{F}$ and 10 s. Applying the principle of linear momentum and impulse Eq. (10. 9) in the x, y direction respectively, yields

$$mv_{x2} - mv_{x1} = F_x \cdot t$$

$$mv_{y2} - mv_{y1} = F_N \cdot t - mg \cdot t + F \cdot t \cdot \sin 45°$$

Substituting the $m=100$ kg, $F=200$ N and $t=10$ s into above equation, we have

$$100v_2 - 0 = 200 \times \cos 45° \times 10$$

$$v_2 = 14.1 \text{ m/s} \qquad \textit{Ans.}$$

$$0-0=F_N\times 10-981\times 10+200\times 10\times \sin 45^\circ$$

$$F_N=840\ \text{N} \qquad Ans.$$

Note: Since no motion occurs in the y direction, direct application of the equilibrium equation gives the same result for F_N.

Example 10-4

The 3 000 kg hammer A falls from rest at a height $h=1.5$ m and strikes the top of the workpiece B, as shown in Fig. 10-5a. If the coupling between the hammer and workpiece takes place in $t=0.01$ s. Determine the average force between them.

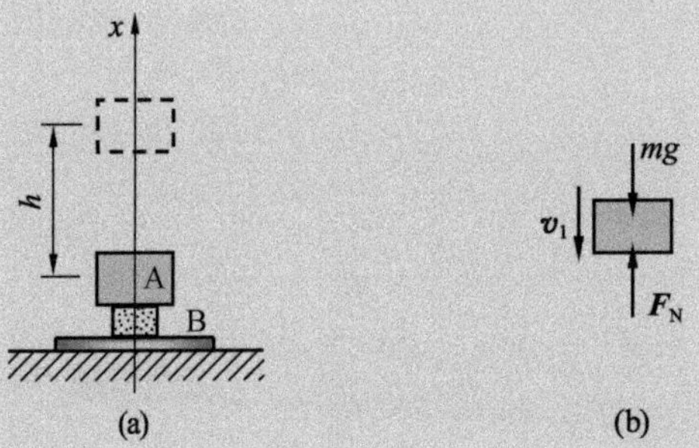

Fig. 10-5

Solution:

The free-body diagram of the hammer is shown in Fig. 10-5b, since the weight and average forces between the hammer and workpiece are constant, the impulses are simply the product of the force and time t. The velocity v_1 of the hammer that contacted initially the workpiece is

$$v_1=-\sqrt{2gh}$$

After deformation of the workpiece, the velocity of the hammer $v_2=0$. Applying the principle of linear momentum and impulse.

$$mv_2-mv_1=(F_N-mg)t$$

Solving, we have

$$F_N=mg\left(1+\frac{1}{t}\sqrt{\frac{2h}{g}}\right)$$

$$=3\ 000\times 9.81\times\left(1+\frac{1}{0.01}\sqrt{\frac{2\times 1.5}{9.81}}\right)$$

$$=1\ 657\ \text{kN} \qquad Ans.$$

Namely the average force between the hammer and workpiece is 1 657 kN.

Example 10-5

A motor body consisting of the rotor m_2 and stator m_1, is mounted on a horizontal foundation, in Fig. 10-6a. Due to an inaccuracy in the production, the rotor is rotating with the uniform angular velocity ω around the axis passes through the mass center O_1 of stator. The eccentric distance from the mass center O_2 of the rotor to O_1 is e. Determine the reaction forces exerted on the mount of the motor by the foundation.

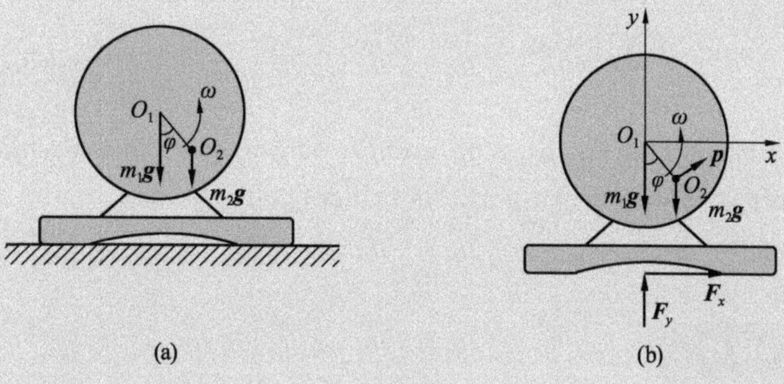

Fig. 10-6

Solution:

The motor will be considered as a system of particles which consists the rotor and stator. The free-body diagram is shown in Fig. 10-6b. The resultant reaction force from the foundation is $\boldsymbol{F}_x$ and $\boldsymbol{F}_y$. Since the stator is at rest, the linear momentum of the rotor is the momentum of the motor system.

$$p = m_2 e\omega$$

$$p_x = p\cos\varphi = m_2 e\omega\cos\omega t$$

$$p_y = p\sin\varphi = m_2 e\omega\sin\omega t$$

Applying the principle of momentum and impulse Eq. (10.14) in the x, y direction respectively, we have

$$\frac{\mathrm{d}p_x}{\mathrm{d}t} = F_x, \quad -m_2 e\omega^2\sin\omega t = F_x$$

$$\frac{\mathrm{d}p_y}{\mathrm{d}t} = F_y, \quad m_2 e\omega^2\cos\omega t = F_y - (m_1 + m_2)g$$

Solving, we get

$$F_x = -m_2 e\omega^2\sin\omega t \qquad \textit{Ans.}$$

$$F_y = (m_1 + m_2)g + m_2 e\omega^2\cos\omega t \qquad \textit{Ans.}$$

This is the dynamic reaction forces coming from the rotating of the rotor. If the motor is at rest, there are only static constraints $(m_1 + m_2)g$ on the foundation.

10.2.3 Law of Conservation of Linear Momentum

When the sum of the external resultant acting on a system of particles is zero ($\boldsymbol{F}=0$), then the linear momentum of the system is a constant, namely

$$\boldsymbol{p}=m\boldsymbol{v}_C=\text{const.} \tag{10.16}$$

This equation is referred to as the *conservation of linear momentum*. It states that *the total linear momentum for a system of particles remains constant during the time period t_1 to t_2.*

Generally this theorem is used in scalar form projecting on the axis of a reference system, for example, if $\sum F_x^{(e)}=0$,

$$p_x=\text{const.}$$

The linear momentum in the x direction is conserved. The *conservation of linear momentum* is often applied when particles collide, an explosion or the striking of one body against another.

Example 10-6

The 10 kg package A is thrown with a speed of $v=3$ m/s into the cart which has a mass of $M=50$ kg, Fig. 10-7a. If the cart is initially at rest, determine the velocity of the cart at the instant the package fully fall into the cart. What is the compression which the cart exerts on the ground when the collision time is $t=0.3$ s? Neglect friction between the ground and cart.

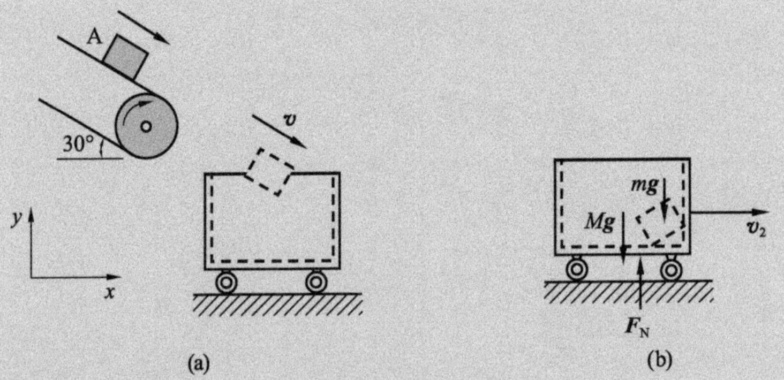

Fig. 10-7

Solution:

Here we will consider the cart and package as a system of particles. The free-body diagram is shown in Fig. 10-7b, the resultant reaction force exerted on the wheel from the ground is $\boldsymbol{F}_N$. Since the external force along the x axis is zero, the linear momentum will be conserved along the x direction. Using the *conservation of linear momentum* Eq. (10.16) in the x direction.

$$mv\cos 30° = (m+M)v_2$$

$$v_2 = \frac{mv\cos 30°}{m+M} = \frac{10 \times 3 \times \cos 30°}{10+50} = 0.443 \text{ m/s} \qquad \textit{Ans.}$$

Applying the *principle of linear momentum and impulse* Eq. (10.15) in the y direction, we have

$$p_2 y - p_1 y = I_y$$

$$0 - (-mv\sin 30°) = (F_N - mg - Mg)t$$

Solving,

$$F_N = \frac{mv\sin 30°}{t} + (m+M)g$$

$$= \frac{10 \times 3 \times 0.5}{0.3} + (10+50) \times 9.81$$

$$= 638.6 \text{ N} \qquad \textit{Ans.}$$

Example 10-7

The 1 200 N cannon shown in Fig. 10-8a fires an 8 N projectile with a muzzle velocity of 1 500 N/s relative to the ground. If firing takes place in 0.03 s, determine (1) the recoil velocity of the cannon just after firing, and (2) the average impulsive force acting on the projectile. The cannon support is fixed to the ground, and the horizontal recoil of the cannon is absorbed by two springs.

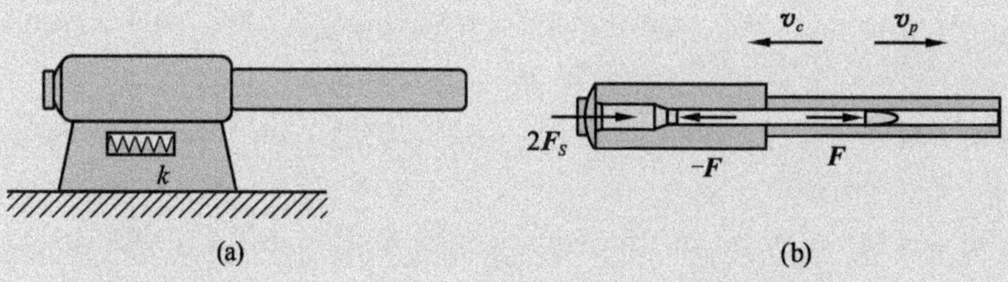

Fig. 10-8

Solution:

(1) As shown in Fig. 10-8b, the projectile and cannon are considered as a system of particles, since the impulsive forces **F**, between the cannon and projectile are internal to the system and will cancel from the analysis. Hence the linear momentum for the system is conserved in the horizontal x direction. Applying the *conservation of linear momentum*, Eq. (10.16).

$$m_c v_{c1} + m_p v_{p1} = -m_c v_{c2} + m_p v_{p2}$$

$$0 = -1\,200 v_{c2} + 8 \times 1\,500$$

$$v_{c2} = 10 \text{ m/s} \qquad \textit{Ans.}$$

(2) The average impulsive force exerted by the cannon on the projectile can be determined by applying the *principle of linear momentum and impulse* to the projectile, applying the Eq. (10. 15).

$$mv_{p2} - mv_{p1} = \int_0^t F\mathrm{d}t$$

$$8\times 1\,500-0=F_{\text{avg}}\times 0.03$$

$$F_{\text{avg}}=12.4\ \text{kN}$$ *Ans.*

10. 3 Motion of the Mass Center

10.3.1 Law of Motion for the Center of Mass

According to the preceding analysis, the total linear momentum $\boldsymbol{p}$ of the system of particles can be expressed using Eq. (10. 5) as

$$\boldsymbol{p}=m\boldsymbol{v}_C$$

Using *the principle of linear momentum* and substituting into Eq. (10. 8) yields the equation of motion for the center of mass:

$$\frac{\mathrm{d}}{\mathrm{d}t}(m\boldsymbol{v}_C) = m\frac{\mathrm{d}\boldsymbol{v}_C}{\mathrm{d}t} = \sum \boldsymbol{F}_i^{(e)} \tag{10. 17}$$

Using definition $\boldsymbol{a}_C=\dfrac{\mathrm{d}\boldsymbol{v}_C}{\mathrm{d}t}$, we obtain

$$m\boldsymbol{a}_C = \sum \boldsymbol{F}_i^{(e)} \tag{10. 18}$$

This equation is referred to as the *law of motion for the center of mass*. It states that *the mass center of a system moves as though it is a particle (with same total mass) subject to the resultant of forces acting on the whole system*. The theorem of motion of the mass center is an equivalent expression to the theorem of linear momentum of a system, being similar in form as the equation of motion Eq. (10. 8) for one particle. Note that only external forces can change the motion of the mass center of system, while internal forces are incapable of changing it. They change only the motion of the individual particles of the system.

The vector Eq. (10. 18) corresponds to three scalar equations one for each component, in rectangular coordinates:

$$\begin{aligned} ma_{Cx} &= \sum F_x^{(e)} \\ ma_{Cy} &= \sum F_y^{(e)} \\ ma_{Cz} &= \sum F_z^{(e)} \end{aligned} \tag{10. 19}$$

or in normal and tangential coordinates:

$$
\begin{aligned}
ma_C^t &= \sum F_t^{(e)} \\
ma_C^n &= \sum F_n^{(e)} \qquad (10.20) \\
0 &= \sum F_b^{(e)}
\end{aligned}
$$

Example 10-8

The crank AB of a mass m_1 and length r is rotating around axis A with the constant angular velocity ω. The I-shape rod BOD (the mass center is at point O) has a mass of m_2. If an external force $\boldsymbol{F}$ acts at point D, Fig. 10-9a, determine the maximum reaction forces at support A in the horizontal direction. Neglect friction and weight of block B.

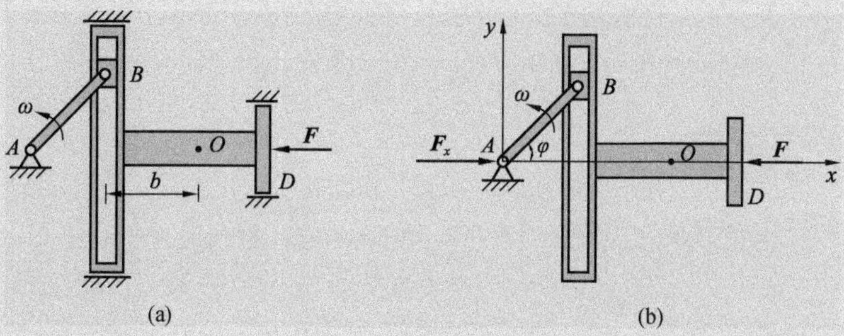

Fig. 10-9

Solution:

We will consider the crank and I-shape rod as a single system, the free-body diagram is shown in Fig. 10-9b. The coordination for mass center of the system in the x direction is

$$x_C = \frac{\sum m_i x_i}{\sum m_i} = \frac{m_1 \frac{r}{2}\cos \omega t + m_2(r\cos \omega t + b)}{m_2 + m_1} \qquad (a)$$

The acceleration of mass center in the x direction is

$$a_C = \frac{d^2 x_C}{dt^2} = \frac{-r\omega^2}{m_1 + m_2}\left(\frac{m_1}{2} + m_2\right)\cos \omega t \qquad (b)$$

From the free-body diagram, only one unknown $\boldsymbol{F}_x$ to be determined. Applying *law of motion for the mass center* in the x direction.

$$(m_1 + m_2)a_{Cx} = F_x - F \qquad (c)$$

Substituting Eq. (b) into Eq. (c), yield

$$F_x = F - r\omega^2\left(\frac{m_1}{2} + m_2\right)\cos \omega t$$

When $\cos \omega t = -1$, F_x is maximum

$$F_{max} = F + r\omega^2\left(\frac{m_1}{2} + m\right) \qquad \textit{Ans.}$$

10.3.2 Principle of Conservation of the Mass Center

In the special case that no external impulses are applied to the system, the external resultant is zero ($\boldsymbol{F}=0$), then Eq. (10.18) gives

$$\boldsymbol{p}=m\boldsymbol{v}_C=\text{const.} \tag{10.21}$$

The linear momentum of the system is a constant and the center of mass moves uniformly and in a straight line. Which indicates that the velocity $\boldsymbol{v}_C$ of the mass center for the system of particles does not change if no external impulses are applied to the system.

This theorem is usually used in scalar form projecting on the axis of a reference system, for example, if $\sum F_x^{(e)}=0$,

$$p_x=mv_{Cx}=\text{const.} \tag{10.22}$$

This indicates that the velocity $\boldsymbol{v}_{Cx}$ of the mass center for the system of particles in the x direction does not change. In the case that the mass center was initially at rest, $\boldsymbol{v}_C^0=0$, then the position of the mass center x_C will remain constant.

Example 10-9

A big triangular column A with mass m_1 is placed on a smooth horizontal plane, a small triangular column B with mass m_2 is at rest on the slope initially, as shown in Fig. 10-10. Determine the distance the big triangular column A moves when the small column B slides down the shape to the bottom of slope. Neglect the frictional force between two triangular columns.

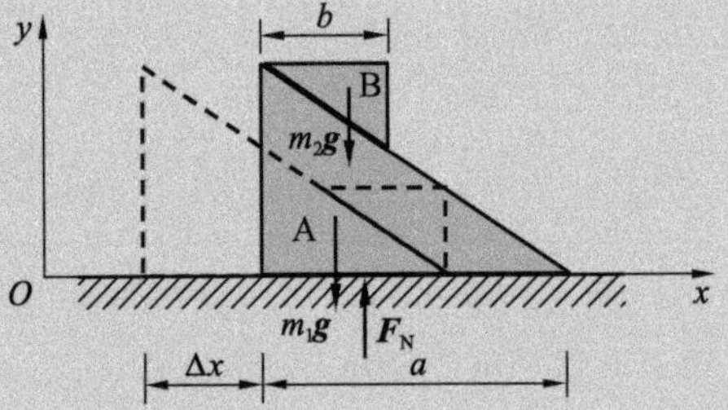

Fig. 10-10

Solution:

Here we will consider both triangular columns as a single system, the free-body diagram is shown in Fig. 10-10, since the external force along the x axis is zero, the linear momentum will be conserved along the x axis. Assume that the displacements of mass centers of every column before motion are x_1, x_2, respectively, then the mass center of system is

$$x_{C1}=\frac{m_1x_1+m_2x_2}{m_1+m_2} \tag{a}$$

Suppose the displacement of the big triangular column A is Δx when the small column B slides down the shape to the end of slope. The mass center of system after motion is

$$x_{C2}=\frac{m_1x_1'+m_2x_2'}{m_1+m_2}=\frac{m_1(x_1-\Delta x)+m_2[x_2-\Delta x+(a-b)]}{m_1+m_2} \quad \text{(b)}$$

According to the *principle of conservation of the mass center* and system of particles is at rest initially, we have

$$x_{C1}=x_{C2} \quad \text{(c)}$$

Substituting Eq. (a), Eq. (b) into Eq. (c), we get

$$\frac{m_1x_1+m_2x_2}{m_1+m_2}=\frac{m_1(x_1-\Delta x)+m_2[x_2-\Delta x+(a-b)]}{m_1+m_2}$$

$$\Delta x=\frac{m_1}{m_2+m_2}(a-b) \quad \textit{Ans.}$$

Exercises

10-1 Determine the linear momentum of the systems, Fig. 10-11. The mass of each body is m.

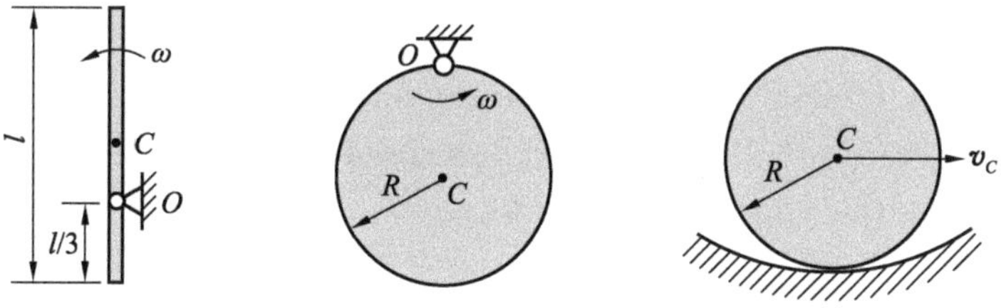

Fig. 10-11

10-2 As shown in Fig. 10-12, block A has a mass of $m_A=1$ kg and block B has a mass of $m_B=2$ kg. At the instant, the velocity of blocks A and B is $v=2$ m/s, determine the linear momentum of the system. Neglect the mass of the pulleys O and cord.

10-3 In Fig. 10-13, the 15 kg block is towed by the force $F=100$ N and starts to move from rest. If the coefficient of kinetic friction between the block and the ground is $\mu_k=0.2$, determine the speed of the block when $t=4$ s.

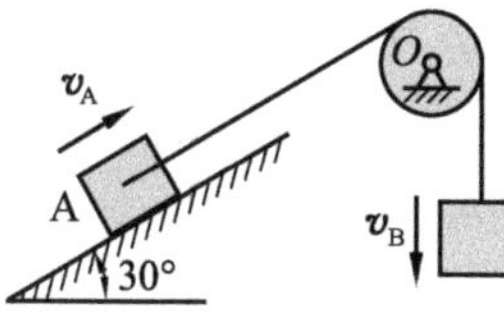

Fig. 10-12

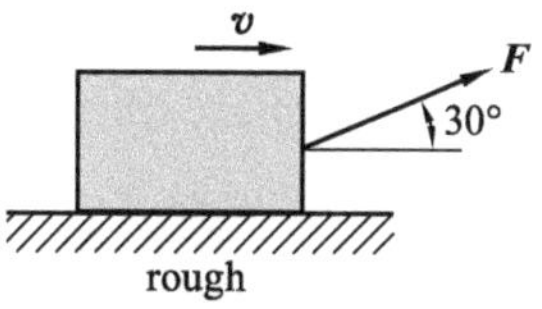

Fig. 10-13

10-4 In Fig. 10-14, the 1 500 kg track tows the 3 000 kg trailer along horizontal surface. If the engine provide a traction force of $F = 3$ kN on the engine wheels, determine the speed of the track when $t = 30$ s, starting from rest. Also, find the horizontal coupling force at C between the engine A and trailer B. Neglect rolling resistance.

10-5 In Fig. 10-15, the platform of mass m_1 is at rest initially on rough horizontal surface, the coefficient of dynamic friction between platform and surface is μ_k. If the trolley of mass m_2 moves on the platform with relative motion $s = bt^2/2$, determine the acceleration of platform. Neglect the mass of winch.

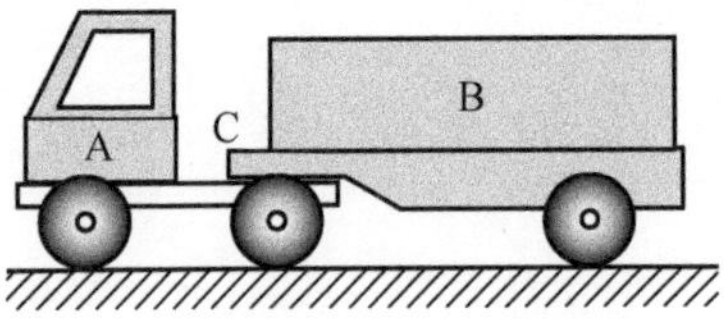

Fig. 10-14

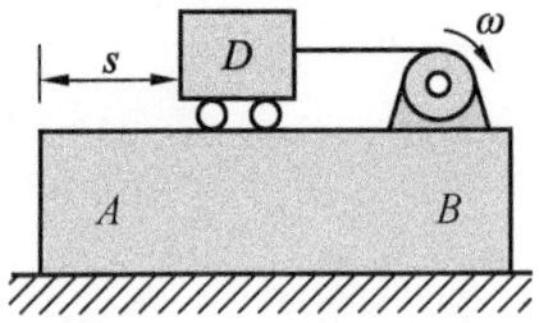

Fig. 10-15

10-6 In Fig. 10-16, the drum has a mass of m_1 and block B has a mass of m_2. Block A has a mass of m_3 and moves on the smooth incline. If the acceleration of block B is $\boldsymbol{a}$, determine the constraint forces of axis O. Neglect the mass of cord.

10-7 The 1 kg cannon ball is fired horizontally by a 50 kg cannon, Fig. 10-17. If the muzzle velocity of the ball is 500 m/s, measured relative to the ground, determine the recoil velocity of the cannon just after firing.

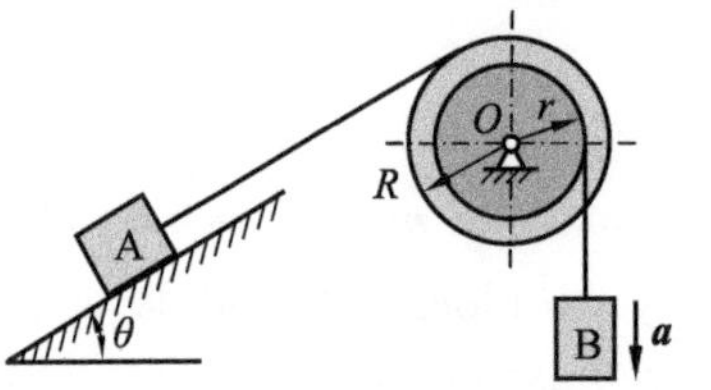

Fig. 10-16

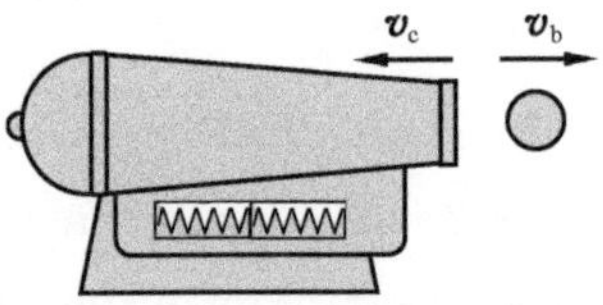

Fig. 10-17

10-8 In Fig. 10-18, the 1 500 kg car A is moving at 1. 5 m/s on the horizontal surface when it encounters a 1 200 kg car B moving at 0. 75 m/s toward it. If the cars collide and couple together, determine (1) the speed of both cars just after the coupling, and (2) the average force between them if the coupling takes place in 0. 8 s.

10-9 The circular plate of mass 1 kg, is at rest initially on a smooth support, Fig. 10-19. A cannon ball of mass 0. 17 kg strikes the plate with the velocity $v_1 = 550$ m/s. When the velocity of ball after shooting through the plate is $v_2 = 275$ m/s, determine the velocity of the circular plate.

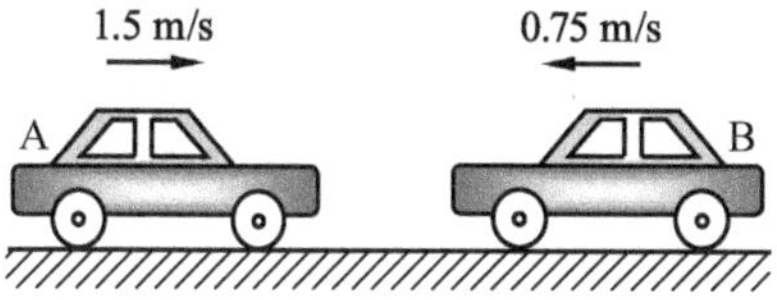

Fig. 10-18

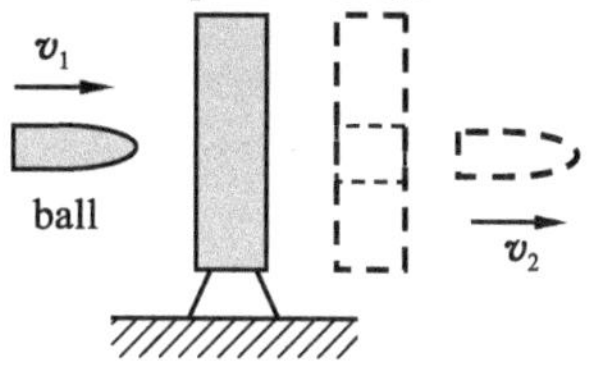

Fig. 10-19

10-10 The cart and crate have a mass of 200 kg and 100 kg, respectively, Fig. 10-20. The cart has a rough surface, both cart and crate move with initial uniform velocity of 3.5 km/h along the smooth horizontal surface. If a 50 kg stone falls into the crate, determine the velocity of the cart. After the stone falling, the crate slides on the cart and stop after 0.2 s, determine the average frictional force between the cart and crate.

10-11 In Fig. 10-21, the crank OA has a mass m_1 and length l, is rotating around O with the constant angular velocity ω. The slider A has a mass of m_2. The T-shape bar BCD whose mass center locates in point C, has a mass of m_3, determine the maximum reaction forces at support O in the horizontal direction. Neglect friction.

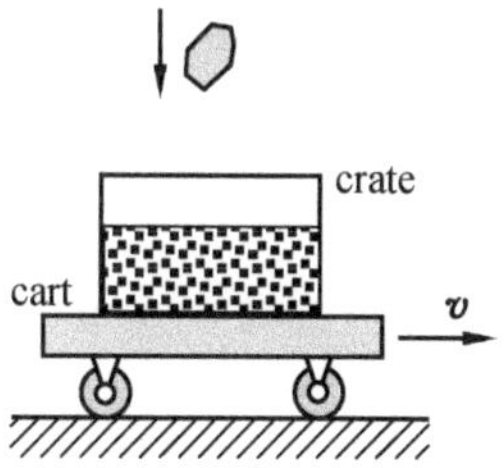

Fig. 10-20

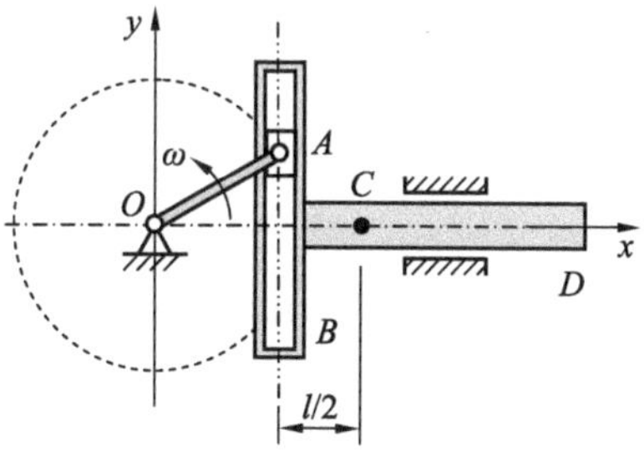

Fig. 10-21

10-12 The homogenous rod AB with length l is erecting on smooth horizontal surface, Fig.10-22. When it falls down from the vertical position with no initial velocity, determine the trajectory of its endpoint A.

10-13 The 20 kg block A is towed up the ramp of the 40 kg cart using the motor M mounted on the cart, Fig. 10-23. Determine how far the cart will move when the block has traveled a distance $s=2$ m up the ramp. Both the block and cart are at rest when $s=0$. Neglect rolling resistance.

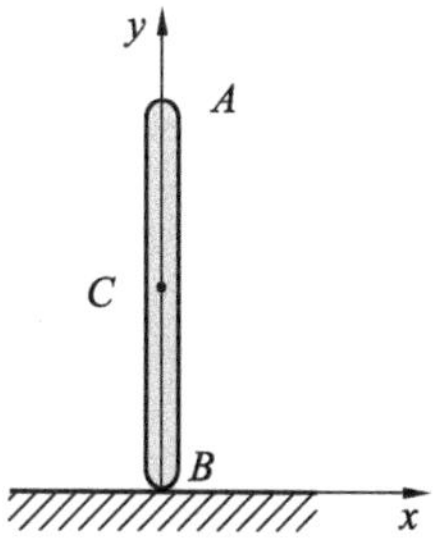

Fig. 10-22

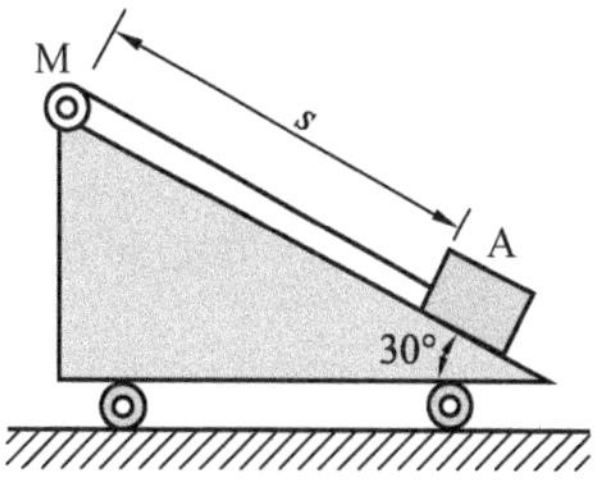

Fig. 10-23

10-14 In Fig. 10-24, the crane has weight of 20 000 kg and floats on the surface of water. The rod OA has a length of 8 m. When the crane lifts the weight of 2 000 kg, the rod OA rotates about the axis O from the rest, and the angle between the rod and vertical line changes from initial 60° to final 30°, determine the displacement of the crane. Neglect the effect of water resistance.

10-15 The 1 500 kg car moves on the 10 000 kg barge from the left to the right side with a constant speed of 4 m/s, measured relative to the barge, Fig. 10-25. Neglect right side water resistance, determine the velocity of the barge when the car reaches point B. Initially, the car and the barge are at rest relative to the water.

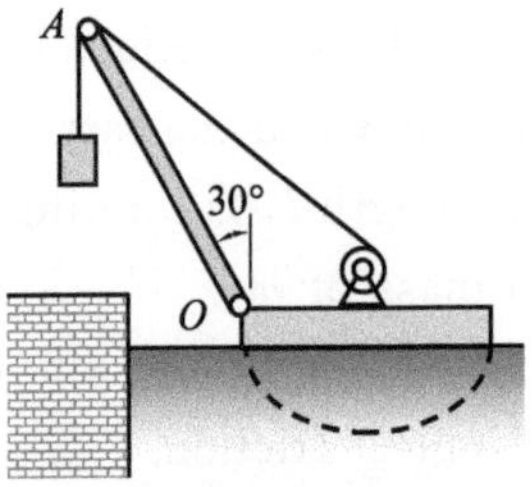

Fig. 10-24

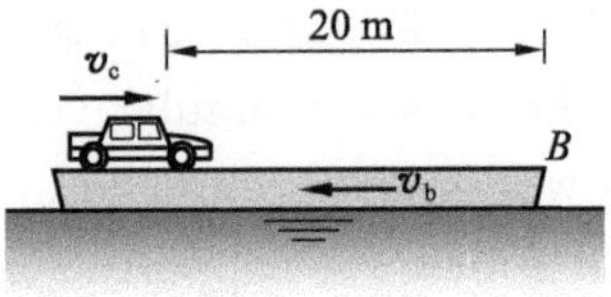

Fig. 10-25

10-16 In Fig. 10-26, three small blocks of mass $m_1=20$ kg, $m_2=15$ kg, $m_3=10$ kg, are connected by cord and put on a pentagon block of mass $m=100$ kg. When the block m_1 falls down 1 m, determine the displacement of pentagon block relative to the smooth surface. Neglect the friction and the mass of the rope and pulleys.

10-17 In Fig. 10-27, the homogenous rod AD and BD have length of $AD=250$ mm, $BD=400$ mm, respectively. They are connected by pin D and put on smooth horizontal surface with $h=240$ mm. When they fall down from rest with no initial velocity, determine the displacement of their endpoint A and B at the instant the A, D, B are in a line.

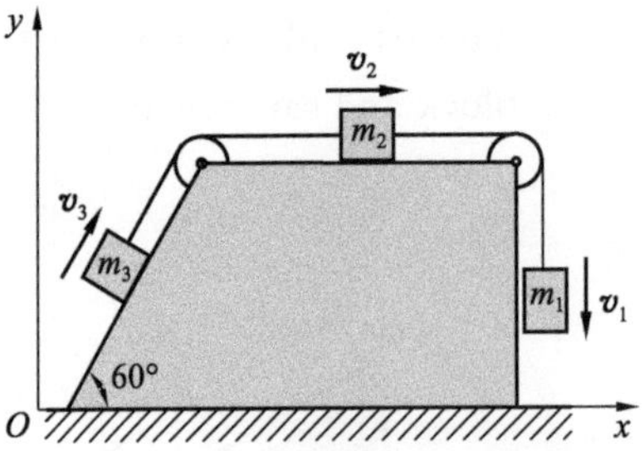

Fig. 10-26

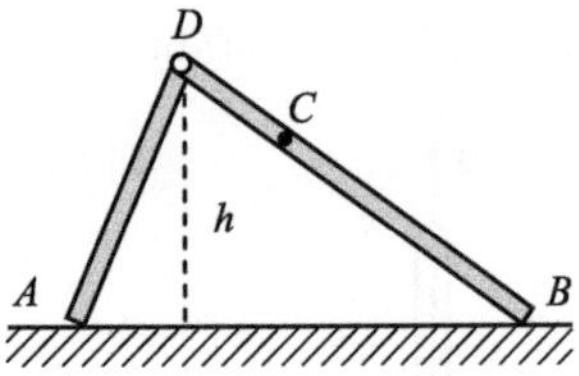

Fig. 10-27

Chapter 11

Principle of Angular Momentum

Objectives

In this chapter, we will first formalize the methods for obtaining a body's angular momentum, then develop the principle of angular momentum and apply it to solve problems that involve moment, angular velocity, and time. And the conservation of angular momentum is discussed. Further, we will apply the principles of angular momentum and the principles of linear momentum to solve rigid-body planar kinetic problems.

11.1 Angular Momentum

11.1.1 Angular Momentum of a Particle

The *angular momentum* of a particle about point O is defined as the "moment" of the particle's linear momentum about O.

(1) Scalar Formulation

If a particle moves along a curve lying in the x-y plane, Fig. 11-1, the angular momentum at any instant can be determined about point O (z axis) by using a scalar formulation. The magnitude of $\boldsymbol{L}_O$ is

$$\boldsymbol{L}_O = mvd \tag{11.1}$$

Here d is the moment arm or perpendicular distance from O to the line of action of $\boldsymbol{mv}$. The units for $\boldsymbol{L}_O$ is kg · m^2/s. The direction of $\boldsymbol{L}_O$ is defined by the right-hand rule. As shown, the curl of the fingers of the right hand indicates the sense of rotation of $\boldsymbol{mv}$ about O, so the thumb (or $\boldsymbol{L}_O$) is directed perpendicular to the x-y plane along the z^+ axis.

(2) Vector Formulation

If the particle moves along a space curve, Fig. 11-2, the angular momentum about O is determined by the *vector cross product*. In this case

$$\boldsymbol{L}_O=\boldsymbol{M}_O(m\boldsymbol{v})=r\times m\boldsymbol{v} \tag{11.2}$$

Here $\boldsymbol{r}$ denotes a position vector drawn from point O to the particle. As shown in the figure, $\boldsymbol{L}_O$ is perpendicular to the shaded plane containing position vector $\boldsymbol{r}$ and $m\boldsymbol{v}$. Its magnitude is given by the product of the orthogonal projection $\boldsymbol{r}$ to the velocity and the linear momentum magnitude $m\boldsymbol{v}$. In the rectangular coordinates, the angular momentum can be determined by evaluating the determinant:

$$\boldsymbol{L}_O=\boldsymbol{r}\times m\boldsymbol{v}=\begin{vmatrix} \boldsymbol{i} & \boldsymbol{j} & \boldsymbol{k} \\ x & y & z \\ mv_x & mv_y & mv_z \end{vmatrix}$$

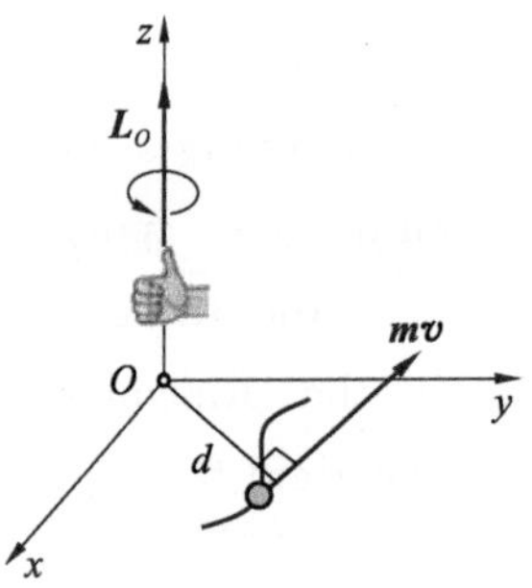

Fig. 11-1

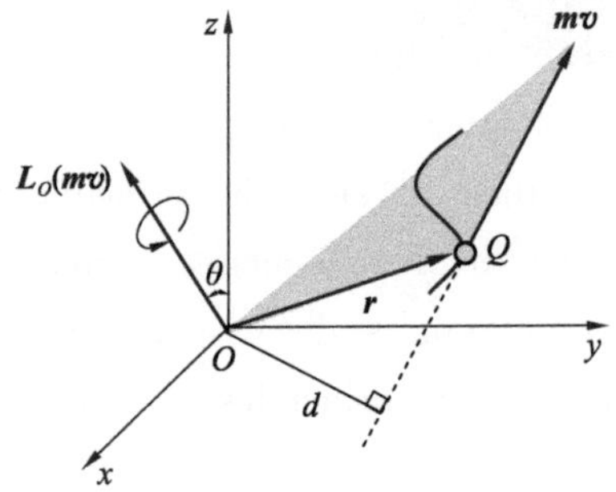

Fig. 11-2

11.1.2 Angular Momentum of a System of Particles

For a system of particles, in Fig. 11-3, each particle has angular momentum $\boldsymbol{L}_O$ about point O, so the angular momentum of the system with respect to the fixed point O is equal to the vector summation of the angular momentum of all the particles in the system:

$$\boldsymbol{L}_O=\sum \boldsymbol{M}_O(m_i\boldsymbol{v}_i)=\sum \boldsymbol{r}_i\times m_i\boldsymbol{v}_i \tag{11.3}$$

Using these results, we will consider two types of motion.

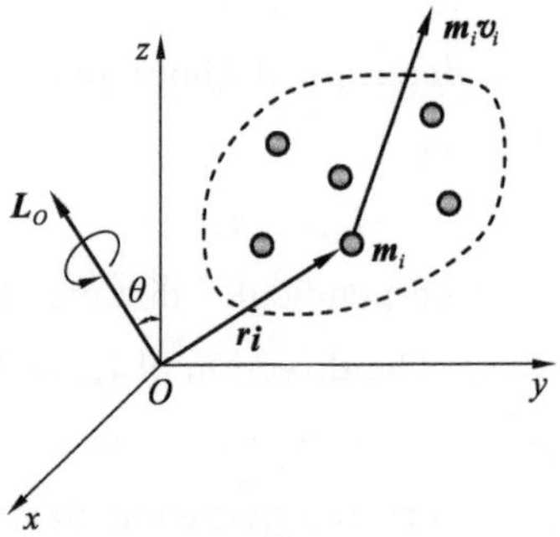

Fig. 11-3

(1) Translation

When a rigid body of mass m is subjected to translation, and its mass center has a velocity of $\boldsymbol{v}_C = v$, Fig. 11-4, then angular velocity $\omega = 0$, and the angular momentum about *mass center* C, is $\boldsymbol{L}_C = 0$. If the angular momentum is computed about *other point* A, the "moment" of the linear momentum $\boldsymbol{L}_A$ can be found about that point.

$$\boldsymbol{L}_A = m\boldsymbol{v}d \tag{11.4}$$

Here d is the "moment arm" as shown in Fig. 11-4.

(2) Rotating about a Fixed Axis

As an important special case, a rigid body rotating about a fixed axis z as shown in Fig. 11-5. Without loss of generality, we place the origin O on the axis of rotation and align the z-axis with it. Following Eq.(11.1), for each particle m_i in the system, the z-component of the angular momentum is given by

$$L_{zi} = m_i v_i r_i = m_i r_i^2 \omega \tag{11.5}$$

Here, r_i is the orthogonal distance of m_i from the axis of rotation. As all the particles move with the same angular velocity ω, summing Eq. (11.5) over all particles gives

$$L_z = \sum m_i r_i^2 \omega = \omega \sum m_i r_i^2 \tag{11.6}$$

Defining $J_z = \sum m_i r_i^2$, the variable J_z is called the *mass moment of inertia* of the system relative to the rotation axis z. Then angular momentum of the rigid body about axis z is

$$L_z = J_z \omega \tag{11.7}$$

This equation states that *angular momentum of the rigid body about axis is the product of the moment of inertia of the system relative to the same rotation axis and angular velocity*.

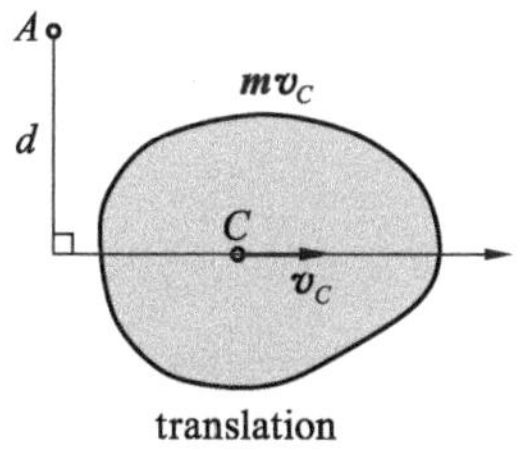

Fig. 11-4

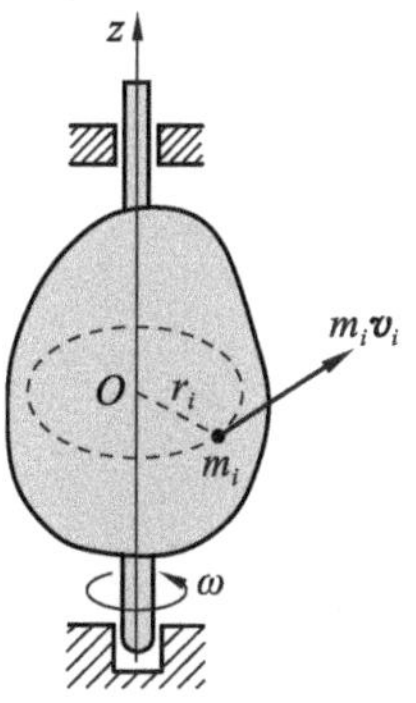

Fig. 11-5

Example 11-1

The drum of radius R has a mass moment of initial J_O about point O, rotates around point O with uniform angular velocity ω as shown in Fig. 11-6, two blocks connected by a rope have same mass m and travel on the smooth surfaces, determine the angular momentum of the system about point O.

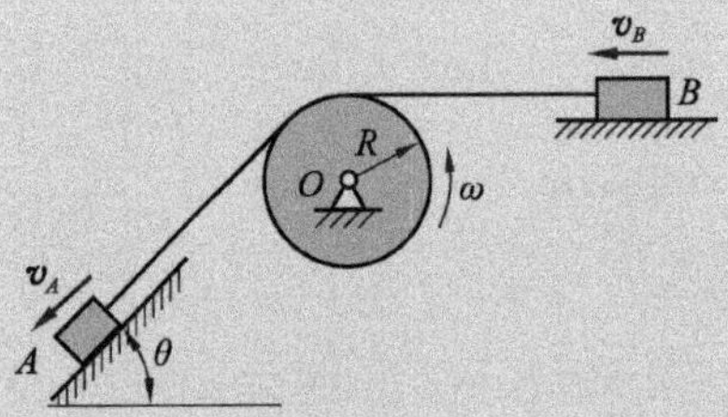

Fig. 11-6

Solution:

Since the velocity of blocks A and B is tangent to the drum, Fig. 11-6, the $\boldsymbol{v}_A$ and $\boldsymbol{v}_B$ can be determined

$$v_A = v_B = \omega \cdot R$$

The angular momentum of the system about point O is

$$L_O = J_O\omega + 2 \cdot mv = J_O\omega + 2m\omega \cdot R$$
$$= (J_O + 2mR)\omega \qquad Ans.$$

11. 2 Principle of Angular Momentum

11.2.1 Angular Momentum Theorem for a Particle

To determine a relation between the particle's angular momentum and moment of all the forces acting on the particle in Fig. 11-7. We perform a cross-product multiplication of each side of Eq. (10. 8) by the position vector $\boldsymbol{r}$. We have

$$\boldsymbol{r} \times \frac{\mathrm{d}}{\mathrm{d}t}(m\boldsymbol{v}) = \boldsymbol{r} \times \boldsymbol{F} \tag{11. 8}$$

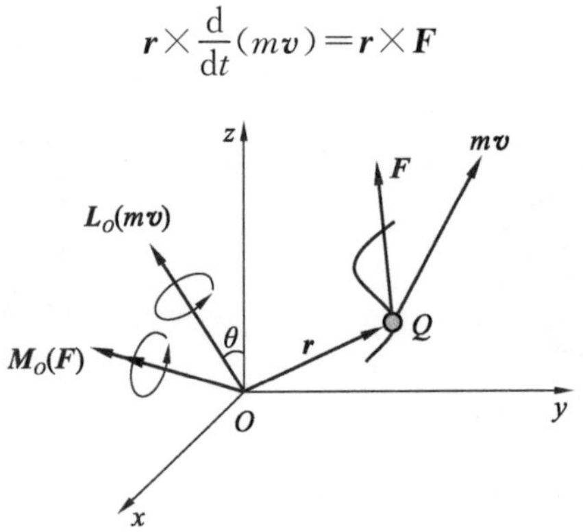

Fig. 11-7

The right side of above equation is the moment $\boldsymbol{M}_O = \boldsymbol{r}\times\boldsymbol{F}$ about point O as defined in Eq. (3.10).

From the chain rule of differentiation, the derivative of Eq. (11.2) can be written as

$$\frac{\mathrm{d}\boldsymbol{L}_O}{\mathrm{d}t} = \frac{\mathrm{d}}{\mathrm{d}t}(\boldsymbol{r}\times m\boldsymbol{v}) = \frac{\mathrm{d}\boldsymbol{r}}{\mathrm{d}t}\times m\boldsymbol{v} + \boldsymbol{r}\times\frac{\mathrm{d}}{\mathrm{d}t}(m\boldsymbol{v})$$

Since the cross product of a vector with itself is zero, namely, $\frac{\mathrm{d}\boldsymbol{r}}{\mathrm{d}t}\times m\boldsymbol{v} = \boldsymbol{v}\times m\boldsymbol{v} = 0$, the first term on the right side vanishes. Using the Eq. (11.8), we obtain

$$\frac{\mathrm{d}}{\mathrm{d}t}(\boldsymbol{r}\times m\boldsymbol{v}) = \boldsymbol{r}\times\frac{\mathrm{d}}{\mathrm{d}t}(m\boldsymbol{v}) = \boldsymbol{r}\times\boldsymbol{F}$$

That is

$$\frac{\mathrm{d}\boldsymbol{L}_O}{\mathrm{d}t} = \boldsymbol{M}_O(\boldsymbol{F}) \tag{11.9}$$

This is the *angular momentum theorem*, which states that *the time rate of change of the particle's angular momentum relative to a fixed arbitrary point O is equal to the moment of the force acting on the particle relative to the same point O.*

In general, the Eq. (11.9) can be expressed in x, y, z component form, three scalar equations can be written to express the motion, namely,

$$\begin{aligned} \frac{\mathrm{d}L_x}{\mathrm{d}t} &= M_x(\boldsymbol{F}) \\ \frac{\mathrm{d}L_y}{\mathrm{d}t} &= M_y(\boldsymbol{F}) \\ \frac{\mathrm{d}L_z}{\mathrm{d}t} &= M_z(\boldsymbol{F}) \end{aligned} \tag{11.10}$$

11.2.2 Angular Momentum Theorem for a System of Particles

For the system of particles shown in Fig. 11-8, the forces acting on the arbitrary ith particle of the system consist of a resultant *external force* $\boldsymbol{F}_i^{(e)}$, and a resultant *internal force* $\boldsymbol{F}_i^{(i)}$. According to the angular momentum theorem Eq. (11.9), for each particle m_i, we have

$$\frac{\mathrm{d}\boldsymbol{L}_{Oi}}{\mathrm{d}t} = \boldsymbol{M}_{Oi}(\boldsymbol{F}_i^{(i)}) + \boldsymbol{M}_{Oi}(\boldsymbol{F}_i^{(e)}) \quad (i = 1, 2, 3, \cdots, n)$$

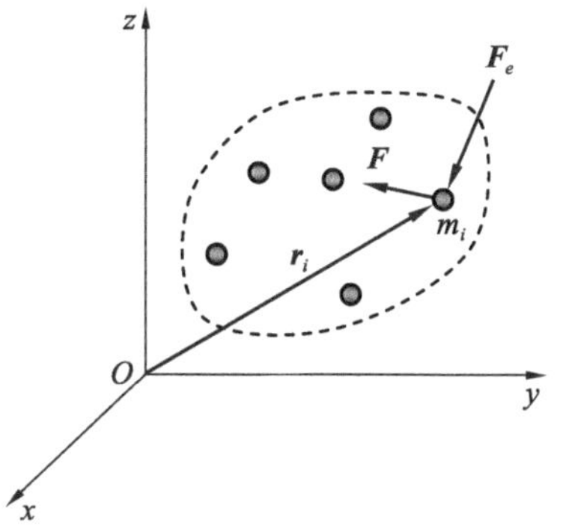

Fig. 11-8

Similar equations can be written for each of the other particles of the system. If we sum the equations of motion over all n particles, then we have

$$\sum \frac{\mathrm{d}\boldsymbol{L}_{Oi}}{\mathrm{d}t} = \sum \boldsymbol{M}_{Oi}(\boldsymbol{F}_i^{(i)}) + \sum \boldsymbol{M}_{Oi}(\boldsymbol{F}_i^{(e)})$$

or

$$\frac{\mathrm{d}}{\mathrm{d}t}\left(\sum \boldsymbol{L}_{Oi}\right) = \sum \boldsymbol{M}_{Oi}(\boldsymbol{F}_i^{(i)}) + \sum \boldsymbol{M}_{Oi}(\boldsymbol{F}_i^{(e)})$$

Since the internal forces occur in equal but opposite collinear pairs, hence the moment of each pair of *internal forces* about point O is zero, the first term of right side, $\sum \boldsymbol{M}_{Oi}(\boldsymbol{F}_i^{(i)}) = 0$, only the total moment of the *external forces* remains, the above equation can be written in a simplified form as

$$\frac{\mathrm{d}\boldsymbol{L}_O}{\mathrm{d}t} = \sum \boldsymbol{M}_{Oi}(\boldsymbol{F}_i^{(e)}) = \sum \boldsymbol{M}_O \tag{11.11}$$

This is the *angular momentum theorem* for a system of particles, which states that *the time rate of change of the total angular momentum of a particle's system relative to a fixed point O is equal to the resultant moment of all the external forces about the same point O.*

In general, the above equation can be expressed in x, y, z component form, three scalar equations can be written to express the motion, namely,

$$\begin{aligned} \frac{\mathrm{d}L_x}{\mathrm{d}t} &= \sum M_x(\boldsymbol{F}) \\ \frac{\mathrm{d}L_y}{\mathrm{d}t} &= \sum M_y(\boldsymbol{F}) \\ \frac{\mathrm{d}L_z}{\mathrm{d}t} &= \sum M_z(\boldsymbol{F}) \end{aligned} \tag{11.12}$$

When applying Eq. (11.9)~Eq. (11.12), we should pay attention to the assumed positive sense of rotation for angular momentum $\boldsymbol{L}$ and moment $\boldsymbol{M}$. For example, if $\boldsymbol{L}$ is taken as *positive* for clockwise rotation, then $\boldsymbol{M}$ should also be measured as *positive* for clockwise moments and vice-versa.

The angular momentum theorem establishes the dependence between the change of the *angular momentum* of a particle or a system with respect to a center (or a fixed axis) and the *moment* given by all external forces acting on the particle or the system with respect to the same center (or axis).

11.2.3 Conservation of Angular Momentum

When the resultant external moment acting on the system of particles is zero ($\boldsymbol{M}_O = 0$), then $\frac{\mathrm{d}\boldsymbol{L}_O}{\mathrm{d}t} = 0$ and the angular momentum about a fixed point O is constant

$$\boldsymbol{L}_O = \boldsymbol{r} \times m\boldsymbol{v} = \text{const.} \tag{11.13}$$

This equation is known as the *conservation of angular momentum*. In some cases, the angular momentum about a fixed *point* may not be conserved and the angular momentum about a fixed *axis* will be conserved, for example, if $\sum M_x^{(e)} = 0$,

$$L_x = \text{const.} \tag{11.14}$$

The angular momentum about axis x is conserved.

An example of *conservation of angular momentum* occurs when the particle of mass m is subjected only to a *central force* (such as the motion of the satellite around earth), as shown in Fig. 11-9, the central force $\boldsymbol{F}$ is always directed toward point O as the particle moves along the path. Hence, the moment created by $\boldsymbol{F}$ about the z (point O in plane) axis is always zero, and therefore angular momentum of the particle is conserved about this axis, namely Eq. (11.13)

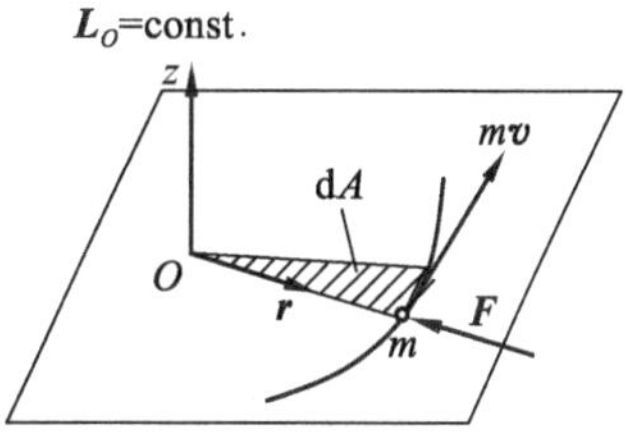

Fig. 11-9

$$\boldsymbol{L}_O = \boldsymbol{r} \times m\boldsymbol{v} = \text{const.}$$

In a time increment dt the position vector $\boldsymbol{r}$ sweeps out an area with magnitude

$$\mathrm{d}\boldsymbol{A} = \frac{1}{2}|\boldsymbol{r} \times \mathrm{d}\boldsymbol{r}|$$

Introduce the corresponding vectorial quantity

$$\mathrm{d}\boldsymbol{A} = \frac{1}{2}|\boldsymbol{r} \times \mathrm{d}\boldsymbol{r}| = \frac{1}{2}|\boldsymbol{r} \times \boldsymbol{v}\mathrm{d}t|$$

Then the time rate of change of the swept out area is

$$\frac{\mathrm{d}\boldsymbol{A}}{\mathrm{d}t} = \frac{1}{2}(\boldsymbol{r} \times \boldsymbol{v})$$

Substituting Eq. (11.13) into above equation results in

$$\frac{\mathrm{d}\boldsymbol{A}}{\mathrm{d}t} = \frac{\boldsymbol{L}_O}{2m} = \text{const.} \tag{11.15}$$

Thus, the time rate of change of the swept out area will be constant. For planetary motion, this result is known as *Kepler's 2nd Law*: the ray from the sun to a planet sweeps out *equal areas* in equal times.

Example 11-2

A pendulum consisting of a massless rope with length l and a ball of mass m is suspended from a frictionless pivot O, as shown in Fig. 11-10a. If the system is displaced from the equilibrium position and released, then it will oscillate under the action of gravity in the indicated plane. Find the equation of motion for the pendulum.

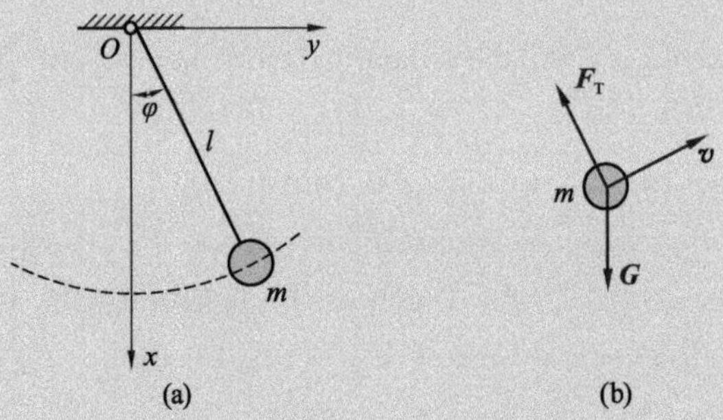

Fig. 11-10

Solution:

The system executes a pure rotation about an axis perpendicular to the page and passing through point O. There are two forces acting on the ball (m), a tension force F_T (pointing towards O) and the gravitational force $G = mg$, Fig. 11-10b. Introducing a positive rotation angle φ to express the angular momentum and moment relative to the fixed point O as

$$L_O = mvl = ml^2 \frac{d\varphi}{dt}$$

$$M_O = -mgl\sin\varphi$$

According to the angular momentum theorem Eq. (11.9), an expression for the equation of motion is

$$\frac{d}{dt}\left(ml^2 \frac{d\varphi}{dt}\right) = -mgl\sin\varphi$$

Re-written above equation, we have

$$\frac{l}{g}\frac{d^2\varphi}{dt^2} = -\sin\varphi$$

$$\frac{d^2\varphi}{dt^2} + \frac{g}{l}\sin\varphi = 0$$

For small angles ($\sin\varphi \approx \varphi$), this equation can be written as

$$\frac{d^2\varphi}{dt^2} + \frac{g}{l}\varphi = 0 \qquad \textit{Ans.}$$

This is the simple harmonic oscillation of an *ideal pendulum*.

Example 11-3

The trolley m is hoisted up the incline by a drum arrangement shown in Fig. 11-11. The drum with radius of R and moment of inertial J about point O, can rotate around an axis perpendicular to the page and passing through point O. If there is a couple M acting on the drum, determine the acceleration a of trolley. Neglect the friction and weight of rope.

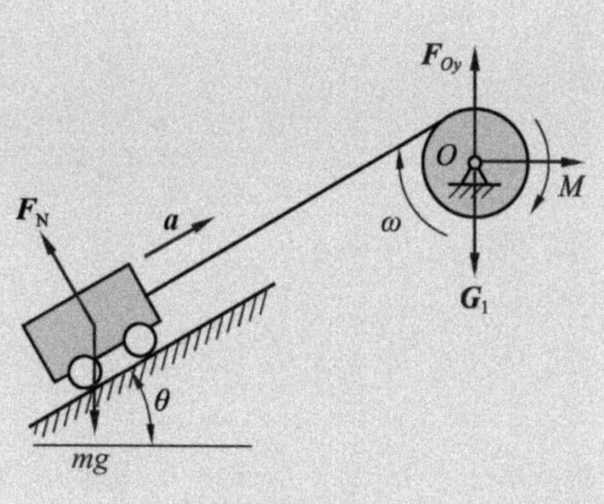

Fig. 11-11

Solution:

Here we will consider the trolley and drum as a single system. The free-body diagramof the system is shown in Fig. 11-11, the moment of all external forces about point O is

$$M_O = M - mgR\sin\theta$$

The angular momentum of the system about same point O is

$$L_O = J_O\omega + mvR$$

Applying the principle of angular momentum about point O. Thus,

$$\frac{\mathrm{d}}{\mathrm{d}t}(J_O\omega + mvR) = M - mgR\sin\theta \qquad \text{(a)}$$

We have the kinematic relation

$$\omega = \frac{v}{R}, \quad \frac{\mathrm{d}v}{\mathrm{d}t} = a$$

Substituting back the Eq. (a) and solving for a, yields

$$a = \frac{MR - mgR^2\sin\theta}{J_O + mR^2} \qquad \textit{Ans.}$$

Example 11-4

Two identical spheres A and B of mass m are attached to the cords of negligible mass. The cords are connected to the rotating shaft, as shown in Fig. 11-12a. Initially two spheres are connected by a cord AB and the shaft is rotating with an angular velocity ω_0 around the axis z. Determine the angular velocity ω of shaft if include angle between the cord and shaft is changed to θ after the connecting cord breaks. Neglect the frictional force.

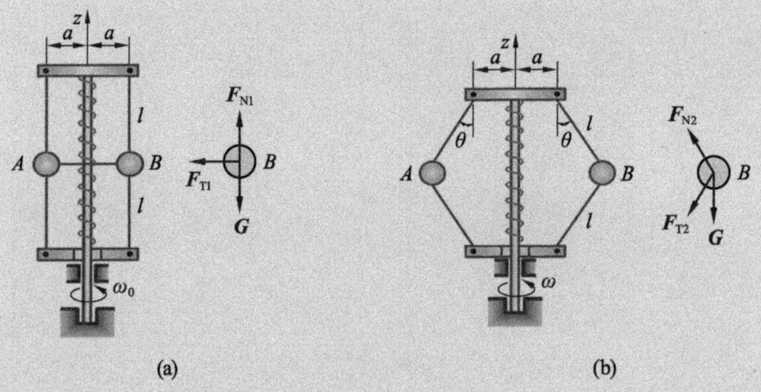

Fig. 11-12

Solution:

Before the connecting cord AB breaks, the free-body diagram of the spheres is shown in Fig. 11-12a, the tension $\boldsymbol{F}_{N1}$ and weight $\boldsymbol{G}$ are always parallel to the z axis, and the tension $\boldsymbol{F}_{T1}$ pass through it. After the cord AB breaks, the free-body diagram of the spheres is shown in Fig. 11-12b, the tension $\boldsymbol{F}_{N2}$ and $\boldsymbol{F}_{T2}$ always pass through the z axis, and the weight $\boldsymbol{G}$ is parallel to it. Hence the moments created by these forces, are all zero about z axis. Therefore, angular momentum is conserved about the z axis. Thus

$$L_{z_1}=L_{z_2}$$

$$2mva=2ma^2\omega_0=2m(a+l\sin\theta)^2\omega=\text{const.}$$

The angular speed of the system is thus

$$\omega=\frac{a^2}{(a+l\sin\theta)^2}\omega_0 \qquad \textit{Ans.}$$

11. 3 Mass Moment of Inertia

11.3.1 Formula of Mass Moment of Inertia

The mass moment of inertia of a body is a measure of the body's resistance to angular acceleration. The mass moment of inertia that is discussed at Section 11. 1. 3 is defined as *the summation of the "second moment" about an axis of all the particles of mass m_i which compose the system*:

$$J_z=\sum m_i r_i^2 \tag{11. 16a}$$

Here the "*moment arm*" r_i is the perpendicular distance from the z axis to the arbitrary particle m_i. From the equation, since the formulation involves r, the value of J is different for each axis about which it is calculated. Since r is squared in Eq.(11. 16a), the mass moment of inertia is always a positive quantity. The units used for its measurement is $kg \cdot m^2$. For example, the flywheel on the engine of the tractor has a large mass *moment of inertia* about its axis of rotation. Once it is set into motion, it will be difficult to stop, and this in turn will prevent the engine from stalling and instead will allow it to maintain a constant power.

For the rigid body, we can consider it as a continuous and non-deformable system of particles by the elementary masses dm. In this way the mass moment of inertia of a rigid body about the z axis is

$$J_z = \rho \int_V r^2 \mathrm{d}V \tag{11.16b}$$

where V is the integral volume.

(1) Slender Homogeneous Rod

A slender homogeneous rod of length l and mass m in Fig. 11-13a. For calculating the mass moment of inertia about the *central* axis z_C that is perpendicular to the rod and passes through the center C of the rod, we choose any point by coordinate x, one infinitesimal element of mass dm and length dx, the linear density is

$$\rho = \frac{\mathrm{d}m}{\mathrm{d}x}$$

With this, the mass moment of inertial about the axis z_C is

$$J_{z_C} = \int_V r^2 \mathrm{d}m = \int_{-\frac{l}{2}}^{\frac{l}{2}} \rho x^2 \mathrm{d}x = \frac{1}{12}\rho l^3 = \frac{1}{12} m l^2 \tag{11.17}$$

If we choose a reference axis z that passes through end point A of rod, Eq. (11.13) leads to

$$J_{z_A} = \int_0^l \frac{m}{l} \cdot x^2 \mathrm{d}x = \frac{1}{3} m l^2 \tag{11.18}$$

(2) Homogeneous Hoop

A homogeneous hoop with radius R and mass m in Fig. 11-13b, the mass moment of inertial about the axis z_C that is perpendicular to the plane of the hoop and passes through its center O is

$$J_z = \int_V r^2 \mathrm{d}m = \int_l R^2 \mathrm{d}m = mR^2 \tag{11.19}$$

(3) Homogeneous Circular Disk

A homogeneous circular disk with mass m, density ρ, and radius R. We choose the reference axis z that is perpendicular to the plane of the disk and passes through its center C, Fig. 11-13c. With the mass element

$$\mathrm{d}m = 2\pi r \mathrm{d}r \cdot \rho = 2\pi \rho r \mathrm{d}r$$

We obtain the mass moment of inertia about the axis z

$$J_C = \int_V r^2 \mathrm{d}m = \int_0^R 2\pi \rho r^3 \mathrm{d}r = \frac{1}{2}\pi \rho R^4 = \frac{1}{2} m R^2 \tag{11.20}$$

The mass moment of inertia J_C depends on the mass m and the radius R, it is independent of the thickness. Therefore, the result Eq. (11.20) is also valid for a *circular cylinder* of arbitrary length.

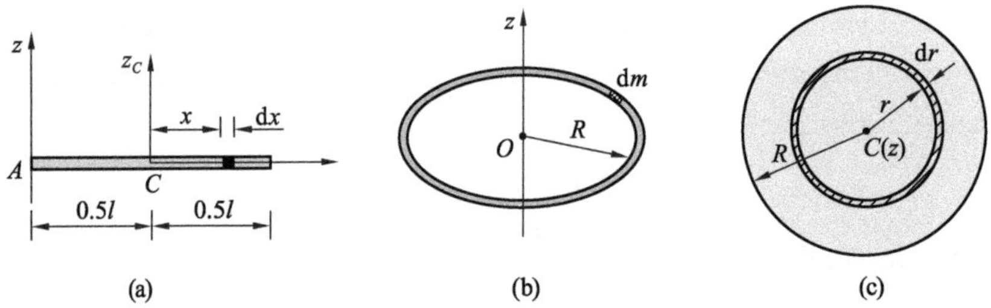

Fig. 11-13

11.3.2 Radius of Gyration

In some cases, it is helpful to use the *radius of gyration* ρ_z. This is a geometrical property which has units of length. When the body's mass m is known, the body's moment of inertia is determined from the equation

$$J_z = m\rho_z^2 \quad \text{or} \quad \rho_z = \sqrt{\frac{J_z}{m}} \tag{11.21}$$

We may interpret ρ_z as the distance from the axis z at which the total mass m can be imagined as being concentrated so that it has the same moment of inertia as the body. The mass moment of inertia and radius of gyration of a body about a specified axis is reported in handbooks.

11.3.3 Parallel-Axis Theorem

If the mass moment of inertia of the rigid body about an axis passing through the body's mass center is known, then the mass moment of inertia about any other parallel axis can be determined by using the *parallel-axis theorem*.

To make the derivation easy to understand, we consider finding the mass moment of inertial about y axis, as shown in Fig. 11-14. Two parallel axes, the axis y_C passes through the mass center C of the rigid body, whereas the corresponding parallel y axis lies at a constant distance d away. We choose a differential element m_i located at an arbitrary distance x_{Ci} from the centroidal y_C axis. With $x_i = x_{Ci} + d$, we obtain the mass moment of inertia of the rigid body about the centroidal axis y_C

$$J_{y_C} = \sum m_i y_{Ci}^2$$

The mass moment of inertia of the body about the y axis is

$$J_y = \sum m_i y_i^2 = \sum m_i (y_{Ci} + d)^2$$
$$= \sum m_i y_{Ci}^2 + 2d \sum m_i y_{Ci} + d^2 \sum m_i$$

The first term represents mass moment of inertia J_{z_C} of the body about the centroidal axis y_C. The second summation equals zero, since the y_C axis passes through the body's mass center, i. e. , $\sum m_i y_{Ci} = m y_C = 0$, since $y_C = 0$. Finally, the third summation represents the total mass m of the body. Hence, the mass moment of inertia about the y axis can be written as

$$J_y = J_{y_C} + md^2 \tag{11.22}$$

where d—perpendicular distance between the parallel z and z_C axes.

A similar expression can be written for J_x, J_z, i,e. ,

$$J_x = J_{x_C} + mc^2$$
$$J_z = J_{z_C} + me^2$$

where c—perpendicular distance between the parallel x and x_C axes;

e—perpendicular distance between the parallel z and z_C axes.

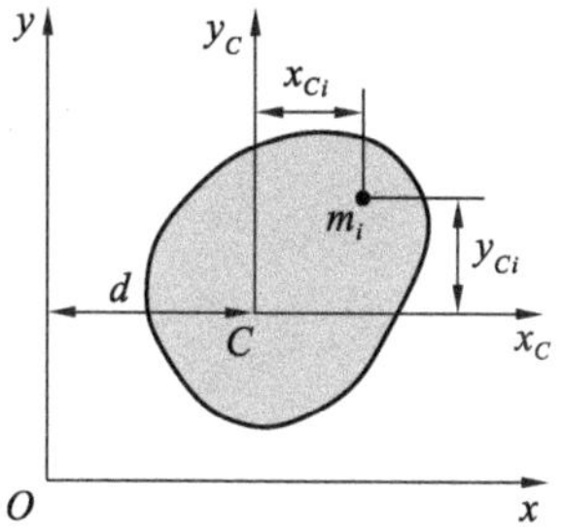

Fig. 11-14

Example 11-5

Determine the mass moment of inertia of the cylinder shown in Fig. 11-15a about the z axis. The density of the material ρ is constant.

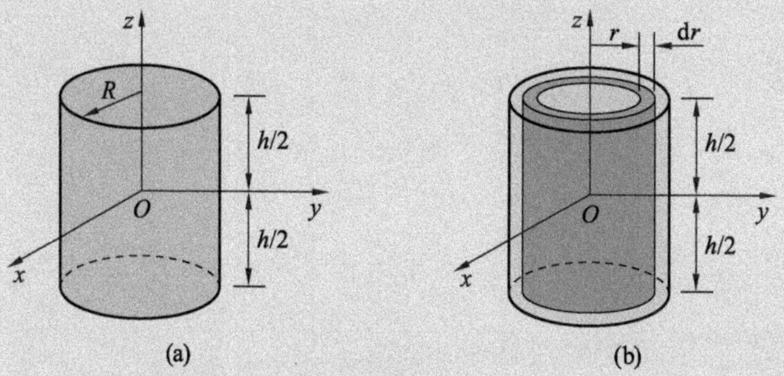

Fig. 11-15

Solution:

The volume of the element is $dV=2\pi r\cdot h\cdot dr$, so that its mass is $dm=\rho\cdot dV=\rho\cdot 2\pi rhdr$. Since the entire element lies at the same distance r from the z axis, the mass moment of inertia of the element is

$$dJ_z=r^2 dm=2\pi\rho hr^3 dr$$

Integrating over the entire region of the cylinder yields

$$J_z=\int_m r^2 dm=2\pi\rho h\int_0^R r^3 dr=\frac{1}{2}\pi\rho R^4 h$$

The mass of the cylinder is

$$m=\int_m dm=2\pi\rho h\int_0^R r dr=\pi\rho hR^2$$

so that

$$J_C=\frac{1}{2}mR^2 \qquad \text{Ans.}$$

The result states that the mass moment of inertia of a circular disk is independent of the thickness.

Example 11-6

The pendulum in Fig. 11-16 is suspended from the pin at O and consists of the slender rod m_1 and the disk m_2. Determine the mass moment of inertia of the pendulum about an axis perpendicular to the page and passing through point O.

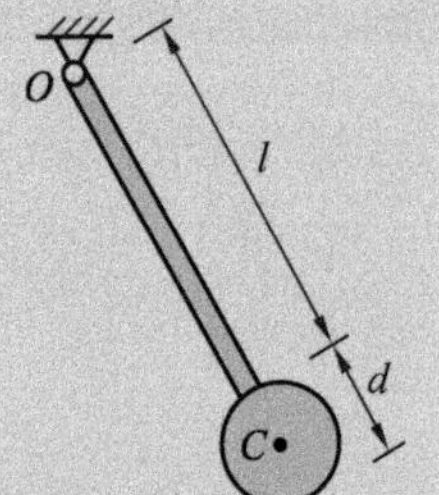

Fig. 11-16

Solution:

Since the pendulum consists of the slender rod and the disk, the mass moment of inertia of rod about an axis perpendicular to the page and passing through point O is

$$(J_{rod})_O=\frac{1}{3}m_1 l^2$$

The mass moment of inertia of disk about an axis O can be obtained using the parallel-axis theorem.

$$(J_{disk})_O=\frac{1}{2}m_2\left(\frac{d}{2}\right)^2+m_2\left(l+\frac{d}{2}\right)^2=m_2\left(\frac{3}{8}d^2+l^2+ld\right)$$

The mass moment of inertia of the pendulum about O is therefore

$$J_O=(J_{rod})_O+(J_{disk})_O=\frac{1}{3}m_1 l^2+m_2\left(\frac{3}{8}d^2+l^2+ld\right) \qquad \text{Ans.}$$

11.4 Rotation about a Fixed Axis

Consider the rigid body shown in Fig. 11-17, which is constrained to rotate about a fixed axis z with angular velocity ω and angular acceleration α. The angular velocity and angular acceleration are caused by the external force and couple moment system acting on the body. The mass moment of inertial of the rigid body about the z-axis is J_z, and M_z is the resultant moment of the external forces with respect to the axis z, the principle of angular momentum of rigid body about axis z is given by

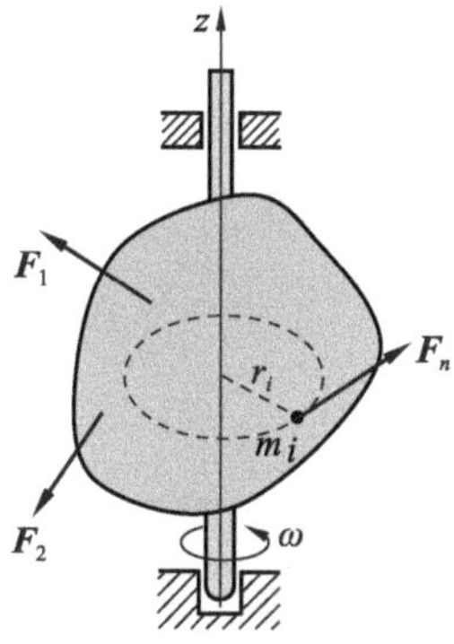

Fig. 11-17

$$\frac{\mathrm{d}}{\mathrm{d}t}(J_z\omega) = \sum M_z(\boldsymbol{F}_i)$$
$$J_z\frac{\mathrm{d}\omega}{\mathrm{d}t} = \sum M_z(\boldsymbol{F}_i) \tag{11.23a}$$

Since $\frac{\mathrm{d}\omega}{\mathrm{d}t}=\alpha=\frac{\mathrm{d}^2\varphi}{\mathrm{d}t^2}$, Eq. (11.23a) can be written in the form

$$J_z\frac{\mathrm{d}^2\varphi}{\mathrm{d}t^2} = \sum M_z(\boldsymbol{F}_i) \tag{11.23b}$$

or

$$J_z\alpha = \sum M_z(\boldsymbol{F}_i) \tag{11.23c}$$

Thus, the product of the moment of initial and angular acceleration is equal to the moment acting by all external forces. By comparison, the moment of inertia is a measure of the resistance of a body to angular acceleration ($M=J\alpha$), in the same way that mass is a measure of the body's resistance to acceleration ($F=ma$). In the special case of $M_z=0$, the angular momentum $L_z=J\omega$ does not change (conservation of angular momentum).

Example 11-7

The pulley shown in Fig. 11-18 has a moment of inertial J_O and a radius R. The belts of both side of pulley are subjected to the tension F_1 and F_2 respectively such that it does not slip at its contacting surfaces. Determine the angular acceleration α of the pulley.

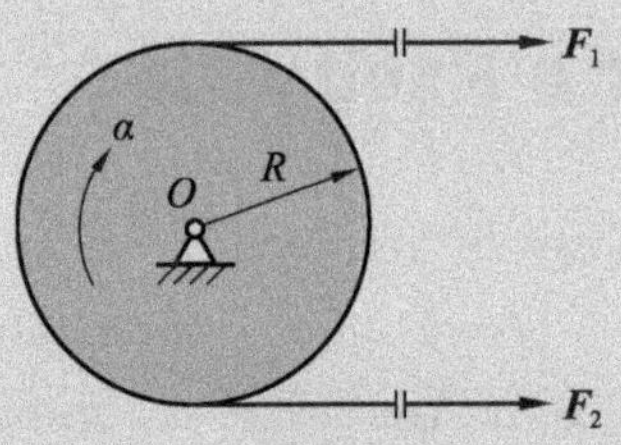

Fig. 11-18

Solution:

The pulley rotates around point O, applying the Eq. (11. 23), we get

$$J_O\alpha=(F_1-F_2)R$$

Solving

$$\alpha=\frac{(F_1-F_2)R}{J_O} \qquad \textit{Ans.}$$

Thus, the condition for the equal tension on both side of pulley is that the pulley rotates with uniform angular velocity.

Example 11-8

The flywheel has a mass moment of inertial J_O and radius R in Fig.11-19a. At beginning, the flywheel rotates around axis O with uniform angular velocity ω_0. If the brake supplies a normal pressure $\boldsymbol{F}_N$ to the flywheel, determine the time required for the wheel to stop. The coefficient of kinetic friction between the flywheel and brake is f.

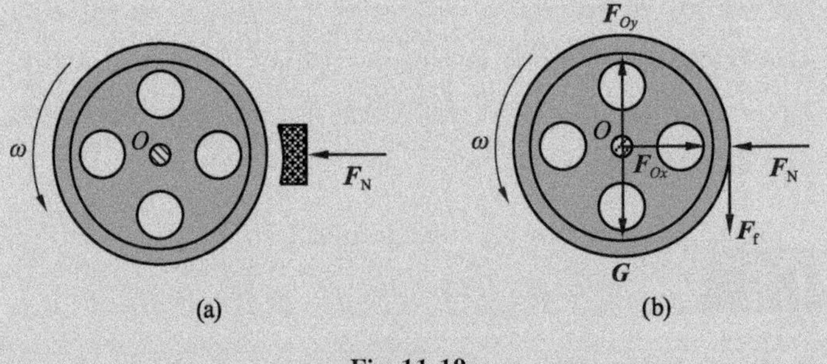

Fig. 11-19

Solution:

The free-body diagram of the flywheel is as shown in Fig. 11-19b, as the flywheel rotates around axis O, the normal pressure $\boldsymbol{F}_N$, the weight of the flywheel and constraints $\boldsymbol{F}_{Ox}$, $\boldsymbol{F}_{Oy}$ pass through the z axis. Hence the moments created by these forces, are all zero about this axis. Applying the Eq. (11. 23), we have

$$J_O\frac{d\omega}{dt}=-F_fR=-fF_NR$$

Rewriting above equation:

$$J_O \mathrm{d}\omega = -fF_N R \mathrm{d}t$$

Integrating both sides,

$$J_O \int_{-\omega_0}^{0} \mathrm{d}\omega = -fF_N R \int_0^t \mathrm{d}t$$

We get

$$J_O \omega_0 = fF_N R t$$

$$t = \frac{J_O \omega_0}{fF_N R} \qquad \textit{Ans.}$$

Example 11-9

The friction clutch coupling as shown in Fig. 11-20, before jointing, rotor Ⅰ rotates with uniform angular velocity ω_1, while the rotor Ⅱ is at rest. When the clutch engages, two rotors rotate together, determine the angular velocity ω_2 of two rotors after coupling. The moments of inertia of two rotors about axis x are J_1 and J_2, respectively, the friction around the rotor is neglected.

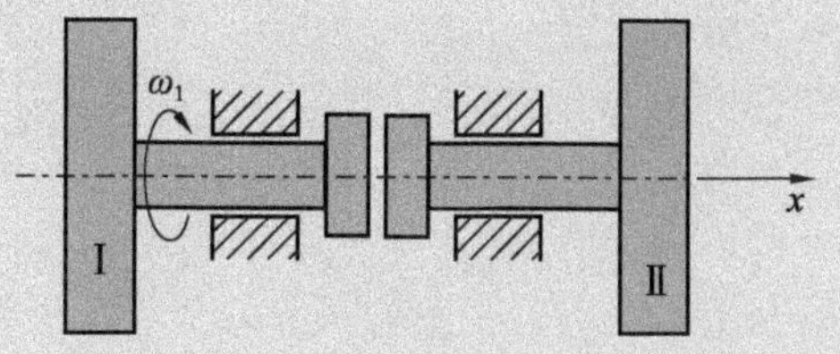

Fig. 11-20

Solution:

Consider two rotors as a single system, Fig. 11-20, since the frictional force between two clutches is internal force. Also, the moments about axis x created by the weights of the rotors and constraints of the bearing are all zero. So the angular momentum is conserved about axis x, thus

$$(J_1 + J_2)\omega_2 = J_1 \omega_1$$

Solving

$$\omega_2 = \frac{J_1}{J_1 + J_2}\omega_1 \qquad \textit{Ans.}$$

Example 11-10

The slender rod AB, as shown in Fig. 11-21a, has a mass m and is at horizontal by a pin connected at end A and suspending at its end B by a cord. If the cord is cut off suddenly, determine the reaction forces of the support A at this instant.

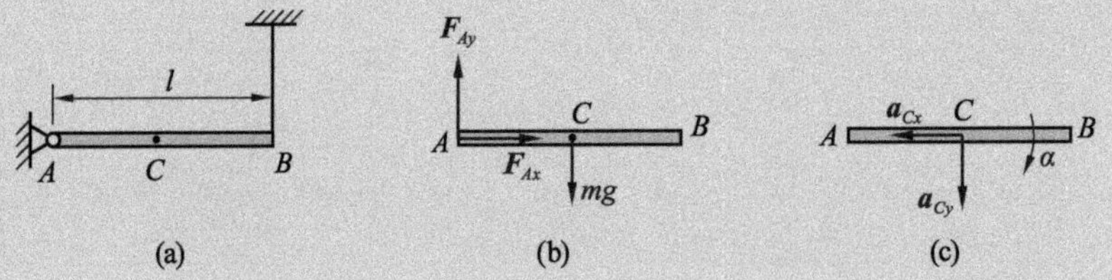

Fig. 11-21

Solution:

The free-body diagram of the rod is shown in Fig. 11-21b. At the instant the cord is cut off, the angular velocity is zero. Thus

$$a_{Cx}=a_{Cn}=0, a_{Cy}=a_{C\tau}=\frac{l}{2}\alpha \tag{a}$$

The rod rotates around the fixed axis, applying the Eq. (11.23), we have

$$J_A\alpha=mg\frac{l}{2} \tag{b}$$

The moment of inertial of rod about point A is $J_A=\frac{1}{3}ml^2$, substituting back above Eq. (b) and solving this equation yield

$$\alpha=\frac{3g}{2l}$$

Substituting the angular acceleration α into the Eq. (a), we have

$$a_{Cy}=\frac{l}{2}\alpha=\frac{3}{4}g \tag{c}$$

Applying the law of motion for the mass center in the x and y direction, respectively.

$$-ma_{Cx}=F_{Ax}$$

$$ma_{Cy}=mg-F_{Ay} \tag{d}$$

Substituting Eq. (a), Eq. (c) into Eq. (d), yield

$$F_{Ax}=0, F_{Ay}=\frac{1}{4}mg \qquad \textit{Ans.}$$

11.5 Angular Momentum Theory about Mass Center

11.5.1 Angular Momentum of a Body in Plane Motion

Consider the body in Fig. 11-22, which is subjected to general plane motion, has an angular velocity ω. The fixed x, y axes are established at an arbitrary point O. The origin of the moving x', y' reference frame will be attached to the mass center C of the body and the axes translate with respect to the fixed frame. At the instant shown, the arbitrary point M of mass m_i has absolute velocity $\boldsymbol{v}_i$, relative velocity $\boldsymbol{v}_{\mathrm{ir}}$ and transport velocity $\boldsymbol{v}_{\mathrm{ie}}$ which is equal to the velocity of mass center $\boldsymbol{v}_C$. Then the velocity of the ith particle of the body is

$$\boldsymbol{v}_i = \boldsymbol{v}_{\mathrm{ie}} + \boldsymbol{v}_{\mathrm{ir}} = \boldsymbol{v}_C + \boldsymbol{v}_{\mathrm{ir}} \tag{a}$$

The position vector $\boldsymbol{r}_i$ in Fig. 11-22 specifies the location of the arbitrary point M, and the relative-position vector $\boldsymbol{r}_i'$ locates point M with respect to origin C of the x', y' reference frame. By vector addition, we have

$$\boldsymbol{r}_i = \boldsymbol{r}_C + \boldsymbol{r}_i' \tag{b}$$

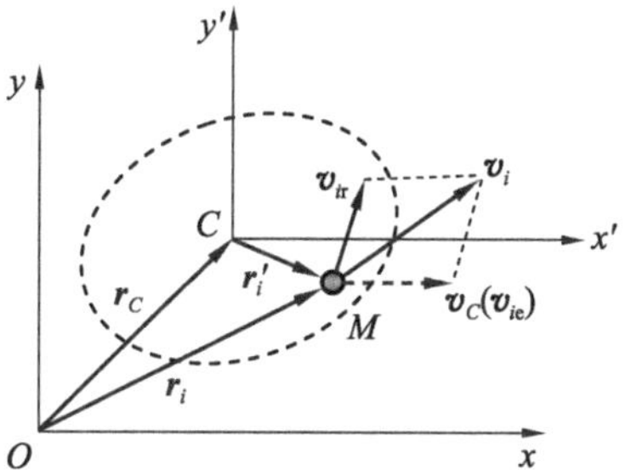

Fig. 11-22

The angular momentum of this particle about point O is equal to the "moment" of the particle's linear momentum about O. Summing the entire mass m of the body, we obtain the angular momentum of the body about point O

$$\begin{aligned} \boldsymbol{L}_O &= \sum \boldsymbol{M}_O(m_i\boldsymbol{v}_i) = \sum \boldsymbol{r}_i \times m_i\boldsymbol{v}_i = \sum (\boldsymbol{r}_C + \boldsymbol{r}_i') \times m_i\boldsymbol{v}_i \\ &= \boldsymbol{r}_C \times \sum m_i\boldsymbol{v}_i + \sum \boldsymbol{r}_i' \times m_i\boldsymbol{v}_i \end{aligned} \tag{c}$$

The first term on the right is the moment of the linear momentum of body about point O and can be equal to

$$\boldsymbol{r}_C \times \sum m_i\boldsymbol{v}_i = \boldsymbol{r}_C \times m\boldsymbol{v}_C \tag{d}$$

Using Eq. (a), the second term on the right of Eq. (c) can be rewritten as

$$\begin{aligned} \sum \boldsymbol{r}_i' \times m_i\boldsymbol{v}_i &= \sum \boldsymbol{r}_i' \times m_i(\boldsymbol{v}_C + \boldsymbol{v}_{\mathrm{ir}}) \\ &= \sum \mathrm{m_i}\boldsymbol{r}_i' \times \boldsymbol{v}_C + \sum \boldsymbol{r}_i' \times m_i\boldsymbol{v}_{\mathrm{ir}} \end{aligned} \tag{e}$$

Since the origin of the moving x', y' reference frame coincides with the mass center C of the body, in which case $\boldsymbol{r}_C' = 0$, hence, the first term on the right of Eq. (e) is

$$\sum m_i \boldsymbol{r}_i' = m\boldsymbol{r}_C' = 0$$

Then, the Eq. (e) can be written as

$$\sum \boldsymbol{r}_i' \times m_i \boldsymbol{v}_i = \sum \boldsymbol{r}_i' \times m_i \boldsymbol{v}_{ir} = \boldsymbol{L}_C \tag{f}$$

Here $\boldsymbol{L}_C$ represents the angular momentum of the body about the mass center C. Substituting Eq. (d) and Eq. (f) into Eq. (c), we have

$$\boldsymbol{L}_O = \boldsymbol{r}_C \times m\boldsymbol{v}_C + \boldsymbol{L}_C \tag{11.24}$$

This result indicates that *the angular momentum of the body about fixed point O is equivalent to the moment of the linear momentum $m\boldsymbol{v}_C$ plus the angular momentum about the mass center of body*. Using these results, we will now consider three types of motion.

(1) Translation

When a rigid body is subjected to either rectilinear or curvilinear translation, Fig. 11-23a, then angular velocity $\omega = 0$ and its mass center C has a velocity of $v_C = v$. Hence, the linear momentum and the angular momentum about C, become

$$\boldsymbol{P} = m\boldsymbol{v}_C$$

$$L_C = 0$$

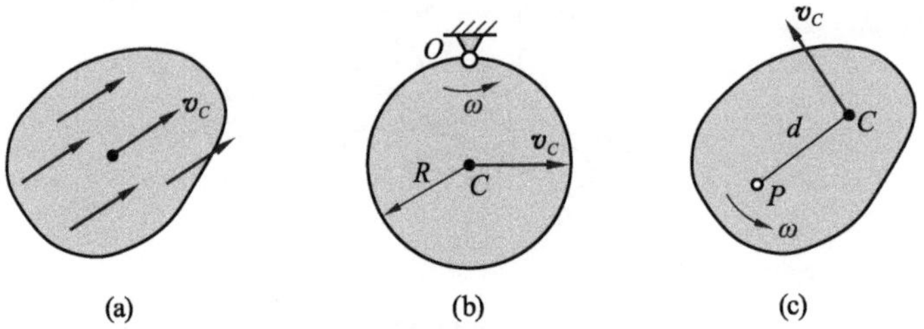

Fig. 11-23

(2) Rotation about a Fixed Axis

When a rigid body is rotating about a fixed axis O, Fig. 11-23b, the linear momentum and the angular momentum about C, are

$$\boldsymbol{P} = m\boldsymbol{v}_C$$

$$L_C = J_C \omega$$

where J_C is the body's moment of inertia about mass center C. Here the angular momentum of the body about C is equal to the product of the moment of inertia of the body about an axis passing through C and the body's angular velocity.

If for the calculation of angular momentum about point O, using the parallel-axis theorem, we have

$$L_O=(J_C+mR^2)\omega=J_O\omega$$

where J_O is the body's moment of inertia about point O.

(3) General Plane Motion

When a rigid body is subjected to general plane motion, Fig. 11-23c, the linear momentum, and the angular momentum about C, become

$$\boldsymbol{P}=m\boldsymbol{v}_C$$

$$L_C=J_C\omega$$

As a special case, if point P is the instantaneous center of zero velocity then, we can write the angular momentum as

$$L_P=(J_C+md^2)\omega=J_P\omega$$

where J_P is the moment of inertia of the body about the P.

Example 11-11

The 5 kg slender bar has a length of $l=4$ m, as shown in Fig. 11-24a. At a given instant the endpoint A of bar has a velocity $\boldsymbol{v}_A=2$ m/s, determine its angular momentum about mass center C and about the instantaneous center at this instant.

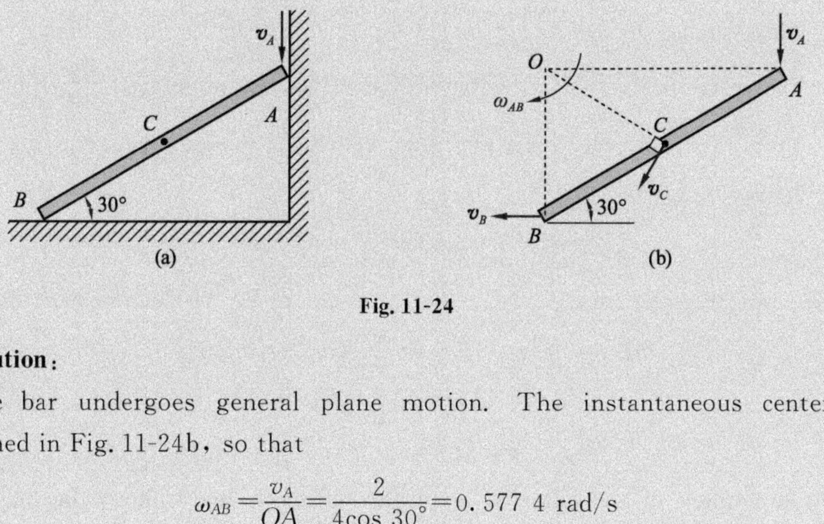

Fig. 11-24

Solution:

The bar undergoes general plane motion. The instantaneous center O is established in Fig. 11-24b, so that

$$\omega_{AB}=\frac{v_A}{OA}=\frac{2}{4\cos 30^\circ}=0.577\,4\ \text{rad/s}$$

$$v_B=OB\cdot\omega_{AB}=0.577\,4\times 2=1.155\ \text{m/s}$$

Thus, angular momentum about point C

$$L_C=J_C\omega=\frac{1}{12}ml^2\omega$$

$$=\frac{1}{12}\times 5\times 4^2\times 0.577\,4$$

$$=3.85\ \text{kg}\cdot\text{m}^2/\text{s} \qquad \textit{Ans.}$$

The angular momentum about instantaneous center O is

$$L_P = J_P\omega = (J_C + m \cdot OC^2)\omega$$
$$= \left(\frac{1}{12}\times5\times4^2+5\times2^2\right)\times0.577\,4$$
$$= 15.4\ \text{kg}\cdot\text{m}^2/\text{s} \qquad \textit{Ans.}$$

11.5.2 Angular Momentum Theorem about the Mass Center

If the rigid body has general plane motion, we have the *angular momentum theorem* Eq. (11.11) about fixed point O,

$$\frac{\mathrm{d}\boldsymbol{L}_O}{\mathrm{d}t} = \sum \boldsymbol{M}_{Oi}(\boldsymbol{F}_i^{(e)}) = \sum \boldsymbol{M}_O$$

Substituting Eq. (11.24) into above equation and using $\boldsymbol{M}_O = \boldsymbol{r}\times\boldsymbol{F}$ about point O as defined in Eq. (3.10) yield

$$\frac{\mathrm{d}\boldsymbol{L}_O}{\mathrm{d}t} = \frac{\mathrm{d}}{\mathrm{d}t}(\boldsymbol{r}_C\times m\boldsymbol{v}_C + \boldsymbol{L}_C') = \sum \boldsymbol{r}_i\times\boldsymbol{F}_i^{(e)} \tag{g}$$

The left side of Eq. (g) can be written as

$$\frac{\mathrm{d}\boldsymbol{L}_O}{\mathrm{d}t} = \frac{\mathrm{d}}{\mathrm{d}t}(\boldsymbol{r}_C\times m\boldsymbol{v}_C + \boldsymbol{L}_C') = \frac{\mathrm{d}}{\mathrm{d}t}(\boldsymbol{r}_C\times m\boldsymbol{v}_C) + \frac{\mathrm{d}\boldsymbol{L}_C'}{\mathrm{d}t}$$
$$= \frac{\mathrm{d}\boldsymbol{r}_C}{\mathrm{d}t}\times m\boldsymbol{v}_C + \boldsymbol{r}_C\times\frac{\mathrm{d}}{\mathrm{d}t}(m\boldsymbol{v}_C) + \frac{\mathrm{d}\boldsymbol{L}_C'}{\mathrm{d}t} \tag{h}$$

Since the cross product of a vector with itself is zero, namely, $\frac{\mathrm{d}\boldsymbol{r}_C}{\mathrm{d}t}\times m\boldsymbol{v}_C = \boldsymbol{v}_C\times m\boldsymbol{v}_C = 0$, the first term on the right side vanishes. Using the principle of linear momentum Eq. (10.17), we have

$$\frac{\mathrm{d}\boldsymbol{L}_O}{\mathrm{d}t} = \boldsymbol{r}_C\times\frac{\mathrm{d}}{\mathrm{d}t}(m\boldsymbol{v}_C) + \frac{\mathrm{d}\boldsymbol{L}_C'}{\mathrm{d}t} = \boldsymbol{r}_C\times\boldsymbol{F} + \frac{\mathrm{d}\boldsymbol{L}_C'}{\mathrm{d}t} \tag{i}$$

Substituting Eq. (b) into Eq. (g), the right side of Eq. (g) can be written as

$$\boldsymbol{M}_O = \sum \boldsymbol{r}_i\times\boldsymbol{F}_i^{(e)} = \sum(\boldsymbol{r}_C + \boldsymbol{r}_i')\times\boldsymbol{F}_i^{(e)}$$
$$= \boldsymbol{r}_C\times\sum\boldsymbol{F}_i^{(e)} + \sum\boldsymbol{r}_i'\times\boldsymbol{F}_i^{(e)}$$
$$= \boldsymbol{r}_C\times\boldsymbol{F} + \boldsymbol{M}_C \tag{j}$$

where $\boldsymbol{M}_C$ is moment of the external resultant acting on body about the mass center. Substituting Eq. (i) and Eq. (j) into Eq. (g), we have

$$\frac{\mathrm{d}\boldsymbol{L}_C'}{\mathrm{d}t} = \boldsymbol{M}_C \tag{11.25}$$

This equation is referred to as the *angular momentum theorem about the mass center*. It states that *the time rate of change of the angular momentum of rigid body relative to the mass center C is equal to the moment of the force acting on the rigid body relative to C*. It has the same mathematical form as that about a fixed point, for the angular momentum to other moving points, this simple relation is not valid, in general.

11.6 Kinetics of a Rigid Body in Plane Motion

Consider the rigid body in Fig. 11-25, which is subjected to general plane motion, has an angular velocity ω. The fixed x, y axes are established at an arbitrary point O. The origin of the moving x', y' reference frame will be attached to the mass center C of the body and the axes translate with respect to the fixed frame. Using linear and angular momentum theorems, it is possible to write two equations which define the body's motion, namely, Eq. (10.18) and Eq. (11.25), restated as

$$m\boldsymbol{a}_C = \sum \boldsymbol{F}_i^{(e)}$$
$$\frac{\mathrm{d}\boldsymbol{L}'_C}{\mathrm{d}t} = \boldsymbol{M}_C \tag{11.26}$$

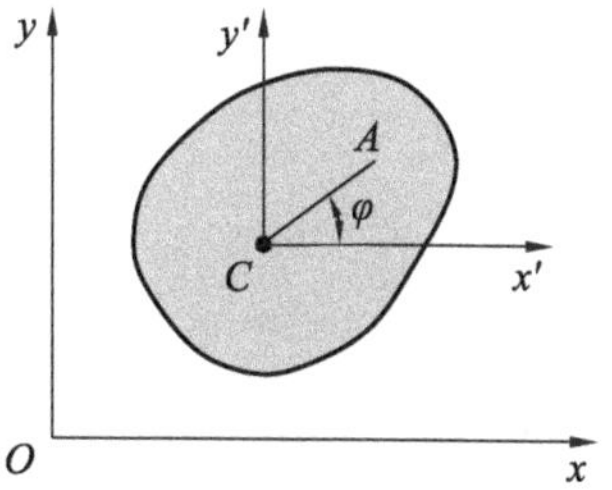

Fig. 11-25

Here the reference point must be the center of mass C. The principles of linear and angular momentum describe the general plane motion of a rigid body. The motion occurs in the x-y plane, and $L'_C = J_C\omega$, then three scalar equations can be written to express the motion, namely, in the rectangular coordinates

$$ma_{Cx} = \sum F_x^{(e)}$$
$$ma_{Cy} = \sum F_y^{(e)} \tag{11.27}$$
$$J_C\alpha = M_C$$

or in the normal and tangential coordinates

$$ma_C^t = \sum F_t^{(e)}$$
$$ma_C^n = \sum F_n^{(e)} \tag{11.28}$$
$$J_C\alpha = M_C$$

The first two of these equations represent the principle of linear momentum and impulse, which has been discussed in Section 10.3, and the third equation represents the principle of angular momentum about the mass center C.

Example 11-12

A homogeneous cylinder has a mass m and radius r, in Fig. 11-26a. An inextensible cord is wrapped around its periphery and the other end of the cord is fastened at A. If the disk releases from rest, determine the tension of cord and the velocity of mass center C of the cylinder after it falls down height h. Neglect the weight of the cord.

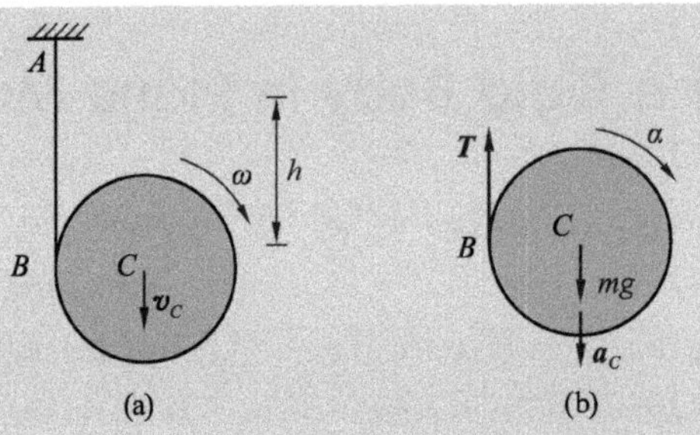

Fig. 11-26

Solution:

The disk is in general plane motion, the free-body diagram is shown in Fig. 11-26b. The tension T of cord is unknown. Assuming the acceleration of the mass center is a_C and angular acceleration of the disk is α. Then the principles of linear and angular momentum yield

$$ma_C = mg - T$$

$$J_C\alpha = Tr$$

where the moment of inertia $J_C = \frac{1}{2}mr^2$, in addition, we have the kinematic relation between $\boldsymbol{a}_C$ and α

$$\alpha = a_C / r$$

Thus, we have three equations for the three unknowns T, $\boldsymbol{a}_C$ and α. Solving these equations yields

$$a_C = \frac{2}{3}g \qquad \textit{Ans.}$$

$$T = \frac{1}{3}mg \qquad \textit{Ans.}$$

Example 11-13

A homogeneous wheel of mass m and radius R, rolls without slipping on the incline track with the angle of θ, in Fig. 11-27a. Determine the acceleration a_C of mass center of the cylinder and the frictional force between wheel and slope.

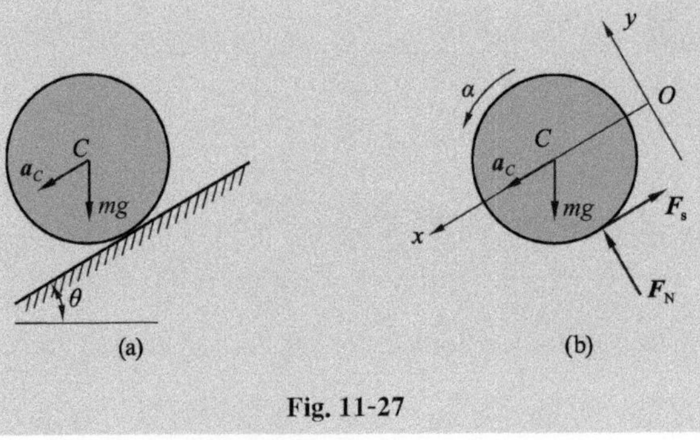

Fig. 11-27

Solution:

The wheel is in general plane motion, the free-body diagram is shown in Fig. 11-27b. The frictional force $\boldsymbol{F}_s$ and reaction force $\boldsymbol{F}_N$ are unknown. Assuming the acceleration of the mass center is $\boldsymbol{a}_C$ and angular acceleration of the wheel is α. Then the principles of linear and angular momentum yield

$$ma_C = mg\sin\theta - F_s$$

$$0 = F_N - mg\cos\theta$$

$$J_C\alpha = F_s R$$

where the mass moment of inertia $J_C = \frac{1}{2}mR^2$, in addition, we have the kinematic relation between a_C and α

$$a_C = R\alpha$$

Thus, we have four equations for the four unknowns F_s, F_N, a_C and α. Solving these equations yields

$$a_C = \frac{2}{3}g\sin\theta,\ \alpha = \frac{2g\sin\theta}{3R},\ F_s = \frac{1}{3}mg\sin\theta \qquad Ans.$$

Example 11-14

The slender rod AB, as shown in Fig. 11-28a, has a mass m and length l. The end A is at rest on smooth horizontal surface and the end B is suspending by a cord. The angle between the surface and rod is $\theta = 60°$ initial. If the cord is cut off suddenly, determine the angular acceleration of rod AB and reaction forces of the support A at this instant.

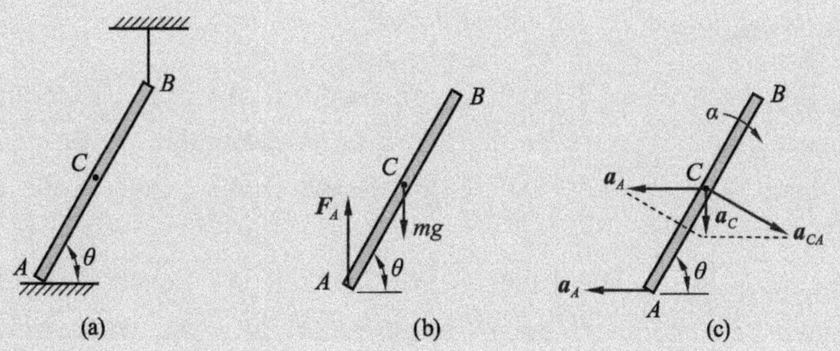

Fig. 11-28

Solution:

The rod is in general plane motion, the free-body diagram of the rod is shown in Fig. 11-28b. The reaction force $\boldsymbol{F}_A$ is unknown. Since the external force along the x axis is zero, the linear momentum will be conserved along the x axis. Hence the mass center C of rod AB drops down vertically. At the instant the cord is cut off, the angular velocity is zero. Assuming the acceleration of the mass center is a_C and angular acceleration of the wheel is α. Apply the principles of linear and angular momentum, we obtain

$$ma_C = mg - F_A$$

$$J_C\alpha = \frac{1}{2}F_A l\cos\theta$$

where the moment of inertia $J_C = \frac{1}{12}ml^2$, we can take the point A as the base point and apply the acceleration equation, the kinematic diagram is drawn in Fig. 11-28c. Then we can obtain the kinematic relation between a_C and α

$$a_C = a_{CA}\cos\theta = \frac{1}{4}l\alpha\cos\theta$$

Thus, we have three equations for the three unknowns $\boldsymbol{F}_A$, a_C and α. Solving these equations yields

$$\alpha = \frac{12g}{7l},\ F_A = \frac{4}{7}mg \qquad \textit{Ans.}$$

Exercises

11-1 Determine the angular momentum J_O about point O, in Fig. 11-29. The mass of each body is m.

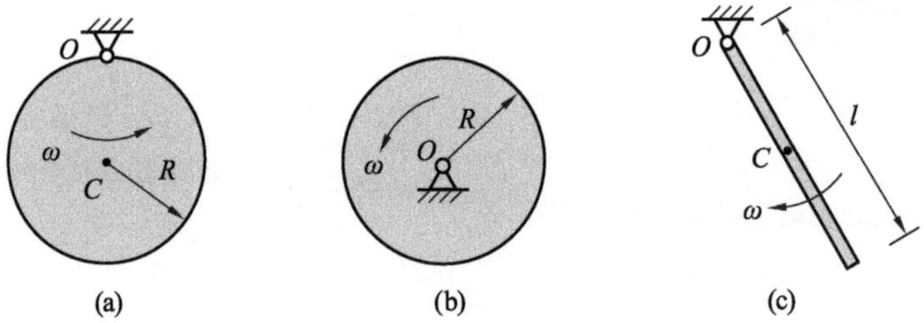

Fig. 11-29

11-2 In Fig. 11-30, the drum has a momentum of initial J_O about point O and different radius R and r, rotates around point O with angular velocity ω. Two blocks have mass m_A and m_B, respectively, determine the angular momentum of the system about point O.

11-3 The mass of the small ball in Fig. 11-31 is 150 kg. If the rod of negligible mass is subjected to a couple moment of $M = 30t^2$ N · m, where t is in seconds, determine the speed of the small ball at the instant $t = 5$ s. The ball starts from rest. Neglect the size of the ball.

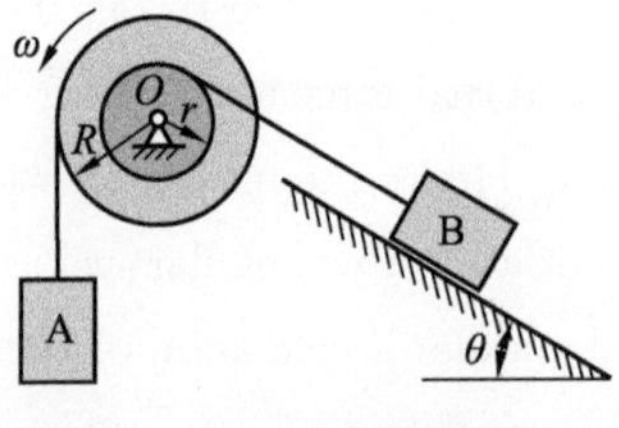

Fig. 11-30

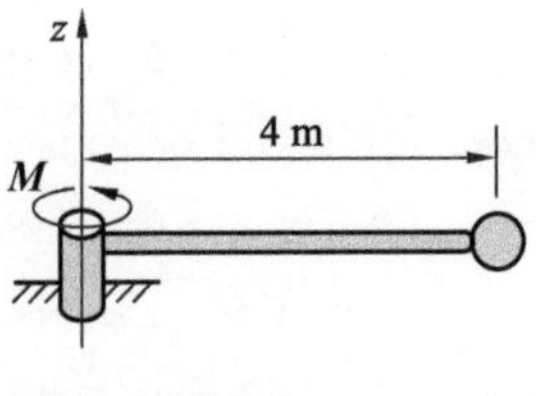

Fig. 11-31

11-4 As shown in Fig. 11-32, an earth satellite M of mass m which is subjected only to the earth's gravitational force $\boldsymbol{F}$, is rotating about the earth with a free-flight trajectory. If the speed of the satellite in the closest point of radius r_1 is $v_1=30$ km/s, determine the speed v_2 of the satellite in the farthest point of radius $r_2=2r_1$.

11-5 The ball M in Fig. 11-33, rotates around a circular path of radius R with angular velocity $\omega_1=12$ rad/s, if the cord length is shortened by pulling the cord through the tube. Determine the angular velocity ω_2 of the ball when it rotates around a radius $r=R/2$ circular path.

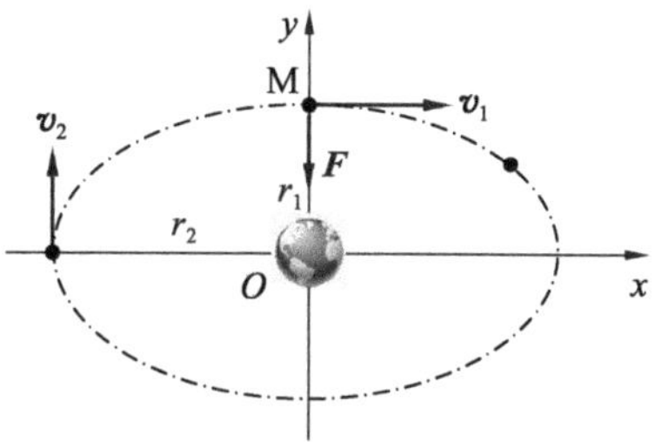

Fig. 11-32

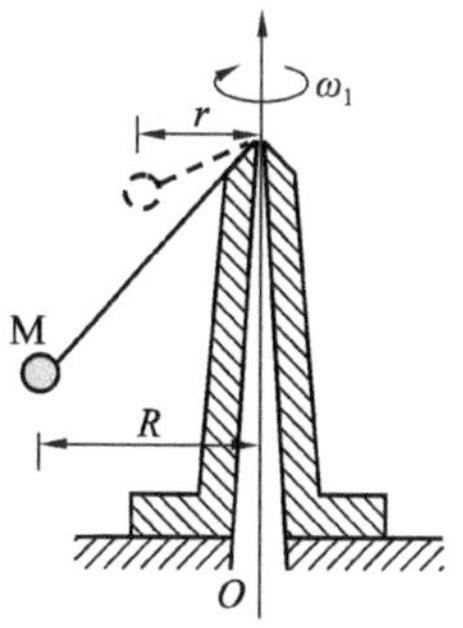

Fig. 11-33

11-6 In Fig. 11-34, two identical 40 N spheres A and B are attached to the light rods of negligible mass. The rods are connected to a shaft as shown. Initially the shaft is rotating with an angular velocity $n=90$ r/min around the axis z when include angle between the light rods and axis is $\theta=60°$. Determine the angular velocity of shaft if include angle is changed to $\theta=30°$. Neglect the frictional force.

11-7 Determine the moment of inertia of a homogeneous solid sphere which has mass m and radius R with respect to an axis a-a that passes through the center C, Fig. 11-35.

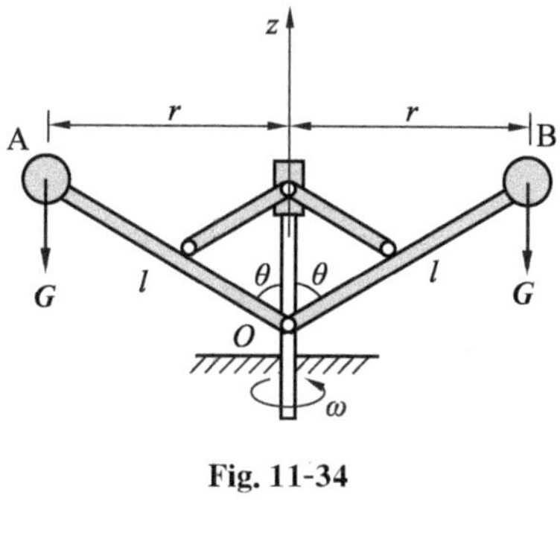

Fig. 11-34

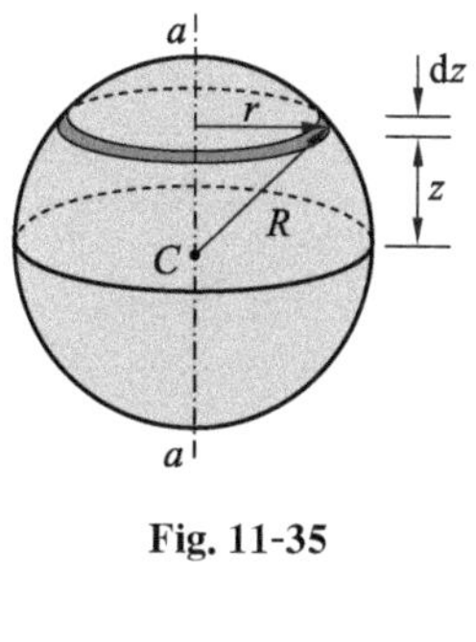

Fig. 11-35

11-8 Determine the mass moment of inertia of the area about the x axis in Fig. 11-36. (mm)

11-9 In Fig. 11-37, the wheel A has radius r and moment of inertial J_1 about its center O_1. Wheel B has radius R and moment of inertial J_2 about its center O_2 and is coupled to the wheel A by means of a belt which does not slip at its contacting surfaces. If a motor supplies a clockwise moment $\boldsymbol{M}$ to the wheel A, determine the angular acceleration of wheel A and B. Neglect the friction and weight of belt.

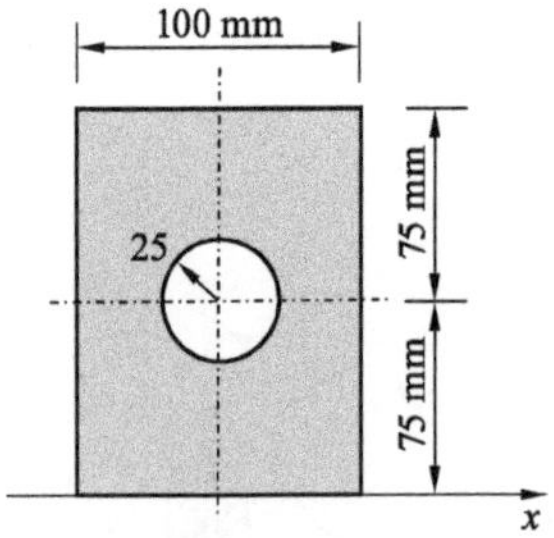

Fig. 11-36

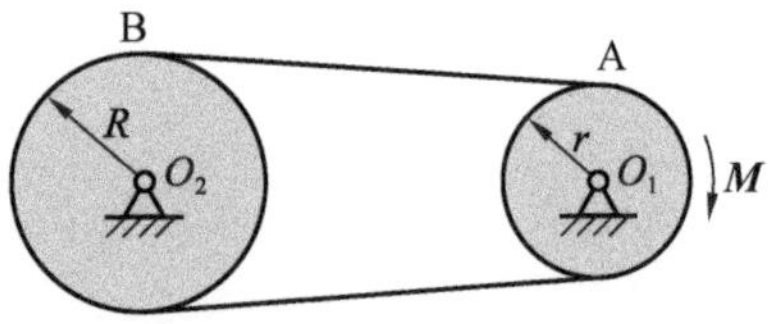

Fig. 11-37

11-10 In Fig. 11-38, the T-shape bar as shown, consisting of two 8 kg homogeneous slender rods, is pin-supported at point O. When it is in horizontal, the bar has an angular velocity $\omega = 4$ rad/s. Determine the angular acceleration and the support reactions of pin O at the instant.

11-11 In Fig. 11-39, two blocks have mass of $m_A = 20$ kg and $m_B = 30$ kg, respectively. They are hanging on two coaxial tub wheels that have radius of R=200 mm and r=100 mm, mass of m_1=5 kg and m_2=2 kg, respectively. If they are released from rest, determine the angular acceleration of the wheels and tension of two side rope. Neglect the mass of the cord and any slipping on the spool.

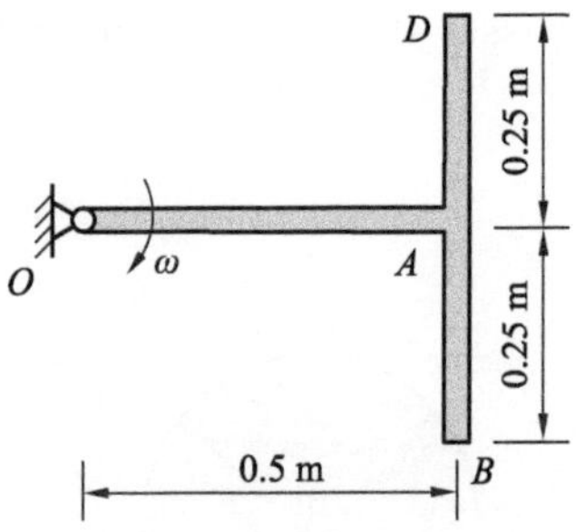

Fig. 11-38

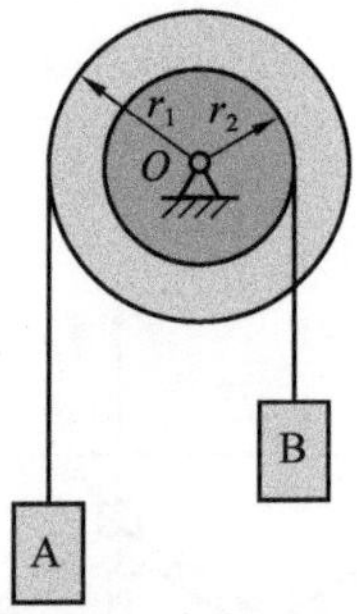

Fig. 11-39

11-12 In Fig. 11-40, the 100 kg disk of radius R=1 m, rotates around the point O with an angular velocity of n=120 r/min. If the brake AB is applied a constant force $\boldsymbol{F}$ on point B and the disk stops after 10 s, determine the magnitude of force $\boldsymbol{F}$. The coefficient of kinetic friction is μ_k=0. 1. Neglect the thickness of the brake.

11-13 In Fig. 11-41, the 5 kg slender rod AB is at horizontal by a pin connected at point O and suspending at its end B by a cord BD. If the cord is cut off suddenly, determine the reaction forces of the support O at this instant.

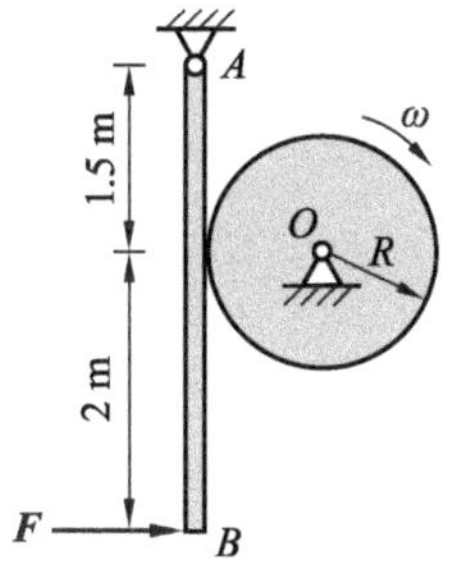

Fig. 11-40

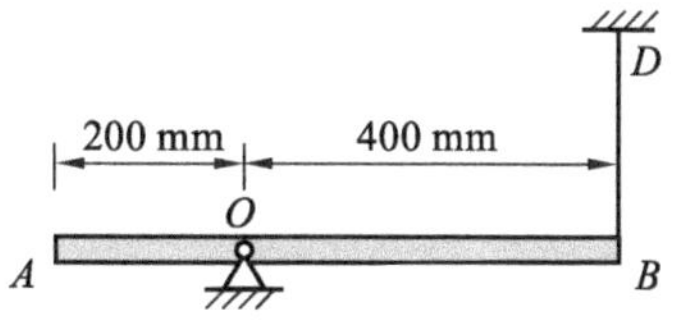

Fig. 11-41

11-14 In Fig. 11-42, the arm OA of negligible mass rotates with constant angular velocities $\omega_O=4$ rad/s about axis O. The homogeneous disk has mass of $m=25$ kg and radius of $R=200$ mm. If the disk rotates with a relative angular velocity $\omega_r=4$ rad/s about the axis A. Determine the angular momentum of the disk about axis O.

11-15 In Fig. 11-43, a homogeneous wheel has a mass m and radius r. If it is subjected to constant couple moment M, determine the acceleration a_C of mass center C of the wheel and the frictional force between the wheel and the surface. The wheel rolls without slipping on the horizontal surface.

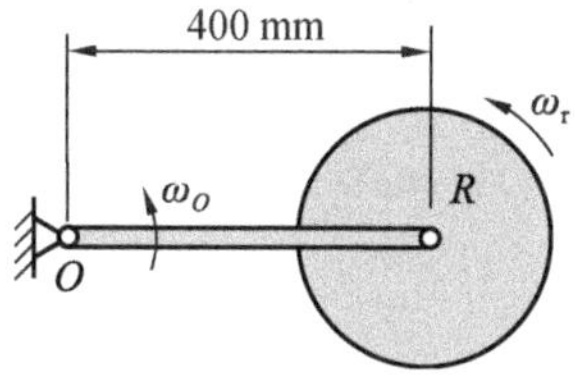

Fig. 11-42

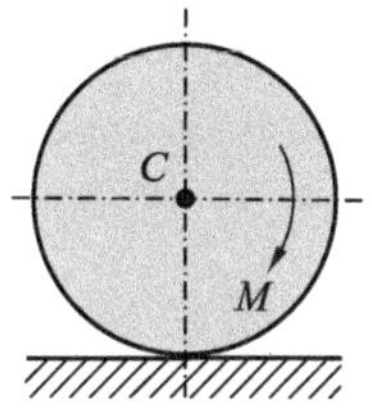

Fig. 11-43

11-16 In Fig. 11-44, the 100 kg spool shown has a radius of gyration $\rho=0.35$ m. A cable is wrapped around the central hub of the spool, and a horizontal force having a variable magnitude of $F=(t+10)$ N is applied, where t is in seconds. If the spool is initially at rest, determine its angular velocity in 5 s. Assume that the spool rolls without slipping. Given: $R=0.75$ m, $r=0.4$ m.

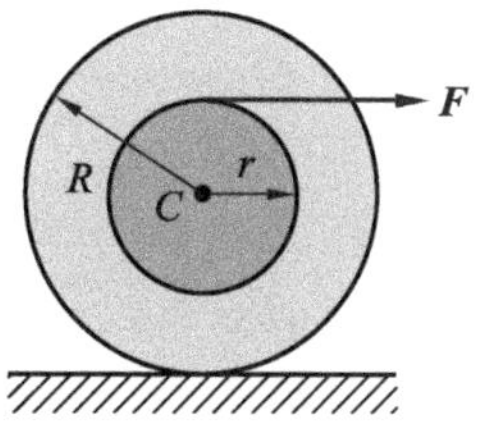

Fig. 11-44

11-17 The homogeneous cylinder has a mass m and radius R, in Fig. 11-45. An inextensible cord is wrapped around its periphery and the other end of the cord is fastened at A. If the cylinder rolls down on the incline track with the angle $\theta = 60°$, determine the acceleration a_C of mass center of the cylinder. The coefficient of the friction between cylinder and slope is $\mu = 0.33$.

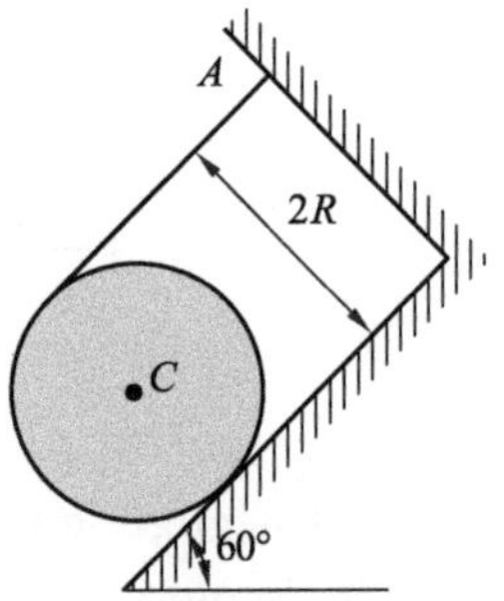

Fig. 11-45

11-18 In Fig. 11-46, the homogeneous cylinder and hoop (thin ring), each has the same mass m and radius r, are coupled using rod AB which is pin connected at both ends. The system is released from rest along an inclined plane, and rolls without sliding, determine the acceleration of mass center of system and internal force of the rod AB. Neglect the weight of the rod AB.

11-19 Two homogeneous cylinders in Fig. 11-47, have same mass m and radius r. An inextensible cable is wrapped on the cylinder A and other end of the cable is wrapped on the cylinder B. If the cylinder B releases from the rest, determine the acceleration of the mass center C of the cylinder B. Neglect the mass of the string and the friction of the axis O.

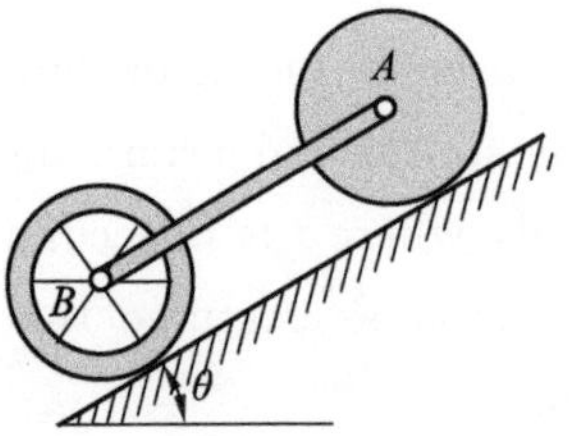

Fig. 11-46

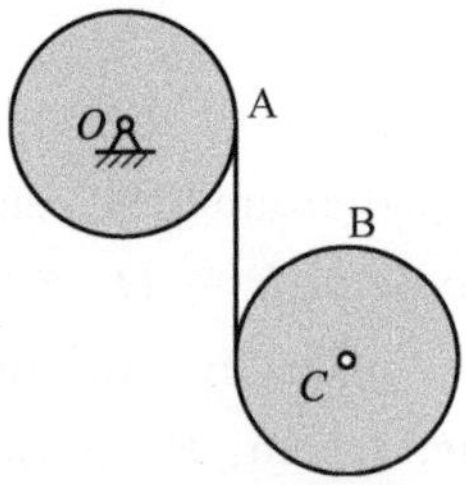

Fig. 11-47

11-20 The block A of mass m_1 shown in Fig. 11-48, is attached to a cord which is cross a negligible pulley and wrapped around the central hub (of radius r) of the spool (of radius R) that has a mass m_2 and radius of gyration ρ about its mass center. If the block A releases from the rest, determine its acceleration. Assume that the spool rolls without slipping.

11-21 The slender rod AB in Fig. 11-49 has a mass m and length l. If the rod releases from the rest when the angle between the normal line and rod is θ initial, determine the angular acceleration of rod AB and reaction forces of the support A at this instant. Neglect the friction between the rod and horizontal surface.

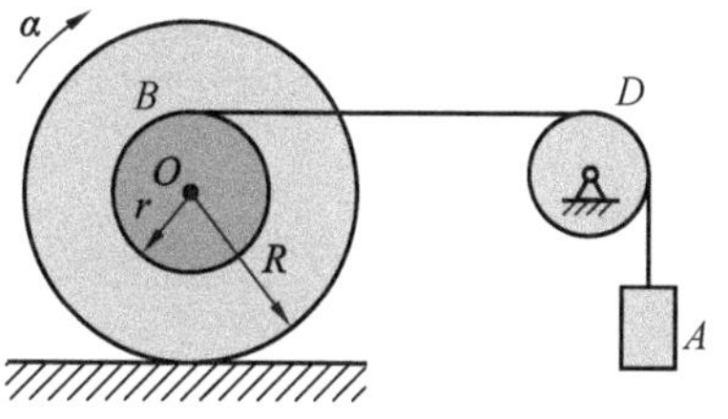

Fig. 11-48

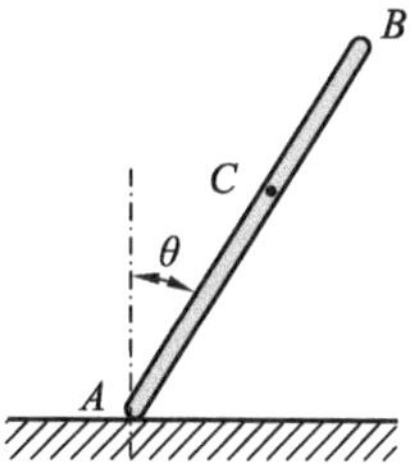

Fig. 11-49

Chapter 12

Work and Kinetics Energy

Objectives

In this chapter, we will first learn the formulations for the kinetic energy, and define the various ways a force and couple do work. Then study the principle of work and energy and apply it to solve problems that involve force, velocity, and displacement. Finally study problems that involve power and efficiency and introduce the concept of a conservative force and apply the theorem of conservation of energy to solve kinetic problems.

12.1 The Work of a Force

12.1.1 Work of a Constant Force

A force $\boldsymbol{F}$ will do work on a particle when the particle undergoes a displacement in the direction of the force. If the force $\boldsymbol{F}$ has a constant magnitude and acts at a constant angle θ from its straight-line path, Fig. 12-1, then the component of $\boldsymbol{F}$ in the direction of displacement is $F \cdot \cos\theta$. The work done by $\boldsymbol{F}$ when the particle moves the displacement s is a scalar quantity, defined by

$$W=\boldsymbol{F}\cdot\boldsymbol{s}=Fs\cos\theta \tag{12.1}$$

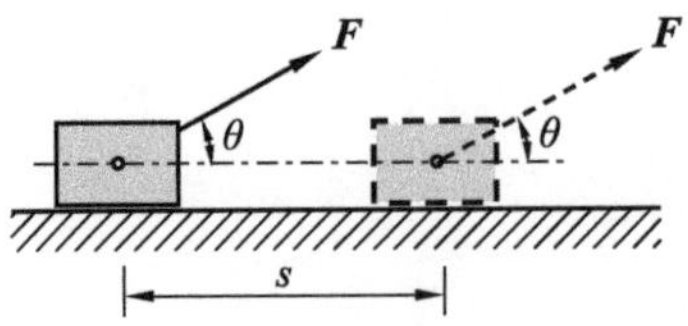

Fig. 12-1

The unit of work in SI units is the *joule* (J). Note that if $0^\circ \leqslant \theta < 90^\circ$, then the force component and the displacement have the same sense so that the work is positive. If $90° < \theta \leqslant 180°$, these vectors will have opposite sense, and therefore the work is negative. If the force is perpendicular to displacement, since $\cos 90° = 0$, or if the force is applied at a *fixed* point, in which case the displacement is zero, the work is zero.

12.1.2 Work of a Variable Force

The variable force $\boldsymbol{F}$ in Fig. 12-2 causes the particle to move along the path s from position $\boldsymbol{r}_1$ to a new position $\boldsymbol{r}_2$, the displacement is then $\mathrm{d}\boldsymbol{r} = \boldsymbol{r}_2 - \boldsymbol{r}_1$. The magnitude of $\mathrm{d}\boldsymbol{r}$ is $\mathrm{d}s$ that is the length of the differential segment along the path. Then the *elementary work* done by $\boldsymbol{F}$ is a scalar quantity, defined by

$$\mathrm{d}W = \boldsymbol{F} \cdot \mathrm{d}\boldsymbol{r} \tag{12.2a}$$

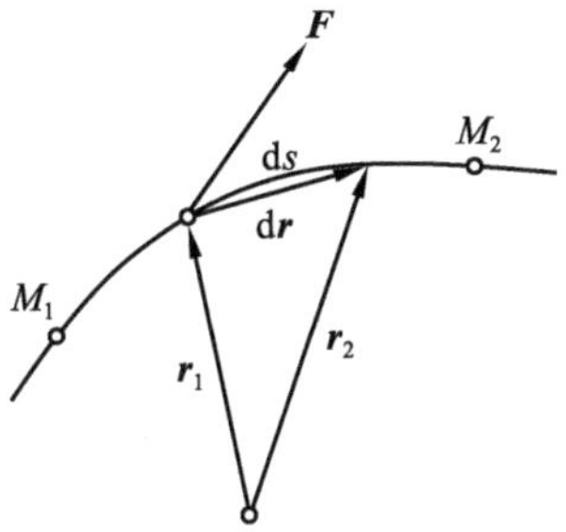

Fig. 12-2

At the intermediate point, displacement $\mathrm{d}\boldsymbol{r}$ and force $\boldsymbol{F}$ can be written as

$$\boldsymbol{F} = F_x \boldsymbol{i} + F_y \boldsymbol{j} + F_z \boldsymbol{k}$$

$$\mathrm{d}\boldsymbol{r} = \mathrm{d}x \boldsymbol{i} + \mathrm{d}y \boldsymbol{j} + \mathrm{d}z \boldsymbol{k}$$

Then the elementary work done by $\boldsymbol{F}$ is given by

$$\mathrm{d}W = \boldsymbol{F} \cdot \mathrm{d}\boldsymbol{r} = F_x \mathrm{d}x + F_y \mathrm{d}y + F_z \mathrm{d}z \tag{12.2b}$$

If the particle acted upon by the force $\boldsymbol{F}$ undergoes a finite displacement along its path from point M_1 to M_2, Fig. 12-2, the work of force $\boldsymbol{F}$ is determined by integration of the Eq. (12.2b), then

$$\begin{aligned} W &= \int_{M_1}^{M_2} \mathrm{d}W = \int_{M_1}^{M_2} \boldsymbol{F} \cdot \mathrm{d}\boldsymbol{r} \\ &= \int_{M_1}^{M_2} (F_x \mathrm{d}x + F_y \mathrm{d}y + F_z \mathrm{d}z) \end{aligned} \tag{12.3}$$

If the particle is subjected to n forces $\boldsymbol{F}_1$, $\boldsymbol{F}_2$, $\cdots$, $\boldsymbol{F}_n$, the resultant is

$$\boldsymbol{F}_{\mathrm{R}} = \boldsymbol{F}_1 + \boldsymbol{F}_2 + \cdots + \boldsymbol{F}_n$$

Substituting into Eq. (12.3), the work done by the resultant is

$$\begin{aligned} W &= \int_{M_1}^{M_2} (\boldsymbol{F}_1 + \boldsymbol{F}_2 + \cdots + \boldsymbol{F}_n) \cdot \mathrm{d}\boldsymbol{r} \\ &= \int_{M_1}^{M_2} \boldsymbol{F}_1 \cdot \mathrm{d}\boldsymbol{r} + \int_{M_1}^{M_2} \boldsymbol{F}_2 \cdot \mathrm{d}\boldsymbol{r} + \cdots + \int_{M_1}^{M_2} \boldsymbol{F}_n \cdot \mathrm{d}\boldsymbol{r} \\ &= W_1 + W_2 + \cdots + W_n \end{aligned} \tag{12.4}$$

This means that the work of the system of forces is equal to the sum of the works of the force components passing the same elementary displacement.

12.1.3 Work of a Weight

Consider a particle of mass m, which moves up along the path s shown in Fig. 12-3 from position M_1 to position M_2. At an intermediate point, the displacement $d\boldsymbol{r}=dx\boldsymbol{i}+dy\boldsymbol{j}+dz\boldsymbol{k}$. Since,

$$F_x=0,\ F_y=0,\ F_z=-mg$$

Applying Eq. (12.3), we have

$$W=\int_{z_1}^{z_2}-mg\,dz=-mg(z_2-z_1) \tag{12.5}$$

Thus, the work of weight is independent of the path and is equal to the magnitude of the particle's weight times its *vertical displacement* Δz. In the case shown in Fig. 12-3, the work is negative, since the weight $G=mg$ is downward and Δz is upward. If the particle is displaced *downward* $(-\Delta z)$, the work of the weight is *positive*.

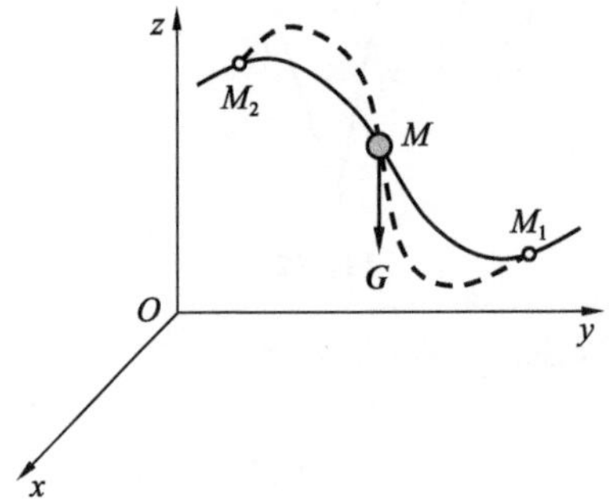

Fig. 12-3

12.1.4 Work of a Spring Force

If an elastic spring is attached to a body, in Fig. 12-4, the elastic force $F_s=-ks$ (k is the stiffness coefficient of the spring, units is N/m) acting on the body does work when the spring either stretches or compresses ds, then the elementary work done by the elastic force that acts on the attached particle is

$$dW=\boldsymbol{F}\cdot d\boldsymbol{r}=-ks\,ds$$

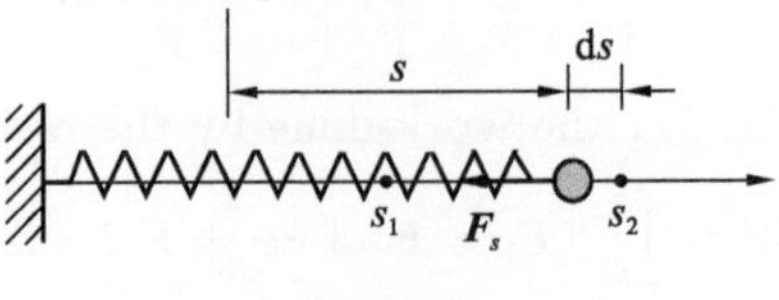

Fig. 12-4

The work is negative since F_s acts in the opposite sense to ds. If the particle displaces from s_1 to s_2, the work of F_s is then

$$W = \int_{s_1}^{s_2} \mathrm{d}W = \int_{s_1}^{s_2} -ks\,\mathrm{d}s$$

$$W = \frac{1}{2}k(s_1^2 - s_2^2) \tag{12.6}$$

Thus, the work of a spring force is also independent of the path. The sign depends on the direction of the spring force acting on the particle and the sense of direction of displacement of the particle, if both are in the *same* sense, *positive* work results; if they are *opposite* to one another, the work is *negative*.

12.1.5 The Work of a Couple Moment

Consider the rigid body in Fig. 12-5, which is subjected to an active force $\boldsymbol{F}$, rotates around axis z. The force can be decomposed into three components, $\boldsymbol{F}_t$, $\boldsymbol{F}_z$ and $\boldsymbol{F}_r$, as shown. If the body undergoes a differential displacement $\mathrm{d}\boldsymbol{r}$, then the work done by the components $\boldsymbol{F}_z$ and $\boldsymbol{F}_r$ is zero since two forces are vertical to the path, hence, the elementary work done is

$$\mathrm{d}W = \boldsymbol{F} \cdot \mathrm{d}\boldsymbol{r} = F_t\mathrm{d}s = F_t r\,\mathrm{d}\varphi = M_z\mathrm{d}\varphi \tag{12.7a}$$

where M_z is the moment of the force about z axis. The work done by the couple can be found by considering the body undergoes a differential rotation $\mathrm{d}\varphi$ about the axis z, Fig. 12-5. The work is positive when M and $\mathrm{d}\varphi$ have the same sense of direction and negative if these vectors are in the opposite sense.

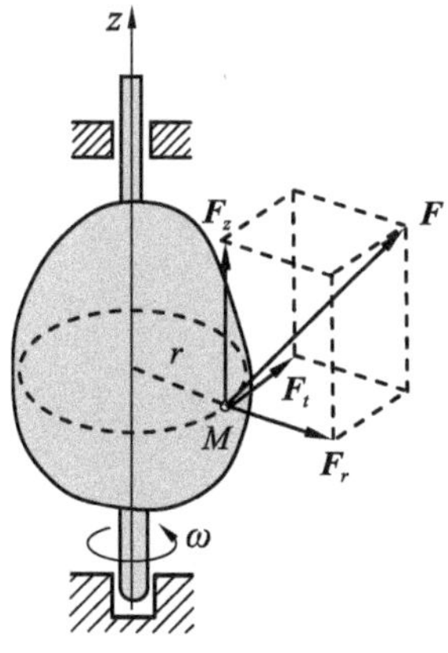

Fig. 12-5

When the body rotates in the plane through a finite angle φ measured in radians, from φ_1 to φ_2, the work of a couple moment is therefore

$$W = \int_{\varphi_1}^{\varphi_2} \mathrm{d}W = \int_{\varphi_1}^{\varphi_2} M_z\mathrm{d}\varphi \tag{12.7b}$$

If the couple moment M has a constant magnitude, then

$$W = M(\varphi_2 - \varphi_1) \tag{12.8}$$

12.1.6 Forces That Do No Work

There are some external forces that do no work when the body is displaced. These forces act either at *fixed* points on the body, or they have a direction *perpendicular* to their displacement. Examples include the reactions at a pin support about which a body rotates, the normal reaction acting on a body that moves along a fixed surface, and the weight of a body when the center of gravity of the body moves in a horizontal plane. Exception, the work of all the internal forces will occur in equal but opposite collinear pairs and so it will cancel out.

A frictional force $\boldsymbol{F}_f$ acting on a round body as it rolls without slipping over a rough surface also does no work, Fig. 12-6. This is because, during any instant of time dt, $\boldsymbol{F}_f$ acts at a point on the body which has zero velocity (instantaneous center) and so the work done by the force on the point is zero. In other words, the point is not displaced in the direction of the force during this instant. Since $\boldsymbol{F}_f$ contacts successive points for only an instant, the work of $\boldsymbol{F}_f$ will be zero.

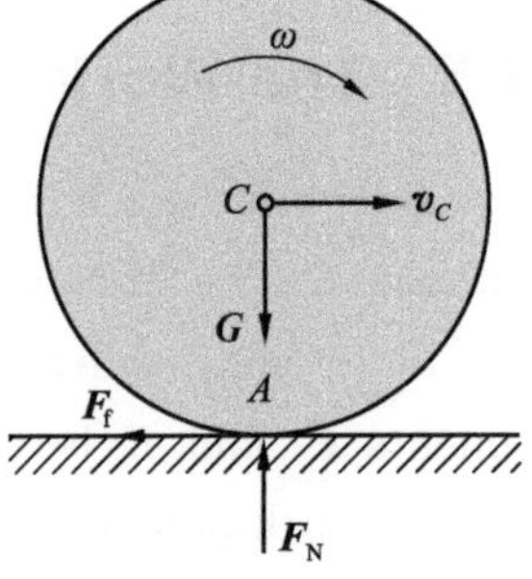

Fig. 12-6

Example 12-1

The collar of mass m slides on a frictionless circular arc of radius R, as shown in Fig. 12-7. The ideal spring attached to the collar has the free length $L_0 = R$ and stiffness k. When the collar moves from A to B, determine (1) the work done by the spring; (2) the work done by the weight.

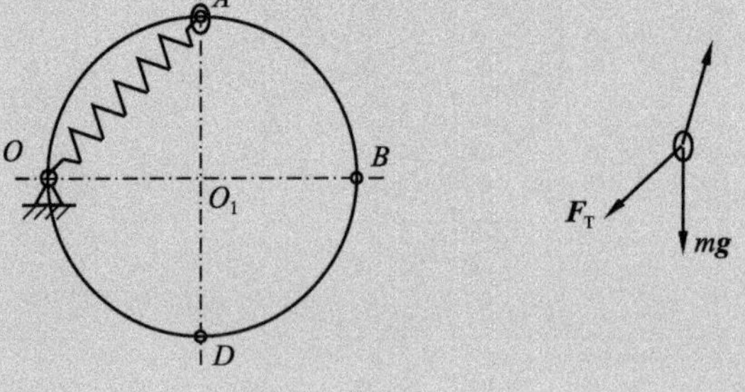

Fig. 12-7

Solution:

In the position A and position B, the spring is stretched, respectively, as

$$\delta_A=\sqrt{2}R-R=(\sqrt{2}-1)R$$

$$\delta_B=2R-R=R$$

When the collar moves from position A to position B, the work of elastic force is thus

$$W_{AB}=\frac{1}{2}k(\delta_A^2-\delta_B^2)=-0.828kR^2 \qquad \textit{Ans.}$$

The work to be negative since the force and displacement are opposite to each other. When the collar moves from position A to position B, the collar undergoes a vertical displace ment R. And the weight acts in the same sense to its vertical displacement, so the work is positive; i. e. ,

$$W_G=mgR \qquad \textit{Ans.}$$

Example 12-2

The 10 kg block shown in Fig. 12-8a is attached to a spring of stiffness $k=120$ N/m on the rough incline of angle $\theta=35°$. If the unstretched position of spring is M_1, determine the total work done by all the forces acting on the block when the block moves down the plane $s=0.5$ m. The coefficient of kinetic friction $\mu_k=0.2$.

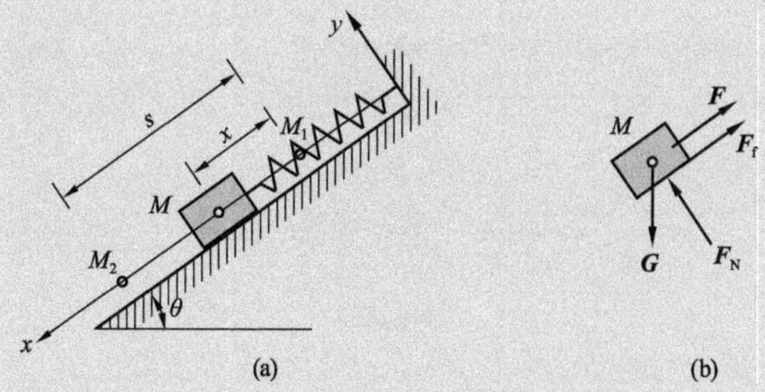

Fig. 12-8

Solution:

The free-body diagram of the block is drawn in Fig. 12-8b. The normal force $\boldsymbol{F}_N$ is always perpendicular to the displacement and does no work, $W_{F_N}=0$. Since the weight $\boldsymbol{G}=mg$ acts in the same sense to its vertical displacement, the work is positive; i. e. ,

$$W_G=mgs\sin\theta=10\times9.8\times0.5\times\sin 35°=28 \text{ J}$$

In the initial position the spring is stretched $s_1=0$ m and in the final position it is stretched $s_2=0.5$ m from the original position. Since the elastic force $\boldsymbol{F}$ and displacement are opposite to each other, the work of $\boldsymbol{F}$ is negative

$$W_F=\frac{1}{2}k(\delta_1^2-\delta_2^2)=-\frac{1}{2}k\delta_2^2=-\frac{1}{2}\times120\times0.5^2=-15 \text{ J}$$

Using the equilibrium conditions in the direction y

$$\sum F_y=0,\ F_N-mg\cos\theta=0$$

$$F_N = mg\cos\theta$$

Then the work of the frictional force $\boldsymbol{F}_f$ is

$$W_{F_f} = -F_f s = -\mu_k F_N s = -0.2\times 10\times 9.8\times \cos 35°\times 0.5 = -8\ \text{J}$$

The work of all the forces when the block is displaced 0.5 m is

$$W = W_{F_N} + W_G + W_{F_f} + W_F$$
$$= 28 - 8 - 15 = 5\ \text{J} \qquad \textit{Ans.}$$

12.2 Kinetic Energy

12.2.1 Kinetic Energy of a Particle

Consider a particle M in Fig. 12-9, which is located on the path defined relative to an inertial coordinate system. If the particle has a mass m and an instantaneous velocity $\boldsymbol{v}$, the kinetic energy T of the particle is defined as:

$$T = \frac{1}{2}mv^2 \tag{12.9}$$

Kinetic energy is a scalar quantity and has units of joules(J) 1 kg · m²/s² = 1 N · m = 1 J. The kinetic energy is always positive, regardless of the direction of motion of the particle.

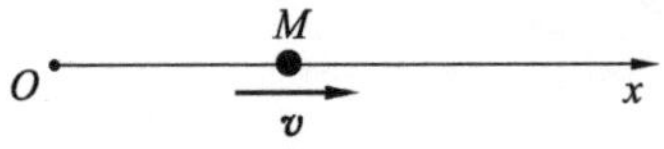

Fig. 12-9

12.2.2 Kinetic Energy of a System of Particles

For a system of particles, the arbitrary ith particle has a mass m_i. If at the instant shown in Fig. 12-10, the particle has a velocity v_i, then the particle's kinetic energy is $\frac{1}{2}m_i v_i^2$. The kinetic energy of the entire system is the scalar quantity and determined by writing similar expressions for each particle of the system and summing the results, i. e.,

$$T = \sum \frac{1}{2}m_i v_i^2 \tag{12.10}$$

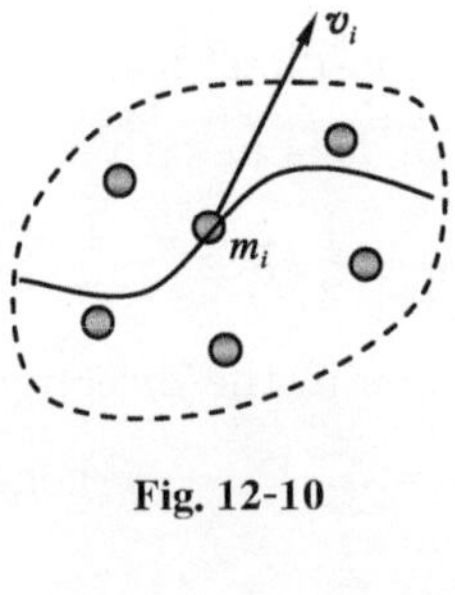

Fig. 12-10

(1) Translation of a Rigid Body

When a rigid body of mass m is subjected to either rectilinear or curvilinear translation, Fig. 12-11, all points in the rigid body move with the same velocity and equal to the velocity of mass center $\boldsymbol{v}_C$. The kinetic energy due to rotation is zero, since $\omega=0$. The kinetic energy of the body is therefore

$$T=\sum\frac{1}{2}m_i v_i^2=\frac{1}{2}\sum m_i {v_C}^2=\frac{1}{2}m v_C^2 \tag{12.11}$$

(2) Rotation about a Fixed Axis

When a rigid body of mass m rotates with an angular velocity ω about a fixed axis passing through point O, Fig. 12-12, each particle m_i in the system has a velocity v_i and the orthogonal distance r_i from the axis of rotation, the body has kinetic energy

$$T=\sum\frac{1}{2}m_i v_i^2=\sum\frac{1}{2}m_i r_i^2\omega^2$$
$$=\frac{1}{2}\left(\sum m_i r_i^2\right)\cdot\omega^2=\frac{1}{2}J_z\omega^2 \tag{12.12}$$

(3) General Plane Motion

When a rigid body is subjected to general plane motion, Fig. 12-13, it has an angular velocity ω and its mass center has a velocity $\boldsymbol{v}_C$. The body has both *translational* and *rotational* kinetic energy, so that

$$T=\frac{1}{2}m v_C^2+\frac{1}{2}J_C\,\omega^2 \tag{12.13}$$

This equation can also be expressed in terms of the body's motion about its *instantaneous center* P of zero velocity, since $v_C=\omega\cdot d$, so the kinetic energy can be written

$$T=\frac{1}{2}m(\omega d)^2+\frac{1}{2}J_C\,\omega^2=\frac{1}{2}(J_C+md^2)\omega^2$$

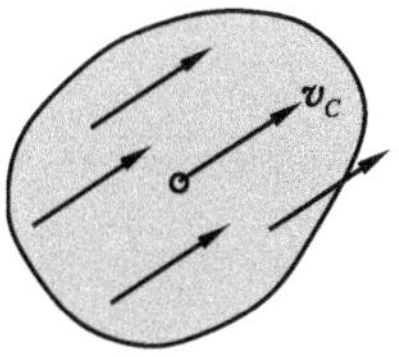

Fig. 12-11

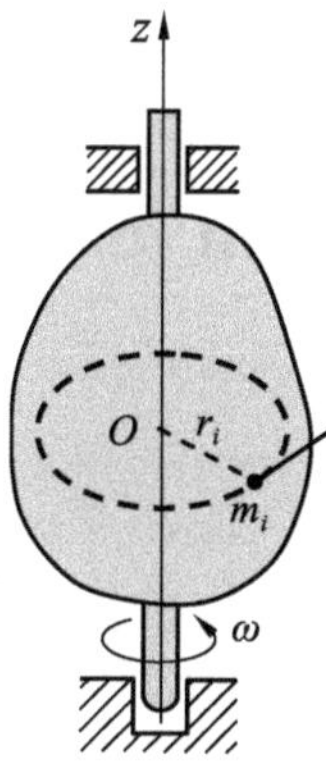

Fig. 12-12

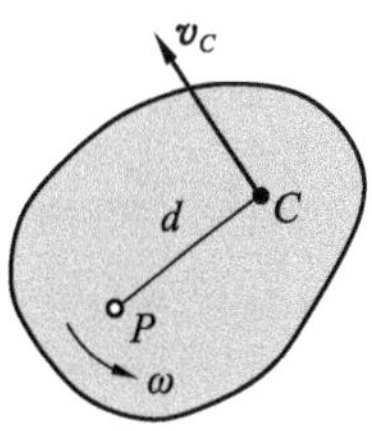

Fig. 12-13

By the parallel-axis theorem, the terms inside the parentheses represent the *moment of inertia* J_P of the body about its instantaneous center P. Hence,

$$T=\frac{1}{2}J_P\omega^2 \tag{12.14}$$

From the derivation, this equation will give the same result as Eq. (12.13).

Example 12-3

Determine the kinetic energy of the rigid body as shown in Fig. 12-14. The mass of each rigid body is m.

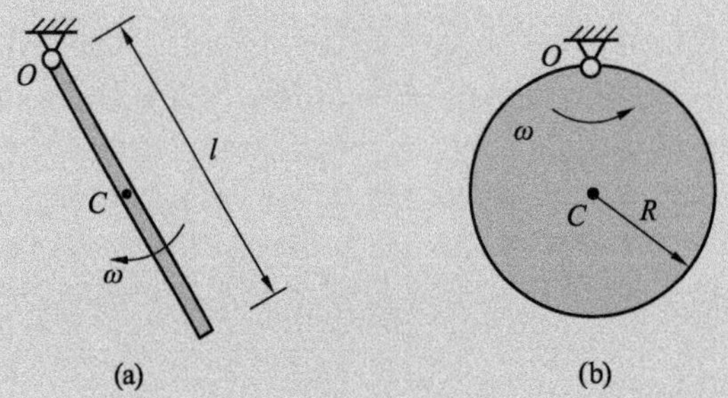

Fig. 12-14

Solution:

① The rod with length l rotates about the fixed end axis with constant angular ω. The kinetic energy of the rod is

$$\boldsymbol{T}=\frac{1}{2}J_O\omega^2=\frac{1}{2}\cdot\frac{1}{3}ml^2\cdot\omega^2=\frac{1}{6}ml^2\omega^2 \qquad \textit{Ans.}$$

② The wheel rotates around the axis passing through the point O with constant angular velocity ω. The kinetic energy of the wheel is

$$\boldsymbol{T}=\frac{1}{2}J_O\omega^2=\frac{1}{2}\cdot\left(\frac{1}{2}mR^2+mR^2\right)\cdot\omega^2=\frac{3}{4}mR^2\omega^2 \qquad \textit{Ans.}$$

12.3 Principle of Work and Kinetics Energy

12.3.1 Work-Energy Theorem for a Particle

Consider the particle M in Fig. 12-15 moving in the path, if the particle has a mass m and is subjected to a system of external forces represented by the resultant $\boldsymbol{F}$, then the equation of motion for the particle is the Newton's 2nd Law

$$m\frac{\mathrm{d}\boldsymbol{v}}{\mathrm{d}t}=\boldsymbol{F}$$

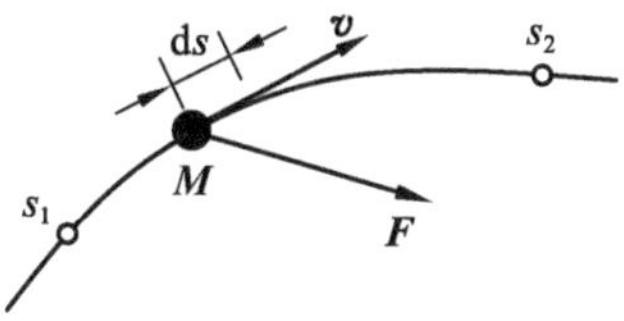

Fig. 12-15

Both sides of the equation are multiplied by d$\boldsymbol{r}$, we obtain

$$m\frac{\mathrm{d}\boldsymbol{v}}{\mathrm{d}t}\cdot \mathrm{d}\boldsymbol{r}=\boldsymbol{F}\cdot \mathrm{d}\boldsymbol{r} \tag{12.15}$$

The right-hand side of Eq. (12.15) is $\boldsymbol{F}\cdot \mathrm{d}\boldsymbol{r}=\delta W$, and its left-hand side is

$$m\frac{\mathrm{d}\boldsymbol{r}}{\mathrm{d}t}\cdot \mathrm{d}\boldsymbol{v}=m\boldsymbol{v}\cdot \mathrm{d}\boldsymbol{v}=\mathrm{d}\left(\frac{mv^2}{2}\right)$$

The Eq. (12.15) can be written as

$$\mathrm{d}\left(\frac{1}{2}mv^2\right)=\delta W \tag{12.16}$$

That represents the *theorem of the kinetic energy* (the differential form of the theorem), "*the variation of the kinetic energy is equal to the elementary work done by the force acting on the particle*".

Assuming initially that the particle has a speed $\boldsymbol{v}=\boldsymbol{v}_1$ at position s_1 and later $\boldsymbol{v}=\boldsymbol{v}_2$ at position s_2, integrating Eq. (12.16), we obtain

$$\int_{v_1}^{v_2}\mathrm{d}\left(\frac{1}{2}mv^2\right)=W_{12}$$

The final result can be written as

$$\frac{1}{2}mv_2^2-\frac{1}{2}mv_1^2=W_{12} \tag{12.17a}$$

The two terms on the left side, which are of the form $T=\frac{1}{2}mv^2$, define the particle's final and initial kinetic energy, respectively.

$$T_2-T_1=W_{12} \tag{12.17b}$$

This equation represents the *principle of work and energy* for the particle, which states that, "*the increment of the kinetic energy of a particle is equal to the work done by the forces acting on the path traveled by the particle*".

Thus, the *kinetic energy* is a measure of the particle's capacity to do work, which is associated with the motion of the particle.

12.3.2 Work-Energy Theorem for a System of Particles

For a system of particles, the arbitrary ith particle has a mass m_i, is subjected to a resultant external force $\boldsymbol{F}_i^{(e)}$ and a resultant internal force $\boldsymbol{F}_i^{(i)}$ which all the other particles exert on the ith particle, in Fig. 12-16. If we apply the principle of work and energy Eq. (12.16) to each of the particles in the system,

$$\mathrm{d}\left(\frac{1}{2}m_i v_i^2\right)=\delta W_i$$

where δW_i is elementary work done by all forces acting on the particle. Since work and energy are scalar quantities, the equations can be summed algebraically, which gives

$$\sum \mathrm{d}\left(\frac{1}{2}m_i v_i^2\right)=\sum \delta W_i \tag{12.18}$$

If the system represents a rigid body, or for rigid constraints, the work of the internal forces is zero. Integrating the Eq. (12.18), we have

$$T_2 - T_1 = \sum W_i \tag{12.19}$$

The sum of the work of all the forces acting on the system is equal to the change in the total kinetic energy of the system.

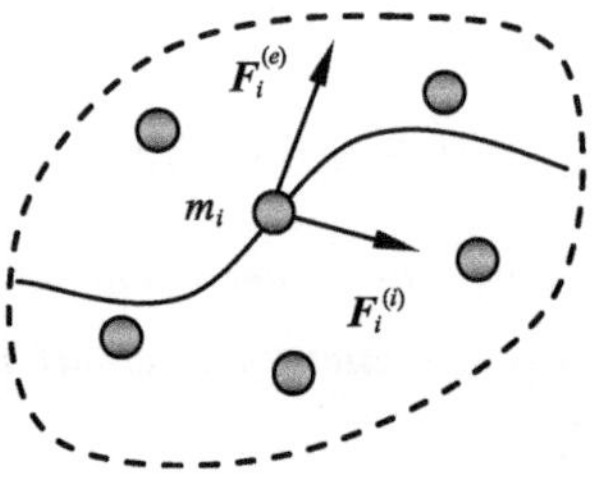

Fig. 12-16

Example 12-4

A block M has mass m is positioned at a height h above the reference position of a spring (spring constant k); shown in Fig. 12-17. The block is released from rest towards the spring. Find the maximum compression of the spring. Neglect the mass of other bodies.

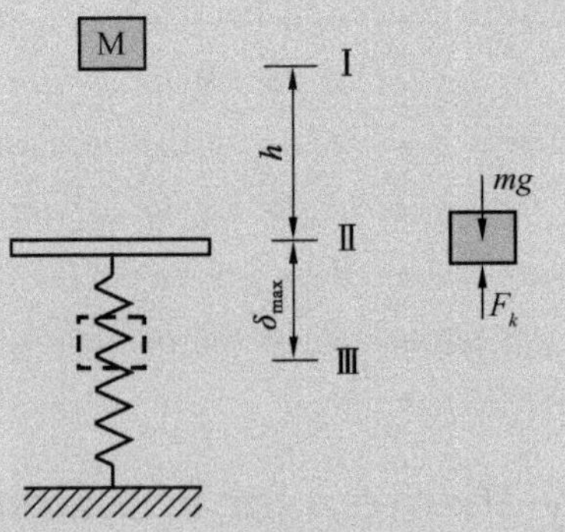

Fig. 12-17

Solution:

Since the block is released from rest and later reaches its maximum height when the spring reaches its maximum deformation, the initial and final velocities are zero, so the change of kinetic energy of the block is zero. The free-body diagram of the block when it is still in contact with the platform is shown in Fig. 12-17. Note that the weight does positive work and the spring force does negative work.

$$W = mg\delta_{max} - \frac{1}{2}k\delta_{max}^2$$

Using principle of work and energy, we have

$$0 - 0 = mg(h + \delta_{max}) - \frac{1}{2}k\delta_{max}^2$$

Solving this quadratic equation gives

$$\delta_{max} = \frac{mg}{k} \pm \frac{1}{k}\sqrt{m^2g^2 + 2kmgh} \qquad \textit{Ans.}$$

Example 12-5

The truck has a weight of $\boldsymbol{G}$ and moves on the straight surface with constant velocity $\boldsymbol{v}$, Fig. 12-18. If the brakes are applied to stop the tires of the truck, how far will the truck skid before stopping? The coefficient of kinetic friction between the tires and the road is μ_k.

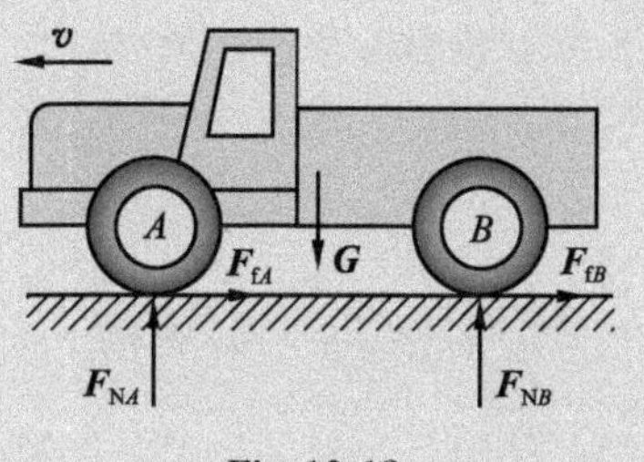

Fig. 12-18

Solution:

As shown in the free-body diagram, Fig. 12-18, the normal force $\boldsymbol{F}_{NA}$, $\boldsymbol{F}_{NB}$ and the weight $\boldsymbol{G}$ do no work since they never undergo displacement along its line of action. The frictional force $\boldsymbol{F}_{fA}$ and $\boldsymbol{F}_{fB}$ do external work when it undergoes a displacement s. This work is negative since it is in the opposite sense of direction to the displacement. Applying the equation of equilibrium normal to the road, we have

$$\sum F_y = 0, \; F_{NA} + F_{NB} - G = 0$$

$$F_{NA} + F_{NB} = G$$

Thus, the frictional force between the tires and the road

$$F_{fA} = \mu_k F_{NA}, \; F_{fB} = \mu_k F_{NB}$$

The work of external force can be written

$$W = -(F_{fA} + F_{fB})s = -\mu_k(F_{NA} + F_{NB})s = -\mu_k Gs$$

The change of kinetic energy of the truck when braking,

$$T_2 - T_1 = -\frac{1}{2}\frac{G}{g}v^2$$

Using principle of work and energy, we have

$$-\frac{1}{2}\frac{G}{g}v^2 = -\mu_k G s$$

Solving for s yields

$$s = \frac{v^2}{2g\mu_k} \qquad \textit{Ans.}$$

Example 12-6

The cylinder C has radius R_2 and mass m_2, is hoisted up the incline without sliding from the rest by a drum shown in Fig. 12-19. The drum has radius of R_1, and its mass m_1 is distributed on the rim, can rotate around an axis perpendicular to the page and passing through point O. If there is a couple M acting on the drum, determine the velocity and the acceleration of the cylinder center C after it travels a distance s up the incline.

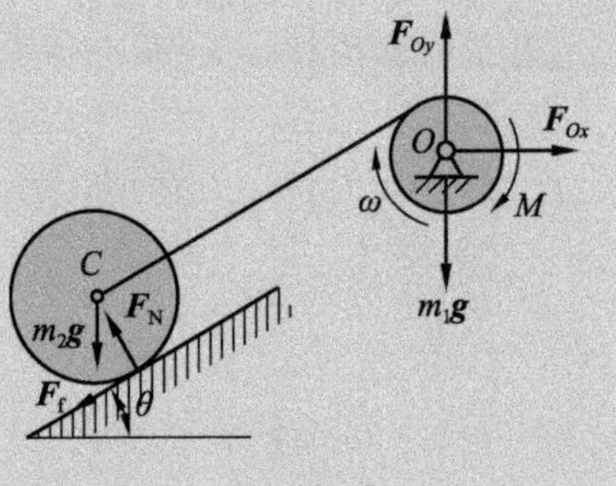

Fig. 12-19

Solution:

Here we will consider the cylinder and drum as a single system. The free-body diagram of the system is shown in Fig. 12-19, The normal force $\boldsymbol{F}_N$ does no work since it never undergoes displacement along its line of action. The weights $m_1\boldsymbol{g}$ does no work since the center of drum is no displacement. The reactions $\boldsymbol{F}_{Ox}$ and $\boldsymbol{F}_{Oy}$ do no work, since these forces represent the reactions at the supports and consequently they do not move while the drum are rotating. A static frictional force $\boldsymbol{F}_f$ also does no work as it rolls without slipping over a rough surface. The weights $m_2\boldsymbol{g}$ do negative work if we assume the cylinder move upward, we have

$$W = M\varphi - m_2 g\sin\theta \cdot s$$

The initial kinetic energy and final kinetic energy are

$$T_1 = 0$$

$$T_2 = \frac{1}{2}J_1\omega_1^2 + \frac{1}{2}m_2 v_C^2 + \frac{1}{2}J_2\omega_2^2$$

Using the methods of kinematics, we have

$$J_1 = m_1 R_1^2,\ J_2 = \frac{1}{2}m_2 R_2^2,\ \omega_1 = \frac{v_C}{R_1},\ \omega_2 = \frac{v_C}{R_2},\ \varphi = \frac{s}{R_1}$$

The final kinetic energy can be written as

$$T_2 = \frac{1}{4}(2m_1 + 3m_2)v_C^2$$

Substituting into principle of work and energy, we have

$$\frac{1}{4}(2m_1 + 3m_2)v_C^2 - 0 = M\varphi - m_2 g\sin\theta \cdot s \qquad \text{(a)}$$

$$v_C = \sqrt{\frac{(M - m_2 g R_1 \sin\theta)s}{R_1(2m_1 + 3m_2)}} \qquad \textit{Ans.}$$

Taking the time derivative of Eq. (a), and note that $a_C = \frac{\mathrm{d}v_C}{\mathrm{d}t}$, $\omega_1 = \frac{\mathrm{d}\varphi}{\mathrm{d}t} = \frac{v_C}{R_1}$, $v_C = \frac{\mathrm{d}s}{\mathrm{d}t}$, yields

$$\frac{1}{2}(2m_1 + 3m_2)v_C a_C = M\omega_1 - m_2 g\sin\theta \cdot v_C$$

$$a_C = \frac{2(M - m_2 g R_1 \sin\theta)}{(2m_1 + 3m_2)R_1} \qquad \textit{Ans.}$$

Example 12-7

The wheel shown in Fig. 12-20a has mass m and radius R. A cord is wrapped around it and the end is fixed on the wall. If the wheel is subjected to a constant horizontal force $\boldsymbol{F}$ on its mass center C and rolls from the rest, determine its angular velocity, angular acceleration and acceleration of mass center C after its center C moves distance s. The coefficient of kinetic friction between the wheel and the horizontal plane is μ_k.

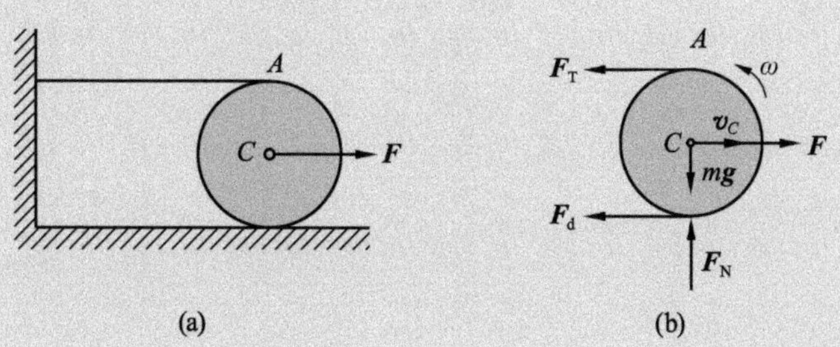

Fig. 12-20

Solution:

The free-body diagram of the wheel is shown in Fig. 12-20b, the normal force $\boldsymbol{F}_N$ and weight mg do no work since their line of action are *perpendicular* to the displacement. The tension force $\boldsymbol{F}_T$ of rope does no work, since the wheel does not slip along the rope as it rolls. Only the kinetics frictional force $\boldsymbol{F}_d$ and the force $\boldsymbol{F}$ do work. Applying the equation of equilibrium normal to the road, we have

$$\sum F_y = 0, \ F_N + mg = 0$$

$$F_N = mg$$

$$F_d = F_N \mu_k = mg\mu_k$$

When the center C moves s, the contact point between the wheel and ground slip $2s$, then the work of external force can be written

$$W = Fs - 2mg\mu_k s$$

Since the wheel is initially at rest, $T_1 = 0$. The wheel is subjected to plane motion, the velocity of mass center $v_C = \omega R$, so the final kinetic energy is determined from

$$T_2 = \frac{1}{2}mv_C^2 + \frac{1}{2}J_C\omega^2 = \frac{1}{2}mv_C^2 + \frac{1}{2}\times\frac{1}{2}mR^2\left(\frac{v_C}{R}\right)^2$$

$$= \frac{3}{4}mv_C^2$$

Using principle of work and energy, we have

$$\frac{3}{4}mv_C^2 - 0 = Fs - 2mg\mu_k s \tag{a}$$

$$v_C = 2\sqrt{\frac{s}{3m}(F - 2mg\mu_k)}$$

$$\omega = \frac{v_C}{R} = \frac{2}{R}\sqrt{\frac{s}{3m}(F - 2mg\mu_k)}$$

Taking the time derivative of Eq. (a), and note that $a_C = \frac{\mathrm{d}v_C}{\mathrm{d}t}$, yields

$$\frac{3}{2}mv_C a_C = Fv_C - 2mg\mu_k v_C$$

$$a_C = \frac{2}{3m}(F - 2mg\mu_k)$$

$$\alpha = \frac{a_C}{R} = \frac{2}{3mR}(F - 2mg\mu_k) \qquad \textit{Ans.}$$

12.4 Power and Efficiency

12.4.1 Power

The term "power" provides a useful basis for choosing the type of motor or machine which is required to do a certain amount of work in a given time. The power generated by a machine or engine that performs an amount of work $\mathrm{d}W$ within the time interval $\mathrm{d}t$, that is, the work done per unit time is known as the *power* P.

$$P=\lim_{\Delta t\to 0}\frac{\Delta W}{\Delta t}=\frac{\mathrm{d}W}{\mathrm{d}t}$$

If the work $\mathrm{d}W$ is expressed as $\mathrm{d}W=\boldsymbol{F}\cdot\mathrm{d}\boldsymbol{r}$, then

$$P=\frac{\mathrm{d}W}{\mathrm{d}t}=\frac{\boldsymbol{F}\cdot\mathrm{d}\boldsymbol{r}}{\mathrm{d}t}=\boldsymbol{F}\cdot\boldsymbol{v} \tag{12.20}$$

Hence, power is a scalar, where in this formulation $\boldsymbol{v}$ represents the velocity of the particle which is acted upon by the force $\boldsymbol{F}$. The basic unit of power used in the SI systems is the watt (W). 1 W=1 J/s=1 N · m/s. Another common unit for power is horsepower (hp); its relation to watt is given as

$$1\ \text{hp}=0.735\ \text{kW}$$

When the body (such as the rotor of the turbine) is subjected to the torque M, the elementary work is $\mathrm{d}W=M\mathrm{d}\varphi$. In this case, the power is calculated as

$$P=\frac{\mathrm{d}W}{\mathrm{d}t}=\frac{M\cdot\mathrm{d}\varphi}{\mathrm{d}t}=M\omega \tag{12.21}$$

12.4.2 Equation of Power

In Section 12.3, we have discussed the differential form of the theorem of kinetic energy.

$$\mathrm{d}T=\sum\mathrm{d}W_i$$

Dividing both sides of the above equation with $\mathrm{d}t$, we obtain

$$\frac{\mathrm{d}T}{\mathrm{d}t}=\sum\frac{\mathrm{d}W}{\mathrm{d}t}=\sum P \tag{12.22}$$

This is known as the *equation of power*, "the rate of change of the kinetic energy of a system of particles is equal to the algebraic sum of power of all forces acting on the system of particles". As the machine is working, the power of the machine is divided into *input power*, *useless power* and *useful power*, then the Eq. (12.22) can be written as

$$\frac{\mathrm{d}T}{\mathrm{d}t}=P_{\text{input}}-P_{\text{useful}}-P_{\text{useless}}$$

or

$$P_{\text{input}}=P_{\text{useful}}-P_{\text{useless}}+\frac{\mathrm{d}T}{\mathrm{d}t}$$

where the *input power* of a machine is the rate at which energy is supplied to the machine. The *useful power* is the rate at which the machine does work. The *useless power* is use to overcome friction.

12.4.3 Mechanical Efficiency

In all machines, energy is lost due to frictional effects in supports and guides. Thus a part of all input or applied work is lost. The *mechanical efficiency* of a machine is defined as the ratio of the output of useful power produced by the machine to the input of power supplied to the machine. Hence,

$$\eta=\frac{P_{\text{useful}}}{P_{\text{input}}}\times 100\% \tag{12.23}$$

Since machines consist of a series of moving parts, frictional forces will always be developed within the machine, and as a result, extra energy or power is needed to overcome these forces. Consequently, power output will be less than power input and so the efficiency of a machine is always less than 1, $\eta<1$.

Example 12-8

The power of the lathe motor shown in Fig. 12-21 is $P=4.5$ kW. When the machine is in stable operation, the rotation speed of the spindle is $n=42$ r/min. Suppose that the power lost due to friction during rotation is 30% of the input power, determine the maximum cutting force at this speed. If rotation speed of the spindle $n=112$ r/min, what is the maximum value of cutting force? The diameter of the workpiece $d=100$ mm.

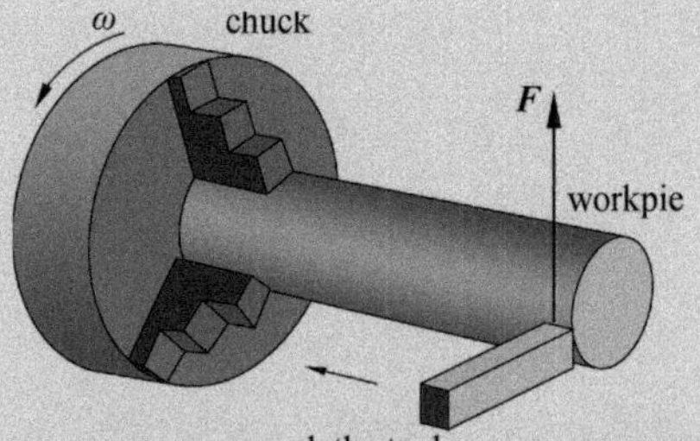

Fig. 12-21

Solution:

When the machine is in stable operation, $\frac{dT}{dt}=0$, then $P_{\text{input}}=P_{\text{useful}}+P_{\text{useless}}$, here the $P_{\text{input}}=4.5$ kW, and the useful power is calculated as

$$P_{\text{useful}}=(1-0.3)P_{\text{input}}=0.7\times 4.5=3.15\ \text{kW}$$

The useful power required to cut the workpiece, therefore

$$P_{useful}=\boldsymbol{F}\cdot\boldsymbol{v}=F\cdot\frac{d}{2}\omega=F\cdot\frac{d}{2}\cdot\frac{2\pi n}{60}$$

$$F=\frac{60}{\pi dn}P_{useful}$$

When $n=42$ r/min, substituting the $P_{useful}=3.15$ kW, and $d=100$ mm to above equation, yield

$$F=\frac{60}{\pi dn}P_{useful}=\frac{60}{\pi\times0.1\times42}\times3.15=14.3\text{ kN} \qquad \textit{Ans.}$$

When $n=112$ r/min, we have

$$F=\frac{60}{\pi dn}P_{useful}=\frac{60}{\pi\times0.1\times112}\times3.15=5.36\text{ kN} \qquad \textit{Ans.}$$

Example 12-9

The truck has a mass of 2 000 kg and is traveling at a speed of 25 m/s as shown in Fig. 12-22. When the brakes to all the wheels are applied, determine the power developed by the frictional force and the truck's velocity after the truck has slid 10 m. The coefficient of kinetic friction is $\mu_k=0.35$.

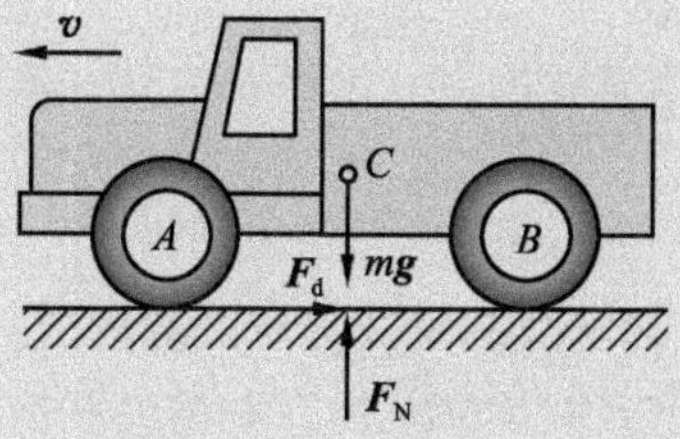

Fig. 12-22

Solution:

The free-body diagram is drawn in Fig. 12-22, the normal $\boldsymbol{F}_N$ and frictional force $\boldsymbol{F}_d$ represent the resultant forces acting on four wheels. Using the equation of equilibrium on y direction.

$$\sum F_y=0,\ F_N-mg=0$$

$$F_N=mg=mg=19.62\text{ kN}$$

The kinetic frictional force $\boldsymbol{F}_d$ is calculated as

$$F_d=F_N\cdot\mu_k=19.62\times0.35=6.876\text{ kN}$$

Using the principle of work and kinetic energy, we have

$$\frac{1}{2}mv_2^2-\frac{1}{2}mv_1^2=\sum W=F_d s$$

$$\frac{1}{2}\times2\,000v_2^2-\frac{1}{2}\times2\,000\times25^2=6.876\times10^3\times10$$

$$v_2 = 26.34 \text{ m/s} \qquad \textit{Ans.}$$

The power of the frictional force at this instant is

$$P = \boldsymbol{F}_d \cdot \boldsymbol{v}_2 = 6.876 \times 10^3 \times 26.34 = 172 \text{ kW} \qquad \textit{Ans.}$$

12.5 Potential Energy and Conservation of Energy

12.5.1 Conservative Force

If the work of a force is independent of the path and depends only on the force's initial and final positions on the path, this force is known as a *conservative force*. Examples of conservative forces are the weight of a particle and the force developed by a spring. The work done by the weight depends only on the vertical displacement of the weight, and the work done by a spring force depends only on the spring's elongation or compression.

In contrast to a conservative force, consider the frictional force exerted on a sliding object by a fixed surface. The work done by the frictional force depends on the path. Consequently, frictional forces are nonconservative. With motion, systems subjected to non-conservative forces dissipate energy which appears as heat.

12.5.2 Potential Energy

Energy is defined as the capacity for doing work. When energy comes from the position of the particle, measured from a fixed datum or reference position, it is called *potential energy*. As a particle or a particle's system moves from a given position M to the reference position M_0, the potential energy, expressed in terms of V, is equal to the work done by the conservative force

$$V = \int_{M_0}^{M} \boldsymbol{F} \cdot \mathrm{d}\boldsymbol{r} \tag{12.24}$$

Thus, potential energy is a measure of the amount of work a conservative force will do when it moves from a given position to the datum. And the potential energy itself is dependent upon a reference position. In mechanics, the potential energy created by gravity (weight) or an elastic spring is important.

(1) Gravitational Potential Energy

If a particle is located a distance y from an arbitrarily selected datum, as shown in Fig. 12-23, the gravitational potential energy of the particle of weight $\boldsymbol{G} = mg$ is

$$V_g = mg(z - z_0) = mgh \tag{12.25}$$

If the distance y is positive upward, the particle's weight $\boldsymbol{G}$ has positive gravitational *potential energy*, V_g, since $\boldsymbol{G}$ has the capacity of doing positive work when the particle is moved back down to the datum. Likewise, if the particle is located a distance y below the datum, V_g is negative since the weight does negative work when the particle is moved back up to the datum. At the datum, the gravitational potential energy, $V_g = 0$.

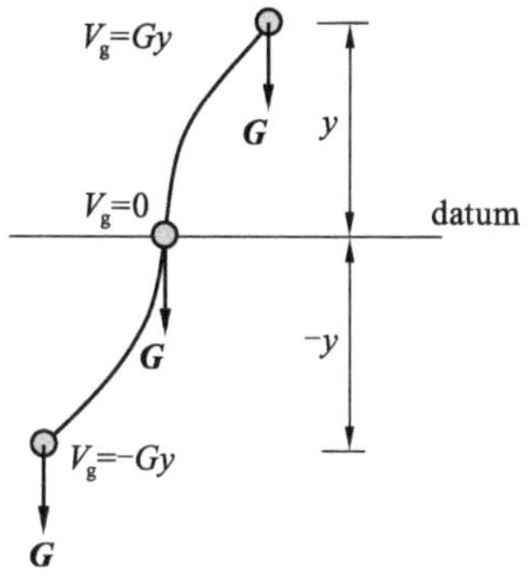

Fig. 12-23

(2) Elastic Potential Energy

When an elastic spring is elongated or compressed a distance s from its unstretched position, elastic potential energy V_e can be stored in the spring. If we select the unstretched position as the reference position, Fig. 12-24, the potential energy is

$$V_e = \frac{1}{2} k s^2 \tag{12.26}$$

Here V_e is always positive, since from the deformed position, the force of the spring always does positive work on the particle when the spring is returned to its unstretched position.

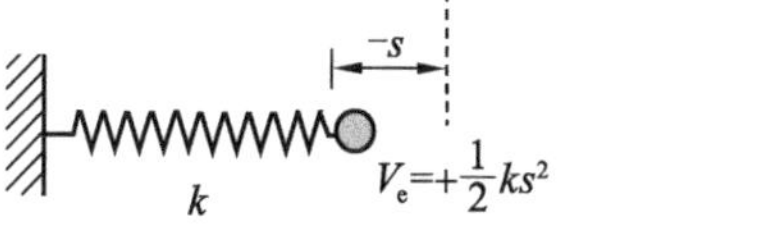

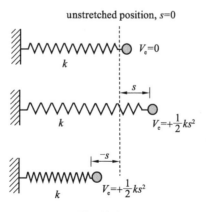

Fig. 12-24

12.5.3 Conservation of Energy

When a system of forces acting on a rigid body consists only of conservative forces, the *conservation of energy theorem* can be used to solve a problem that otherwise would be solved using the principle of work and energy. This theorem is often easier to use since the work of a conservative force is independent of the path and depends only on the initial and final positions of the body.

The work done by a conservative force in the moving particle from one point to another point is expressed as the difference in the body's potential energy,

$$W_{12}=V_1-V_2$$

Using the principle of work and energy

$$T_2-T_1=W_{12}$$

If only conservative forces do work then we have

$$T_2-T_1=V_1-V_2$$

or

$$T_1+V_1=T_2+V_2 \tag{12.27}$$

This equation is referred to as the *conservation of energy*. It states that *during the motion the sum of the particle's kinetic and potential energies remains constant.*

$$T+V=\text{const.} \tag{12.28}$$

For this to occur, kinetic energy must be transformed into potential energy, and vice versa. If the force acting on the system is nonconservative, the *conservation of energy* is no longer valid with such forces. In this case, one needs to use the work-energy theorem.

Example 12-10

The homogeneous rod has mass of $m=10$ kg and length of $l=120$ cm, can rotate about the axis O, shown in Fig. 12-25. The spring is unstretched when the rod is in the initial vertical position ($\theta=0°$). Determine the angular velocity ω_0 initially at which the rod can reach the horizontal position ($\theta=90°$). The spring has a stiffness of $k = 200$ N/m.

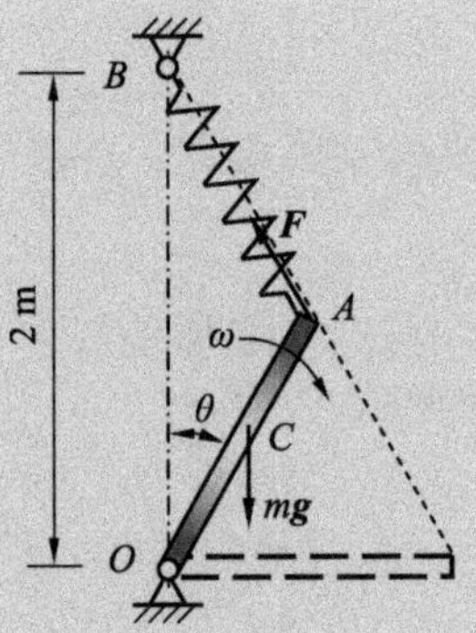

Fig. 12-25

Solution:

For convenience, the datum is established through rod OA. The kinetic energy at initial and finial position is

$$T_1=\frac{1}{2}J_O\omega^2=\frac{1}{6}ml^2\omega^2=2.4\omega^2,\ T_2=0$$

The potential energy at initial and finial position is

$$V_1^r=\frac{1}{2}mgl=\frac{1}{2}\times10\times9.8\times1.2=58.8\ \text{J}$$

$$V_1^e=0,\ V_2^r=0$$

$$V_2^e=\frac{1}{2}k\delta^2=\frac{1}{2}\times200\times[\sqrt{2^2+1.2^2}-(2-1.2)]^2=234.7\ \text{J}$$

Using conservation of energy, we have

$$T_2+V_2^r+V_2^e=T_1+V_1^r+V_1^e$$

$$0+0+234.7=2.4\omega^2+58.8+0$$

Solving, yield

$$\omega=\sqrt{\frac{234.7-58.8}{2.4}}=8.56\ \text{rad/s} \qquad Ans.$$

12.6 Application of the General Theorems of Dynamics

In the previous chapters we have presented the general theorems of dynamics: *theorem of the linear momentum*, *theorem of the angular momentum* and *theorem of the kinetic energy*. The first two theorems are vector theorems using vector notions therefore they can be projected on the axes of a reference system, but the third theorem is a scalar theorem giving us only one equation. These theorems make the join between the change in motion of a body and the forces that cause this change. We can use them to study the motions of the particle, rigid body and systems of particles. For some complex problems, it is sometimes necessary to use several theorems of dynamic to solve all the unknown quantities.

Example 12-11

Both cylinder A and pulley B have radius r and mass m_1, the cylinder is rolling down the incline without sliding and the pulley can rotate around an axis O, shown in Fig. 12-26a. The block C of mass m_2 shown is attached to a cord which is cross the pulley B and attached to the center A of the cylinder. Determine the acceleration of the cylinder center A, tension of the cord AB and reaction force of constraint O. Neglect the mass of the cord.

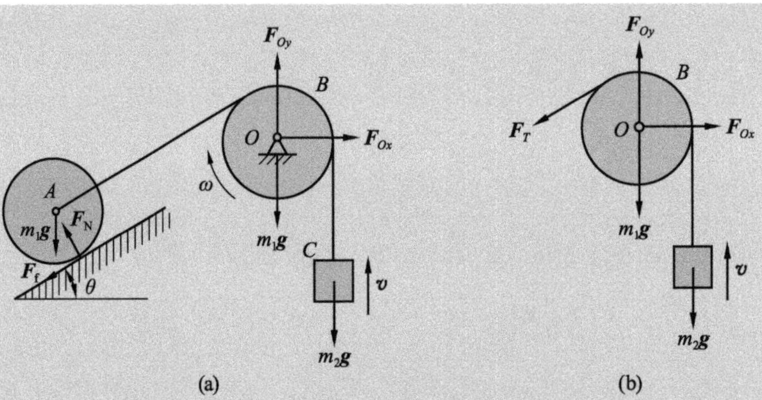

Fig. 12-26

Solution:

Firstly, we consider the cylinder, pulley and block as a single system. The free-body diagram of the system is shown in Fig. 12-26a, the normal force $\boldsymbol{F}_N$ does no work since it never undergoes displacement along its line of action. The weight $m_1\boldsymbol{g}$ of pulley, reactions $\boldsymbol{F}_{Ox}$ and $\boldsymbol{F}_{Oy}$ do no work since the center of drum is no displacement while the drum is rotating. The static frictional force $\boldsymbol{F}_f$ also does no work as it rolls without slipping over a rough surface. The weight $m_1\boldsymbol{g}$ of cylinder do positive work and weight $m_2\boldsymbol{g}$ do negative work when the cylinder moves downward, then the power of external force acting on the system can be calculated as

$$P = m_1 g\sin\theta \cdot v - m_2 g \cdot v$$

The kinetic energy of system at any instant is

$$T = \frac{1}{2}m_1 v_A^2 + \frac{1}{2}J_A\omega_A^2 + \frac{1}{2}J_O\omega_B^2 + \frac{1}{2}m_2 v_C^2 \tag{a}$$

Using the methods of kinematics, we have

$$v_A = v_C = v = r\omega_B,\ \omega_A = \frac{v_A}{r} = \frac{v}{r} = \omega_B = \omega$$

Substituting to the kinetic energy Eq. (a), we have

$$T = \frac{1}{2}m_1 v^2 + \frac{1}{2}\frac{1}{2}m_1 r^2\left(\frac{v}{r}\right)^2 + \frac{1}{2}\frac{1}{2}m_1 r^2\left(\frac{v}{r}\right)^2 + \frac{1}{2}m_2 v^2$$
$$= \frac{1}{2}(2m_1 + m_2)v^2$$

Using the equation of power $\frac{\mathrm{d}T}{\mathrm{d}t} = P$, we have

$$(2m_1 + m_2)v \cdot a_C = m_1 g\sin\theta \cdot v - m_2 g \cdot v$$

$$a_A = a_C = \frac{(m_1\sin\theta - m_2)g}{2m_1 + m_2} \qquad \textit{Ans.}$$

Secondly, we consider the pulley and block as a single system. The free-body diagram of the system is shown in Fig. 12-26b, the moment of all external forces about point O is

$$M_O = F_T r - m_2 g r$$

The angular momentum of the system about same point O is

$$L_O = J_B \omega + m_2 v r = \frac{1}{2}(m_1 + 2m_2) r v$$

Applying the principle of angular momentum about point O, $\frac{dL_O}{dt} = M_O$, we have

$$\frac{1}{2}(m_1 + 2m_2) r a_A = F_T r - m_2 g r$$

$$F_T = \frac{1}{2}(m_1 + 2m_2) a_A + m_2 g$$

$$= \frac{3m_1 m_2 + (m_1^2 + 2m_1 m_2)\sin\theta}{2(2m_1 + m_2)} g \qquad \textit{Ans.}$$

Thirdly, from the free-body diagram of the system in Fig. 12-26b, the linear momentum of the system is

$$p_x = 0, p_y = m_2 v$$

Applying the principle of linear momentum and impulse in the x, y direction respectively, we have

$$\frac{dp_x}{dt} = \sum F_x, 0 = F_{Ox} - F_T \cos\theta$$

$$\frac{dp_y}{dt} = \sum F_y, \ m_2 a = F_{Oy} - F_T \sin\theta - (m_1 + m_2) g$$

Solving, we get

$$F_{Ox} = \frac{3m_1 m_2 + (m_1{}^2 + 2m_1 m_2)\sin\theta}{2(2m_1 + m_2)} g\cos\theta \qquad \textit{Ans.}$$

$$F_{Oy} = \frac{(m_1^2 + 2m_1 m_2)\sin^2\theta + 5m_1 m_2 \sin\theta - 2m_1^2}{2(2m_1 + m_2)} g + (m_1 + m_2) g \qquad \textit{Ans.}$$

Notes: We can obtain the acceleration of the cylinder center A using the principle of work and kinetic energy.

Example 12-12

The homogenous rod AB has a mass m and length l, shown in Fig. 12-27a, is erecting on smooth horizontal surface. When it falls down from the vertical position with no initial velocity, determine the angular velocity of the rod and reaction force of the surface when the rod falls to the ground.

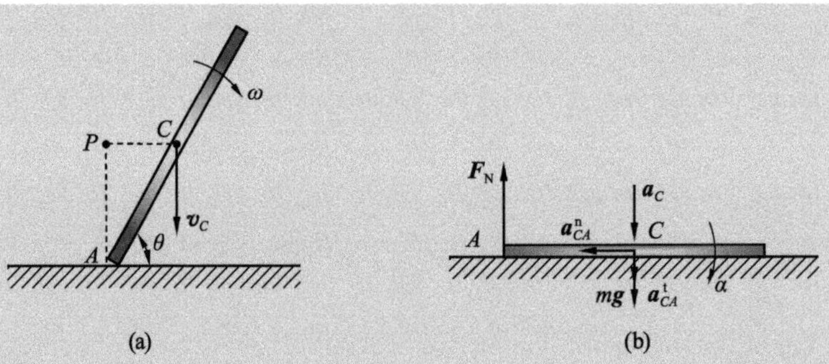

Fig. 12-27

Solution:

The free-body diagram of the rod in arbitrary position is shown in Fig. 12-27a, since the external force along the x axis is zero, the linear momentum will be conserved along the x axis. The mass center C of rod falls vertically. So the instantaneous center of zero velocity for rod AB is located at the intersection P, by the kinematic, we have

$$\omega=\frac{v_C}{CP}=\frac{2v_C}{l\cos\theta}$$

The kinetic energy of rod is

$$T=\frac{1}{2}mv_C{}^2+\frac{1}{2}J_C\omega_2=\frac{1}{2}m\left(1+\frac{1}{3\cos^2\theta}\right)v_C^2$$

Using the principle of work and kinetic energy, we have

$$\frac{1}{2}mv_2^2-\frac{1}{2}mv_1^2=W_{12}$$

$$\frac{1}{2}m\left(1+\frac{1}{3\cos^2\theta}\right)v_C^2=mg\,\frac{l}{2}(1-\sin\theta)$$

When $\theta=0°$, yields

$$v_C=\frac{1}{2}\sqrt{3gl},\ \omega=\sqrt{\frac{3g}{l}} \qquad \textit{Ans.}$$

When the rod falls to the ground, the free-body diagram of the rod is shown in Fig. 12-27b. Applying the principles of linear and angular momentum, we obtain

$$mg-F_N=ma_C$$

$$F_N\,\frac{l}{2}=J_C\alpha=\frac{ml^2}{12}\alpha$$

According to the kinematic diagram drawn in Fig. 12-27b, we can obtain the kinematic relation between a_C and α.

$$a_C=a_{CA}^t=\alpha\,\frac{l}{2}$$

Solving these equations yields

$$F_N=\frac{1}{4}mg \qquad \textit{Ans.}$$

Exercises

12-1 The 10 kg block shown in Fig. 12-28 rests on the smooth incline. If the spring is originally stretched 0.5 m, determine the total work done by all the forces acting on the block when a horizontal force $F = 400$ N pushes the block up the plane $s=2$ m. Set $k=30$ N/m.

12-2 In Fig. 12-29, the pulley of radius $r=0.5$ m can rotate about axis O. The block A of mass $m_A=3$ kg is attached to a cord which is cross pulley O and attached to block B of mass $m_B=2$ kg. If a torque of $M=4\varphi$ (φ is in rad) is acting on the pulley C, determine the total work done by all the forces acting on the system when the angle φ changes from 0 to 2π. Neglect the mass of the cord and any slipping on the spool.

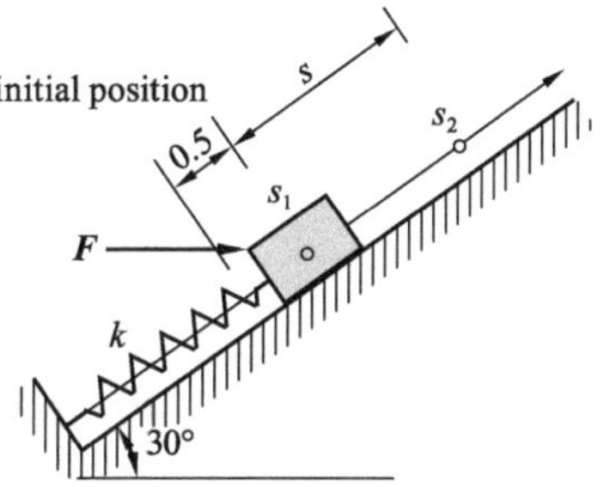

Fig. 12-28

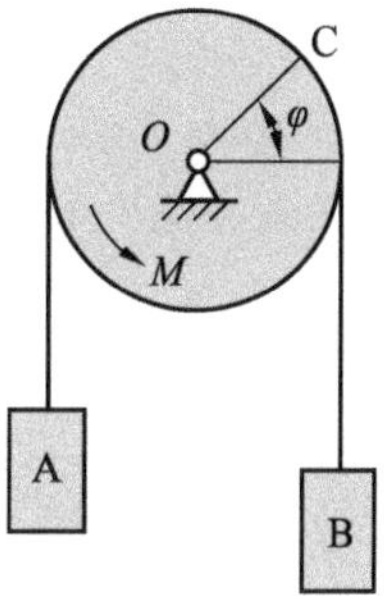

Fig. 12-29

12-3 Determine the kinetic energy of the bodies in Fig. 12-30. The mass of each body is m.

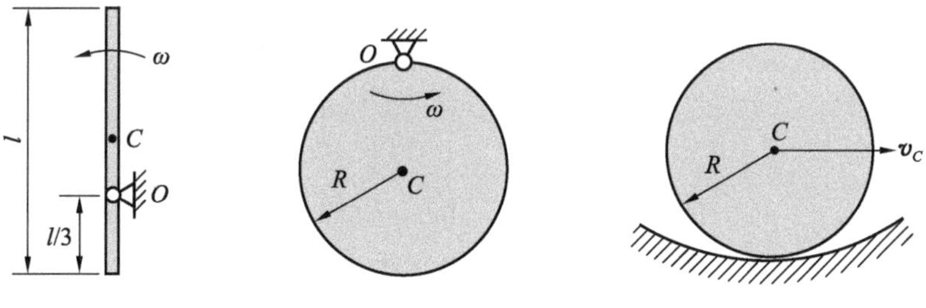

Fig. 12-30

12-4 The tank in Fig. 12-31 moves with velocity of v_0. The track of tank has a mass of m, two homogeneous wheels of tank has the mass of m_1 and the radius of R. The distance between the center of two wheels is πR. Determine the kinetic energy of this system.

12-5 The 100 kg cart in Fig. 12-32 moves along a smooth plane and strikes a linear spring with a speed of $v_0=2$ m/s. Determine the compression δ_{max} of the spring. The spring constant $k=900$ N/cm.

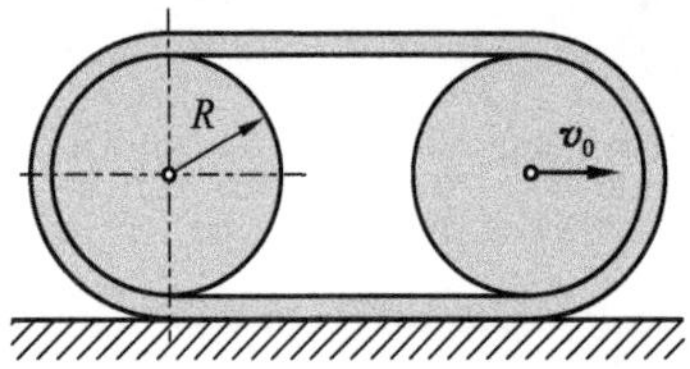

Fig. 12-31

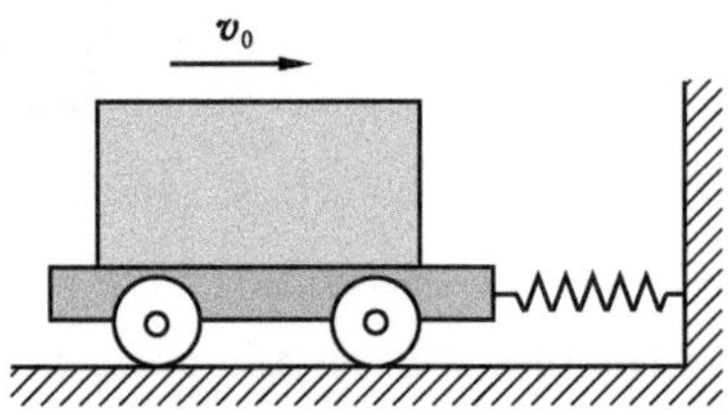

Fig. 12-32

12-6 As shown in Fig. 12-33, the spring in the catapult has an unstretched length of 200 mm. The spring can be compressed 10 mm under the action of force 2 N. When it is compressed 100 mm and releases from rest, the 30 g ball is "fired" from the catapult shown. Determine the speed of the ball when it leaves the catapult barrel.

12-7 The 2 000 kg car in Fig. 12-34 has a velocity of v=100 km/h when the driver sees an obstacle in front of the car. If he locks the brakes, causing the car to skid, determine the distance the car travels before it stops. The coefficient of kinetic friction between the tires and the road is $\mu_d = 0.25$.

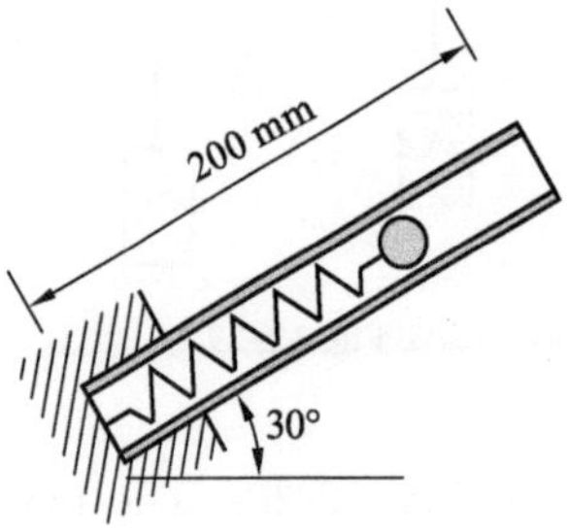

Fig. 12-33

Fig. 12-34

12-8 In Fig. 12-35, the 100 kg block A is released from rest and slides down the rough plane. If the coefficient of kinetic friction between the block and the plane is $\mu_d = 0.25$, determine the compression x of the spring when the block momentarily stops. Set k=2 kN/m.

12-9 In Fig. 12-36, the uniform 5 kg slender rod of length 0.5 m is subjected to a couple moment of M=20 N · m. If the rod is at rest when $\theta=0°$, determine its angular velocity when $\theta=90°$.

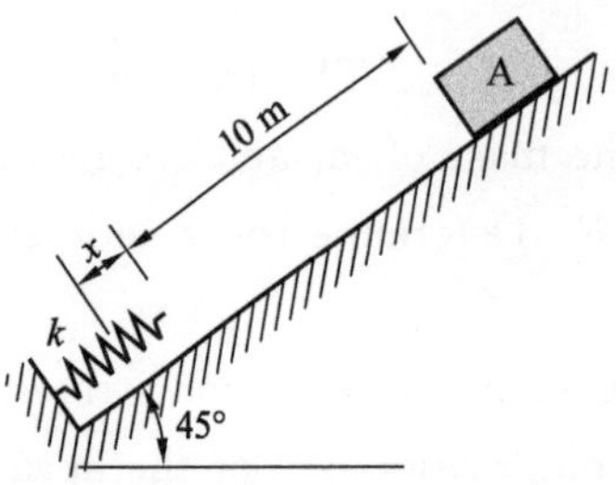

Fig. 12-35

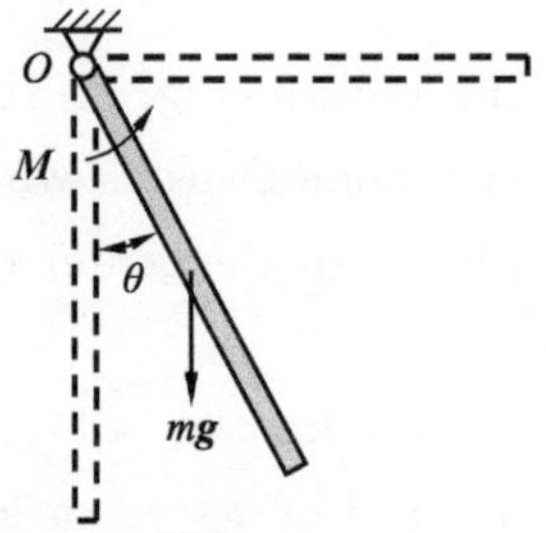

Fig. 12-36

12-10 In Fig. 12-37, the wheel weighs 40 kg and has a radius of $R=0.8$ m. If it is subjected to a clockwise couple moment of $M=15$ N • m and rolls from rest without slipping, determine its angular velocity after its center C moves 0.5 m. The spring has a stiffness $k=10$ N/m and is initially unstretched when the couple moment is applied.

12-11 In Fig. 12-38, the 10 kg rod of length $l=0.8$ m is constrained so that its ends move along the grooved slots. The rod is initially at rest when $\theta=0°$. If the slider block at B is acted upon by a horizontal force $F=50$ N, determine the angular velocity of the rod at the instant $\theta=45°$. Neglect friction and the mass of blocks A and B.

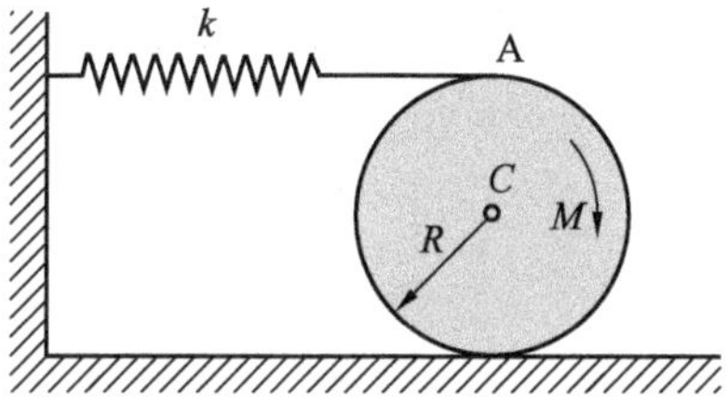

Fig. 12-37

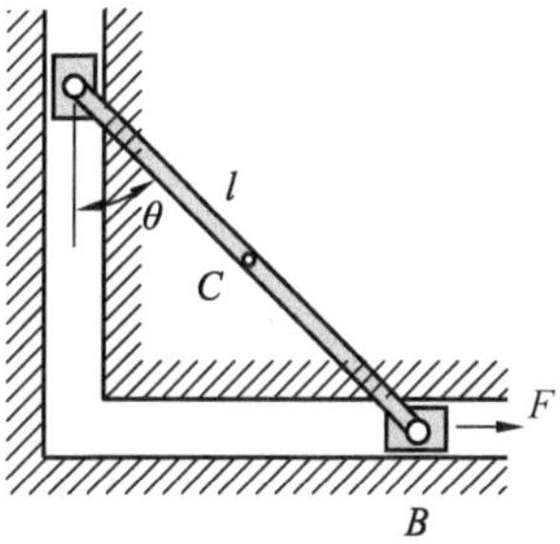

Fig. 12-38

12-12 A smooth 30 kg collar fits loosely on the vertical shaft, Fig. 12-39. The spring with $k=89.6$ N/m is unstretched when the collar is in the horizontal position A. If a constant force $F=250$ N acted on the collar, determine the speed of collar when it moves to the position B from rest at A.

12-13 Two homogeneous pulleys in Fig. 12-40 have same mass m_3 and radius r. The block A has mass m_1, the block B has mass m_2, and $m_2>2m_1-m_3$, determine the velocity of the block B after it fall down height h. Neglect the weight and friction of rope.

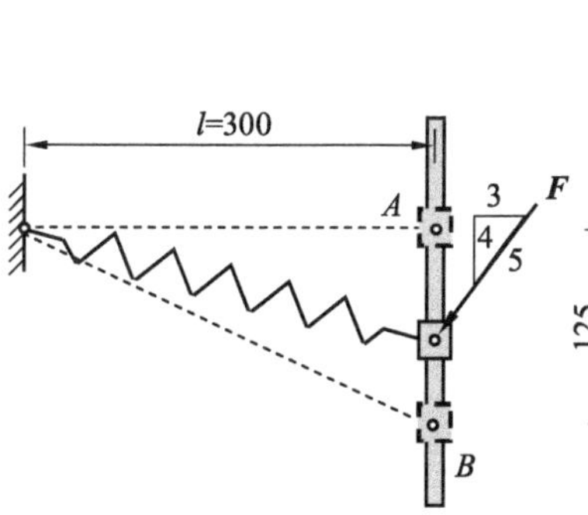

Fig. 12-39

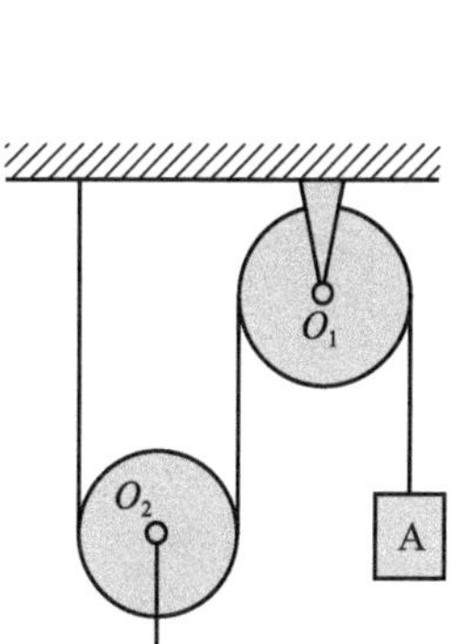

Fig. 12-40

12-14 In Fig. 12-41, the homogeneous wheels B and C have same mass m and radiuses of r. If a constant couple M acted on the wheel B and the conveyor moves from the rest, determine the velocity and acceleration of block A after it moves distance s.

12-15 In Fig. 12-42, the homogeneous wheel A, has mass m_1 and radius R, rolls without slipping on the surface. The pulley C has mass m_2 and radius r. The block B, has mass m_3, is attached to an inextensible cord which is cross the pulley C and attached to the center of the wheel A. If the block B falls a height h, determine the velocity and the acceleration of the center of the wheel A. Neglect the mass of the cord. At the initial instant the system is at rest.

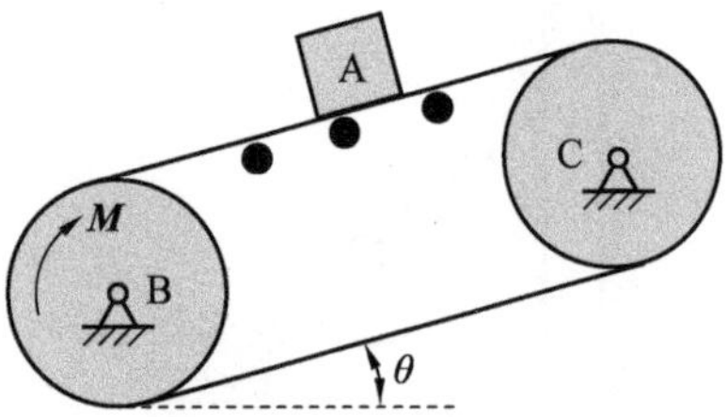

Fig. 12-41

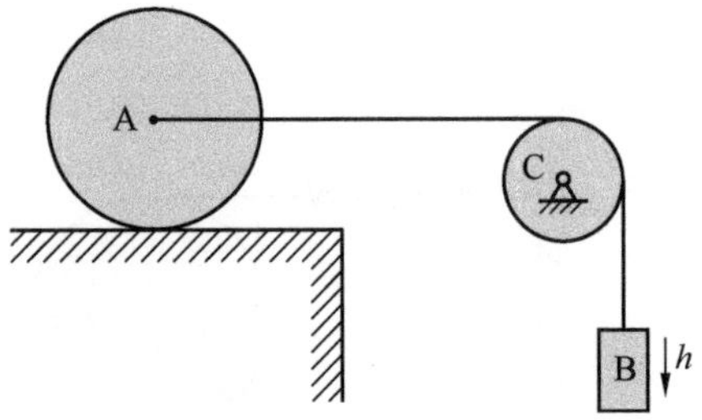

Fig. 12-42

12-16 In Fig. 12-43, the block A has mass m_2, is attached to cord which is wrapped on a drum that has a mass m_1 and radius r. If a constant couple M is acting on the drum, the block moves up the incline from the rest, determine the angular velocity and angular acceleration of the drum after it rotates an angle φ. The coefficient of kinetic friction between the incline and the block is μ_k. Neglect the mass of cord.

12-17 As shown in Fig. 12-44, a force of $F=150$ N is acting on the 50 kg crate. Determine the power supplied by the force when $t=4$ s. The coefficient of kinetic friction between the floor and the crate is $\mu_k=0.2$. Initially the crate is at rest.

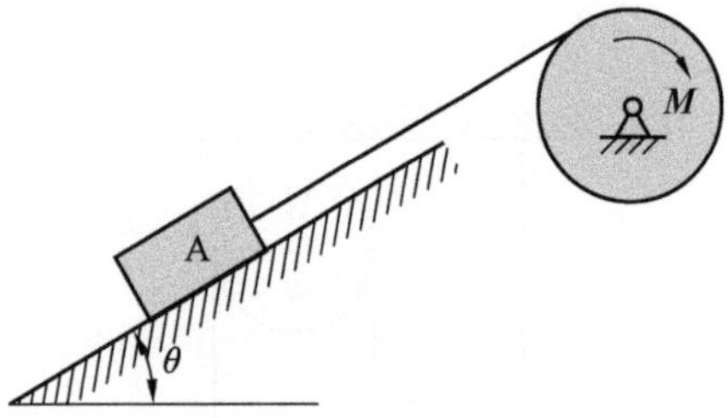

Fig. 12-43

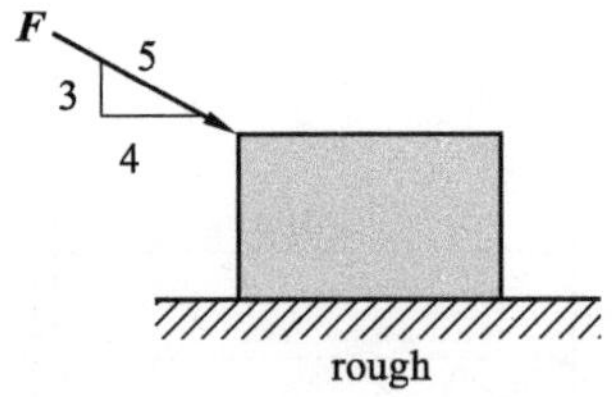

Fig. 12-44

12-18 The car has a mass of 2 000 kg and is traveling at a speed of 25 m/s shown in Fig. 12-45. When the brakes to all the wheels are applied, if the coefficient of kinetic friction is $\mu_k=0.35$, determine the power developed by the frictional force when the car slides. Then find the car's speed after it has slid 10 m.

12-19 In Fig. 12-46, two homogeneous wheels A and B both have mass m_1 and *radius r*. An inextensible cable is wrapped on both wheels, a block C of mass m_2 is attached in the middle of cable and can move on the smooth surface. If the wheel A is subjected to a constant couple $\boldsymbol{M}$, determine the tension of the cable between the wheel A and block C. Neglect the weight and friction of rope.

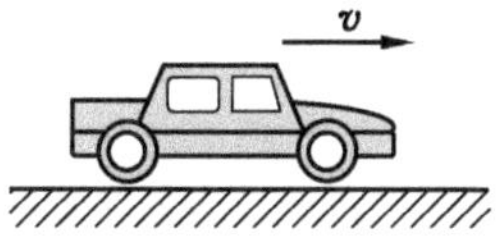

Fig. 12-45

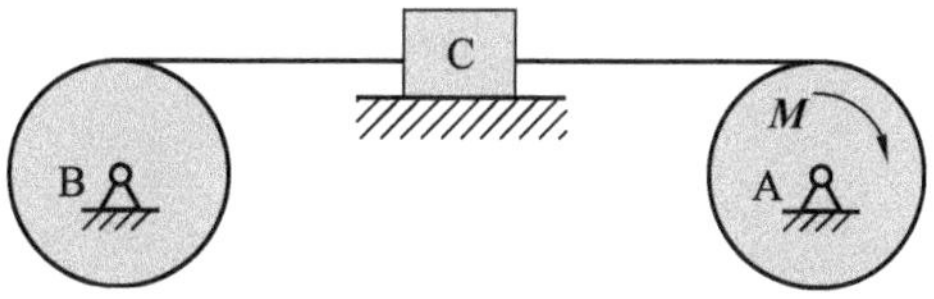

Fig. 12-46

12-20 As shown in Fig. 12-47, the block A of mass m_1 is attached to a cord which is cross pulley C and attached to block B of mass m_2. If the block A slid down the smooth incline, determine the constraint force between the triangle D and barrier E. Neglect the friction and mass of the cord and pulley C.

12-21 The homogeneous rod OA in Fig.12-48, has mass m_1 and length l, can rotates about axis O. A homogeneous disk, has mass m_2 and radius R, can rotate about endpoint A of rod. If the rod and disk is released from the rest of the horizontal position, determine the angular velocity and angular acceleration of the rod at the instant that the include angle between the rod and horizontal is θ.

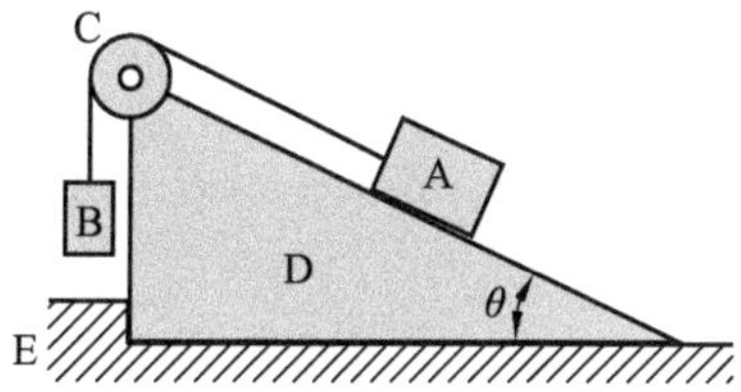

Fig. 12-47

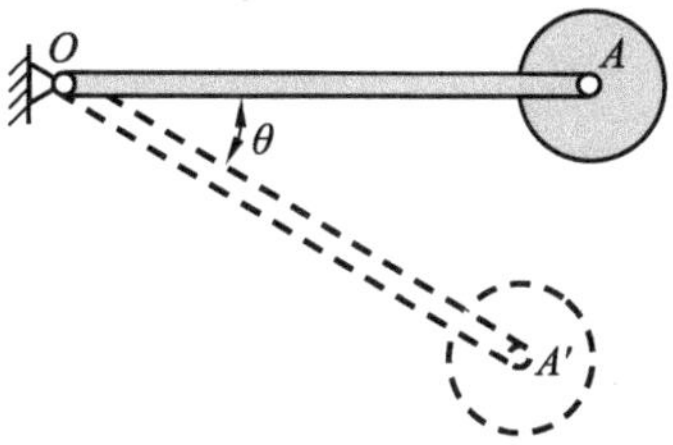

Fig. 12-48

Chapter 13

D'Alembert's Principle

Objectives

In this chapter, we will study the principle of D'Alembert that transforms the dynamical problems into a static problem. Using this theorem, we can obtain a method of studying the motion of the mechanical systems through equilibrium equations. In many cases it is of advantage to use these principles instead of Newton's laws when formulating the equations of motion.

13.1 Inertial Force and D'Alembert's Principle

The dynamical problems can be transformed formally into a static problem. Then they can be solved by the theorem of equilibrium. This method to solve the dynamical problems is called the *dynamic-static method*.

13.1.1 The Inertial Force

Suppose a free particle P of mass m is subjected to the force $\boldsymbol{F}$ and performing a motion with the instantaneous acceleration $\boldsymbol{a}$. Based on the Newton's second law of motion we may write:

$$\boldsymbol{F}=m\boldsymbol{a} \tag{13.1}$$

where $\boldsymbol{F}$ is the resultant of all forces acting on the particle. In accordance with the Newton's third law (principle of action and the reaction). The particle, on which is acting the force $\boldsymbol{F}$, will act on the system that has produced the force $\boldsymbol{F}$ with a force equal in magnitude but with opposite sense. This force is called ***inertia force*** and defined as

$$\boldsymbol{F}_{\mathrm{I}}=-m\boldsymbol{a} \tag{13.2}$$

We can see that this *inertia force* does not act on the particle P, it acts on the mechanical system that have produced the force $\boldsymbol{F}$. For example, a small ball with a mass of m attached to the end of the rope moves in a circle in the horizontal plane, as shown in Fig. 13-1. This ball in the horizontal plane is only affected by the tension force $\boldsymbol{F}$ of the rope. It is this force that changes the ball's motion state, resulting in centripetal acceleration $\boldsymbol{a}_n$. The force $\boldsymbol{F}=m\boldsymbol{a}_n$ is called the *centripetal force*. The force exerted by the ball on the rope, $\boldsymbol{F}_I=-\boldsymbol{F}=-m\boldsymbol{a}_n$, is due to the ball's inertia, which is trying to maintain its original state of motion. This *inertial force* is opposite to the direction of the acceleration $\boldsymbol{a}$, and deviates from the center of the circle. In conclusion the inertia force for the particle is a *fictional* force.

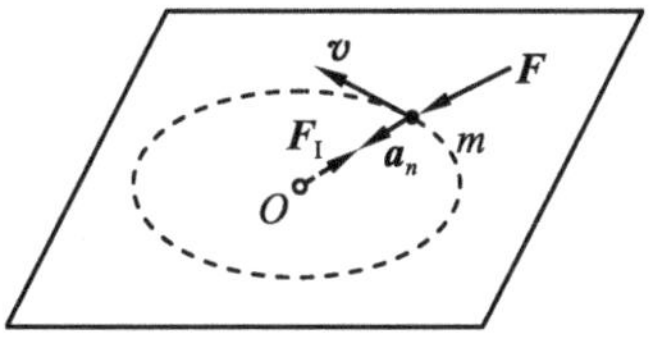

Fig. 13-1

13.1.2 D'Alembert's Principle for a Particle

Suppose a particle P of mass m, having simple ideal constraints and moving with the instantaneous acceleration $\boldsymbol{a}$, is acted by active resultant force $\boldsymbol{F}$ and the reaction resultant force with $\boldsymbol{F}_N$ as shown in Fig. 13-2. According to the Newton's second law,

$$m\boldsymbol{a}=\boldsymbol{F}+\boldsymbol{F}_N \tag{13.3}$$

Binging all the terms in the left part, we now rewrite Eq. (13. 3) in the form

$$\boldsymbol{F}+\boldsymbol{F}_N-m\boldsymbol{a}=0 \tag{13.4}$$

By introducing the *inertial force* Eq. (13. 2), we can formally transform the law of motion Eq. (13. 3) to the equilibrium condition.

$$\boldsymbol{F}+\boldsymbol{F}_N+\boldsymbol{F}_I=0 \tag{13.5}$$

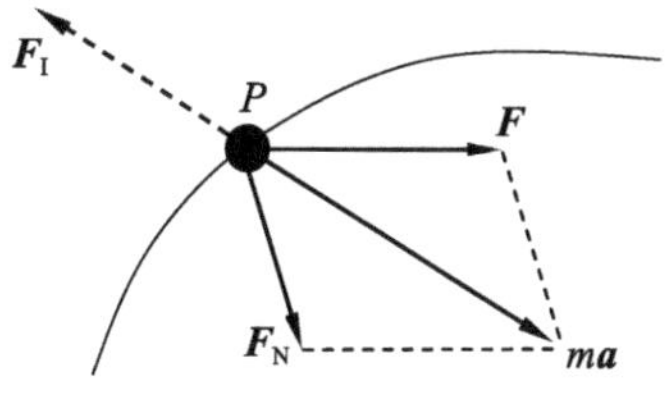

Fig. 13-2

This relation expresses the D'Alembert's principle: *The active forces, the reaction forces and the inertia forces form a system in equilibrium*. Since the particle is not at rest but undergoes a motion, this is often referred to as a state of dynamic equilibrium. Note: the *inertia force* is **not** a real force for the particle P because as we have seen in the previous section this force acts on the bodies that have produced the active forces and not on the particle P. This means that the equilibrium expressed by the *D'Alembert's* *principle* is also a *fictional* equilibrium called *dynamic equilibrium*. And thereby converting problems of dynamics to problems of statics.

D'Alembert's principle provides a new method for studying the dynamics problems, which is to use the static method to deal with the dynamics problem. This procedure may be advantageous when setting up the equations of motion. If we want to apply this method to a problem, the inertial force $\boldsymbol{F}_\mathrm{I}$ must be drawn into the free-body diagram in addition to the real forces. The equations of motion then are given by the condition "sum of all forces is equal to zero".

Example 13-1

A single pendulum is hung in a carriage moving in a horizontally straight line. When the train is moving at uniform acceleration $\boldsymbol{a}$, the pendulum has a deflecting angle θ with the vertical line, as shown in Fig. 13-3a. Find the relation between the acceleration $\boldsymbol{a}$ of the train and the deflecting angle θ.

Fig. 13-3

Solution:

The free-body diagram of the sphere is shown in Fig. 13-3b, the sphere is subjected to the weight $\boldsymbol{G}=mg$, and tensile force $\boldsymbol{F}_\mathrm{T}$. The inertia force $\boldsymbol{F}_\mathrm{I}$ having the direction of the acceleration $\boldsymbol{a}$ of the train but with opposite sense, its magnitude is

$$F_\mathrm{I}=ma$$

The dynamic equilibrium equation on axis x will be:

$$\sum F_x = 0, \quad mg\sin\theta - F_\mathrm{I}\cos\theta = 0$$

Substituting $F_\mathrm{I}=ma$ into above equation, solving yield

$$a=g\tan\theta \qquad \textit{Ans.}$$

We can see that the angle θ varies with the acceleration $\boldsymbol{a}$ of the train. This is the working principle of the pendulum accelerometer.

13.1.3 D'Alembert's Principle for the System of Particles

Let us consider a system of particles, every particle m_i, is subject to both active resultant force $\boldsymbol{F}_i$ as well as reaction resultant force $\boldsymbol{F}_{Ni}$. *For each particle we may write the same kind of relation expressing the dynamic equilibrium,*

$$\boldsymbol{F}_i + \boldsymbol{F}_{Ni} + \boldsymbol{F}_{Ii} = 0 \quad (i = 1, 2, \cdots, n) \tag{13.6}$$

If each particle can be subjected to the action of external forces $\boldsymbol{F}^{(e)}$ and internal forces $\boldsymbol{F}^{(i)}$ coming from other (even all) particles of the considered system of particles, the dynamic equilibrium can be written as,

$$\boldsymbol{F}_i^{(e)} + \boldsymbol{F}_i^{(i)} + \boldsymbol{F}_{Ii} = 0 \quad (i = 1, 2, \cdots, n) \tag{13.7}$$

which follow directly from Newton's second law. Let us introduce the notions of main force vector of external forces, inertia forces and internal forces, and the main moment of the force vector in the following form,

$$\begin{aligned} &\sum \boldsymbol{F}_i^{(e)} + \sum F_i^{(i)} + \sum \boldsymbol{F}_{Ii} = 0 \\ &\sum \boldsymbol{M}_O(\boldsymbol{F}_i^{(e)}) + \sum \boldsymbol{M}_O(\boldsymbol{F}_i^{(i)}) + \sum \boldsymbol{M}_O(\boldsymbol{F}_{Ii}) = 0 \end{aligned} \tag{13.8}$$

For a system of particles, pairs of internal forces $\boldsymbol{F}_i$ can be canceled out each other, and thus do not appear in the equilibrium equation. Eq. (13.8) is reduced to

$$\begin{aligned} &\sum \boldsymbol{F}_i^{(e)} + \sum \boldsymbol{F}_{Ii} = 0 \\ &\sum \boldsymbol{M}_O(\boldsymbol{F}_i^{(e)}) + \sum \boldsymbol{M}_O(\boldsymbol{F}_{Ii}) = 0 \end{aligned} \tag{13.9}$$

We can note that the sums of vector products occurring above represent the main moment of force vectors of the system of external forces $\boldsymbol{F}^{(e)}$ and of the system of inertia forces $\boldsymbol{F}_I$. In this way the system of particles remains in equilibrium under the action of inertia forces and external forces, and the moments of forces due to the aforementioned forces. The obtained result Eq. (13.9) is summarized in the following principle,

A system of vectors consisting of inertia forces, external forces, and their torques is equivalent to zero.

Using the principle of D'Alembert we can obtain a method to study the motion of the mechanical systems through equilibrium equations.

Example 13-2

The pulley of radius r, the mass m is uniformly distributed on the rim, can rotate about axis O, shown in Fig. 13-4. The block A of mass m_1 is attached to a cord which is cross pulley and attached to block B of mass m_2 ($m_1 > m_2$). Determine the acceleration of the block. Neglect the mass of the cord and any slipping on the pulley.

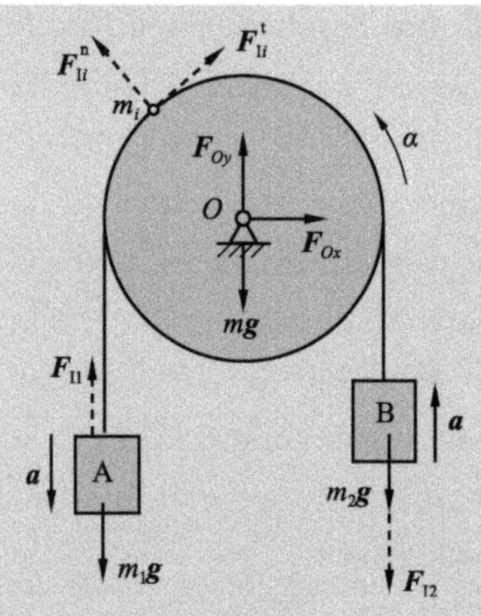

Fig. 13-4

Solution:

We consider the pulley and block as a single system. The free-body diagram of the system is shown in Fig. 13-4, the system is subjected to the weights mg, $m_1 g$, and $m_2 g$. Since the weight mg and constraints F_{Ox}, F_{Oy} pass through the O axis, the moments created by these forces are zero about this axis. The direction of inertia force F_{I1}, F_{I2}, and F^n_{Ii}, F^t_{Ii} of each particle are shown in figure. Their magnitudes are calculated as:

$$F_{I1} = m_1 a,\ F_{I2} = m_2 a$$

$$F^t_{Ii} = m_i a_t = m_i r\alpha = m_i a$$

$$F^n_{Ii} = m_i a_n = m_i \omega^2 r$$

Since the inertia force F^n_{Ii} pass through the O axis, the moments created by this force is zero about this axis. The dynamic equilibrium equation about axis O will be:

$$\sum M_O = 0, (m_1 g - F_{I1} - m_2 g - F_{I2})r - \sum F^t_{Ii} r = 0$$

$$(m_1 g - m_1 a - m_2 g - m_2 a)r - \sum m_i a r = 0$$

Solving yield

$$a = \frac{m_1 - m_2}{m_1 + m_2 + m} g \qquad Ans.$$

13. 2 Simplification of the System of Inertial Forces

Any rigid body can be considered as a continuous and rigid system of particles, so the dynamic equilibrium will be expressed by the same kinds of relations, but the inertia forces are acting in each point of the body and consequently we have to determine the force-couple system of the inertia forces acting about a rigid body. We have obtained the main force and main moment of the inertia force,

$$\sum F_i^{(e)} + \sum F_I = 0$$

$$\sum M_O(F_i^{(e)}) + \sum M_O(F_I) = 0$$

If we express the motion of the mass center C of the rigid body using the theorem of the linear momentum and the theorem of the angular momentum, we shall obtain the following two vector equations,

$$m a_C = \sum F_i^{(e)}$$

$$\frac{dL_C}{dt} = \sum M_C(F_i^{(e)})$$

Comparing the two pairs of vector equations (those resulted from D'Alembert's principle and those from the two general theorems), we obtain the two vectors of the force couple system of the inertia forces,

$$m a_C = -\sum F_I = F_{IR}$$

$$\frac{dL_C}{dt} = -\sum M_C(F_I) = M_{IC} \quad (13.10)$$

13.2.1 Translation

When the rigid body is doing translation, Fig. 13-5a taking center of mass C as simplified center, the acceleration of any point in the body at any instant is the same as the acceleration of the center of mass, a_C. Using the Eq. (13.10), we have

$$F_{IR} = -m a_C$$

$$M_{IC} = J_C \alpha = 0 \quad (13.11)$$

When the rigid body is in translation, the simplified result of the inertial force system is through the center of mass C, which is equal to the product of the mass of the rigid body and the acceleration of the center of mass, and the direction is opposite to that of the acceleration of the center of mass C, as shown in Fig. 13-5b.

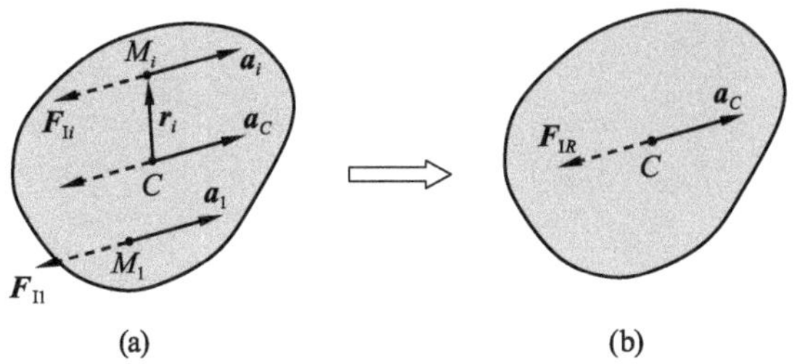

Fig. 13-5

13.2.2 Rotating about a Fixed Axis

For a rigid body rotating around a fixed axis, we study only the case that a homogeneous rigid body has a symmetric plane and the rotation axis is perpendicular to the symmetric plane, Fig. 13-6a. If we take the rotation axis O as the simplified center, the main vector of the inertial force system is

$$\boldsymbol{F}_{IR} = -m\boldsymbol{a}_C = -m(\boldsymbol{a}_C^t + \boldsymbol{a}_C^n)$$

The main moment of the inertial force system to the point O is

$$M_{IO} = \sum M_O(\boldsymbol{F}_I) = \sum M_O(\boldsymbol{F}_I^t) = -\sum m_i r_i \alpha \cdot r_i$$
$$= -\left(\sum m_i r_i^2\right)\alpha = -J_O \alpha$$

The negative sign expresses that it is opposite to the direction of α. Therefore, when the rigid body rotates about a fixed axis, the simplified result of the inertial force system is

$$\boldsymbol{F}_{IR} = -m\boldsymbol{a}_C = -m(\boldsymbol{a}_C^t + \boldsymbol{a}_C^n)$$
$$M_{IO} = -J_O \alpha \tag{13.12}$$

This states that when a rigid body has a mass symmetric plane and rotates about a fixed axis perpendicular to the symmetry plane, the result of the simplification of the inertial force system towards the rotation axis is a force and a moment. The force is equal to the product of the mass of the rigid body and the acceleration of the center of mass. The moment of the force couple is equal to the product of the moment of inertia of the rigid body about the axis of rotation and the angular acceleration, and the rotation is opposite to the angular acceleration, as shown in Fig. 13-6b.

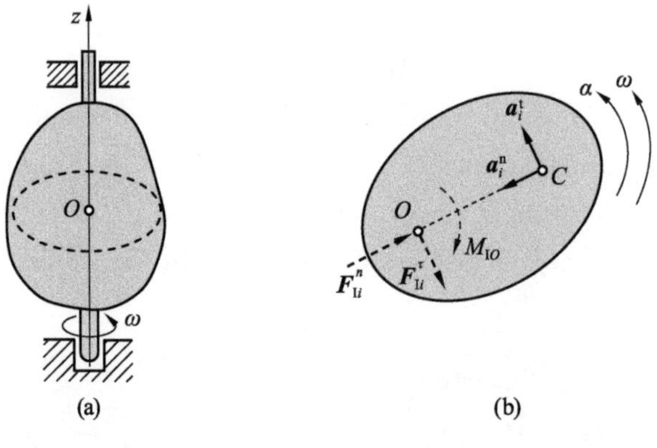

Fig. 13-6

The following special cases can further simplify the problem:

① The rotation axis passes the center of mass, whose angular acceleration is α, and the inertia force system is simplified as a couple, as shown in Fig. 13-7a.

$$F_{IR} = 0,\ M_{IC} = -J_C\alpha$$

② When the rigid body rotates at the uniform angular velocity (angular acceleration $\alpha = 0$), but the rotation axis does not pass the center of mass C, the inertial force system is simplified as a resultant force, as shown in Fig. 13-7b.

$$F_{IR} = -ma_C^n,\ M_{IC} = 0,$$

③ When the rotation axis passes through the center of mass and the angular acceleration equals 0, the main vector and main moment of the inertial force system are both zero, that is, the inertial force system is the equilibrium force system, as shown in Fig. 13-7c.

$$F_{IR}=0,\ M_{IC}=0$$

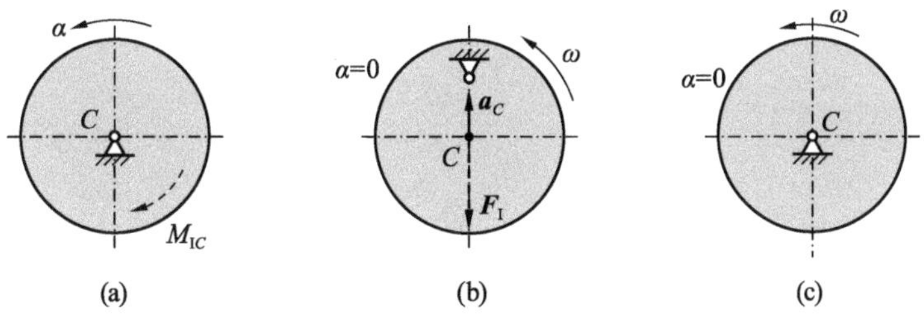

Fig. 13-7

13.2.3 Plane Motion

When a rigid body is in general plane motion, we study only the case that the rigid body has a symmetric plane, the special system of inertial forces can be simplified as the system of coplanar force in this plane, shown in Fig. 13-8. We take the mass center C as simplified center, at this point, the simplified main forces and main moment of the inertial force system toward the center of mass C is

$$\begin{aligned} \boldsymbol{F}_{IR} &= -m\boldsymbol{a}_C \\ M_{IC} &= -J_C\alpha \end{aligned} \tag{13.13}$$

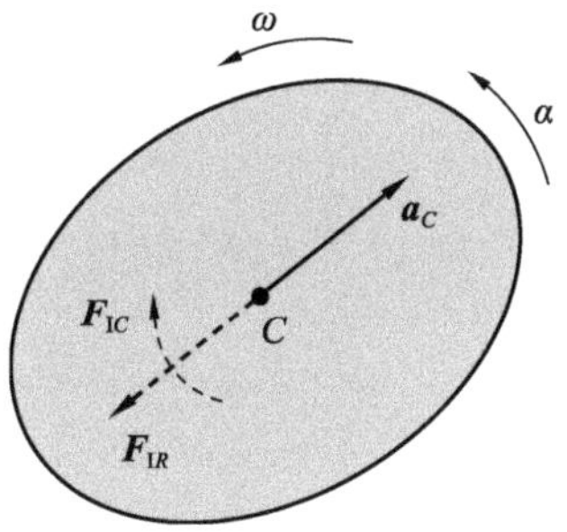

Fig. 13-8

The inertial force system of a rigid body with a symmetric plane which is plane motion can be simplified as a force and a couple in the symmetric plane. The force, which passes through the center of mass, is equal to the product of the mass of the rigid body and the acceleration of the center of mass, and the direction is opposite to the acceleration of mass center. The moment of the couple is equal to the product of the moment of inertia about the center axis and the angular acceleration and the rotation is opposite to the angular acceleration.

Example 13-3

A homogeneous rod with length l and mass m is rotating about the end axis O with angular velocity ω and angular acceleration α, shown in Fig. 13-9. Determine the results of the simplification of the inertial forces about rotation point O.

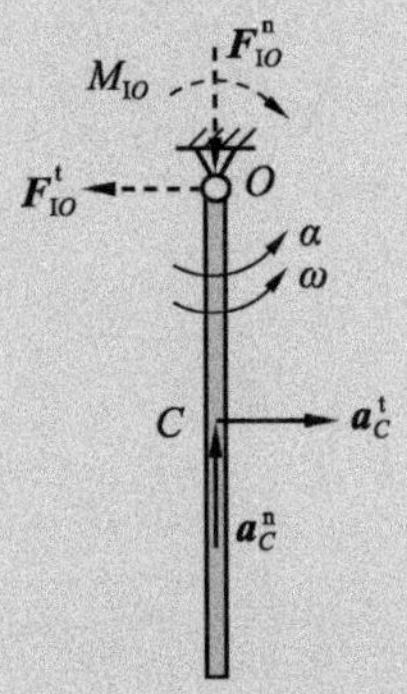

Fig. 13-9

Solution:

The rod rotates about a fixed axis perpendicular to the page and passing through point O. The direction of inertia force F_{IO}^n and F_{IO}^t are opposite to that of the acceleration of the center of mass C, shown in Fig. 13-9. The inertia couple M_{IO} is opposite to the angular acceleration of the rod. Their magnitudes are calculated as:

$$F_{IO}^t = ma_C^t = m\frac{l}{2}\alpha \qquad Ans.$$

$$F_{IO}^n = ma_C^n = m\frac{l}{2}\omega^2 \qquad Ans.$$

$$M_{IO} = J_O\alpha = \frac{1}{3}ml^2\alpha \qquad Ans.$$

Example 13-4

The slender rod AB, has mass m and length $l=1$ m, is at horizontal by a pin connected at endpoint A and suspending at its end B by a cord, as shown in Fig. 13-10a. If the cord is cut off suddenly, determine the angular acceleration of rod AB and reaction forces of the support A at this instant.

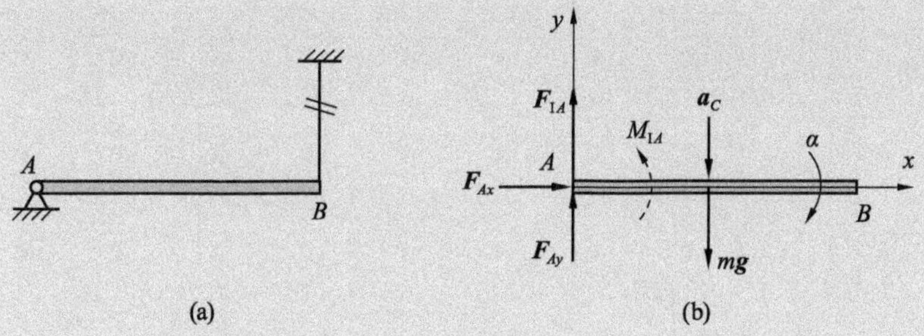

Fig. 13-10

Solution:

When the rope is cut off suddenly, the angular velocity of the rod AB is equal to zero at this instant, and the angular acceleration is α. The free-body diagram of the rod AB is shown in Fig. 13-10b, the rod is subjected to the weights mg, and constraint forces $\boldsymbol{F}_{Ax}$, $\boldsymbol{F}_{Ay}$. The direction of inertia force $\boldsymbol{F}_{IA}$ is opposite to the acceleration of the center of mass C, the inertia couple M_{IA} is opposite to the angular acceleration of the rod, shown in Fig. 13-10b. Their magnitudes are calculated as:

$$F_{IA}=m\cdot\frac{l}{2}\alpha$$

$$M_{IA}=J_A\cdot\alpha=\frac{1}{3}ml^2\alpha$$

The dynamic equilibrium equation will be:

$$\sum F_x=0,\ F_{Ax}=0$$

$$\sum F_y=0,\ F_{Ay}+F_{IA}-mg=0$$

$$\sum M_A=0,\ M_{IA}-mg\frac{l}{2}=0$$

Solving yield

$$F_{Ax}=0,\ F_{Ay}=\frac{1}{4}mg,\ \alpha=\frac{3}{2l}g \qquad Ans.$$

Example 13-5

The block of mass m_1 is attached to cord which is wrapped around the central hub of spool that has a mass m_2 and radius of gyration ρ about its mass center, shown in Fig. 13-11a. If a driving moment M is acting on the spool, determine the acceleration of the block and the constraint forces of support O.

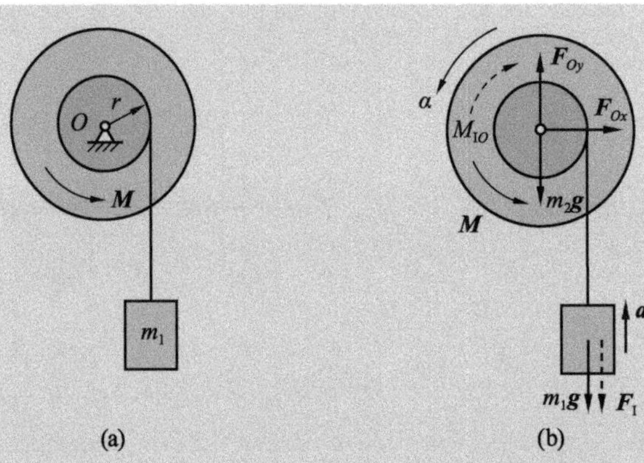

Fig. 13-11

Solution:

Here we will consider the spool and block as a single system. The free-body diagram of the system is shown in Fig. 13-11b, the system is subjected to moment M, the weights $m_1\boldsymbol{g}$, $m_2\boldsymbol{g}$ and constraint forces $\boldsymbol{F}_{Ox}$, $\boldsymbol{F}_{Oy}$. The direction of inertia force $\boldsymbol{F}_I$ is opposite to the acceleration $\boldsymbol{a}$, the inertia couple M_{IO} is opposite to the angular acceleration α of the spool, shown in Fig. 13-11b. Their magnitudes are calculated as:

$$F_{IA}=m_1 a$$

$$M_{IO}=J_O\alpha=m_2\rho^2\frac{a}{r}$$

All forces form a system in dynamic equilibrium, the dynamic equilibrium equation will be:

$$\sum F_x=0,\ F_{Ox}=0$$

$$\sum F_y=0,\ F_{Oy}-m_1 g-m_2 g-F_I=0$$

$$\sum M_O=0,\ M-M_{IO}-m_1 gr-F_I r=0$$

Solving yield

$$a=\frac{(M-m_1 gr)r}{m_1 r^2+m_2\rho^2} \qquad Ans.$$

$$F_{Ox}=0 \qquad Ans.$$

$$F_{Oy}=(m_1+m_2)g+m_1 a=(m_1+m_2)g+\frac{(M-m_1 gr)m_1 r}{m_1 r^2+m_2\rho^2} \qquad Ans.$$

Example 13-6

A homogeneous disk has a mass m and radius r. An inextensible cord is wrapped around its periphery and the other end of the cord is fastened at A, in Fig. 13-12a. If the disk releases from rest, determine the tension of cord and the acceleration of mass center of the disk at an any instant of the motion. Neglect the weight of the cord.

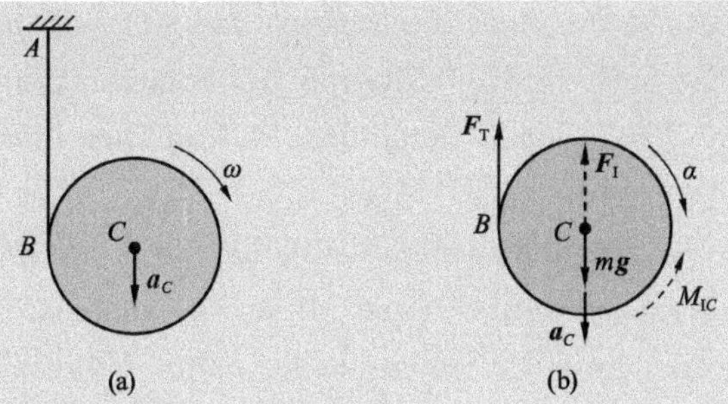

Fig. 13-12

Solution:

The free-body diagram of the disk is shown in Fig. 13-12b, the system is subjected to the weight mg and tensile force $\boldsymbol{F}_T$. The direction of inertia force $\boldsymbol{F}_I$ is opposite to the acceleration $\boldsymbol{a}_C$ of the center of mass C, the inertia couple M_{IC} has opposite sense as the angular acceleration α of the disk. Their magnitudes are calculated as,

$$\boldsymbol{F}_{IR}=-m\boldsymbol{a}_C$$

$$M_{IC}=-J_C\alpha$$

Because the vertical cord is fixed in the point A, consequently the point B is the instantaneous center of the disc. We have the kinematic relation,

$$\alpha=a_C/r$$

The dynamic equilibrium equations will be:

$$\sum F_y=0,\ F_T-mg+F_I=0$$

$$\sum M_C=0,\ M_{IC}-F_Tr=0$$

Solving these equations yields

$$a_C=\frac{2}{3}g \qquad \textit{Ans.}$$

$$F_T=\frac{1}{3}mg \qquad \textit{Ans.}$$

Exercises

13-1 Determine the simplified result of the inertial force system of the member, Fig. 13-13. The mass of each body is m.

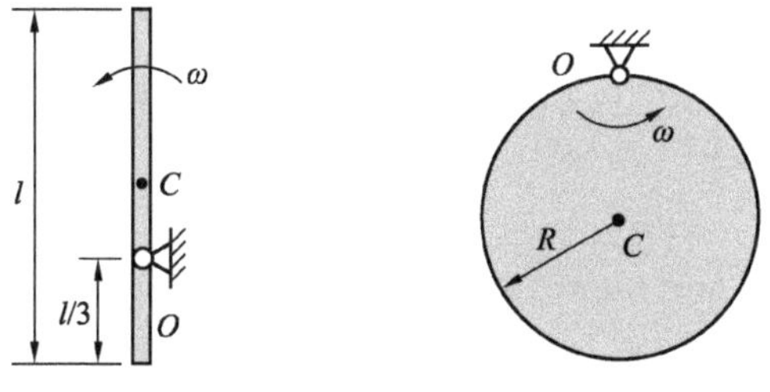

Fig. 13-13

13-2 In Fig. 13-14, in the plane mechanism, rod $AD /\!/ BC$ and $AD=BC=l$, the homogenous rod AB has mass of m. If the rod AD rotates about axis D with angular velocity ω and angular acceleration α, determine the simplified result of the inertial force system of rod AB.

13-3 In Fig. 13-15, the homogeneous disk has radius R and mass m, is rolling without slipping along the straight surface. If at any instant, the disk has angular velocity ω and angular acceleration α, determine the simplified results of the inertial force system about the center of mass C.

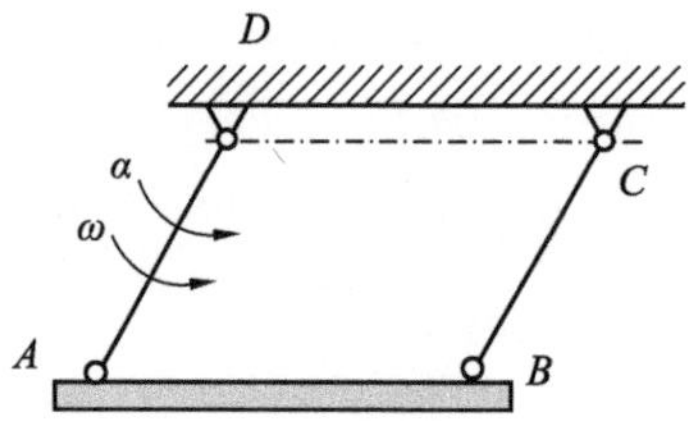

Fig. 13-14

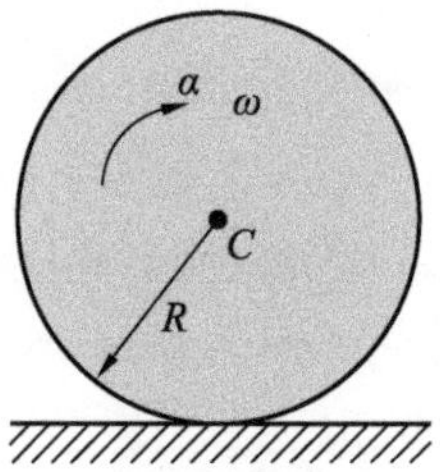

Fig. 13-15

13-4 The truck in Fig. 13-16 has a mass m. The height of mass center C is h, the distance between the mass center and front wheel and back wheel are b and c, respectively. If the truck moves along the straight rough surface with acceleration $\boldsymbol{a}$, determine the vertical reaction force of front wheel and back wheel, respectively.

13-5 A small ball in Fig. 13-17 has the mass $m=0.1$ kg and is attached to the cord of length $l=0.3$ m. The end of cord is connected at the point O. If the ball travels around the circular path and makes the cord have an angle $\theta=60°$ with the vertical line OC, determine the velocity of the small ball and the tensile force of the cord. Neglect air resistance and the size of the ball.

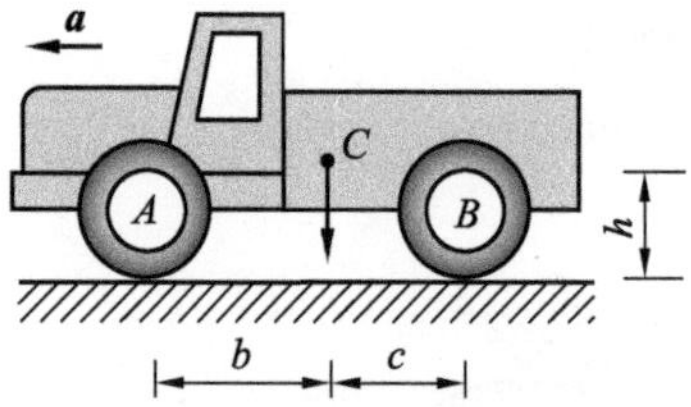

Fig. 13-16

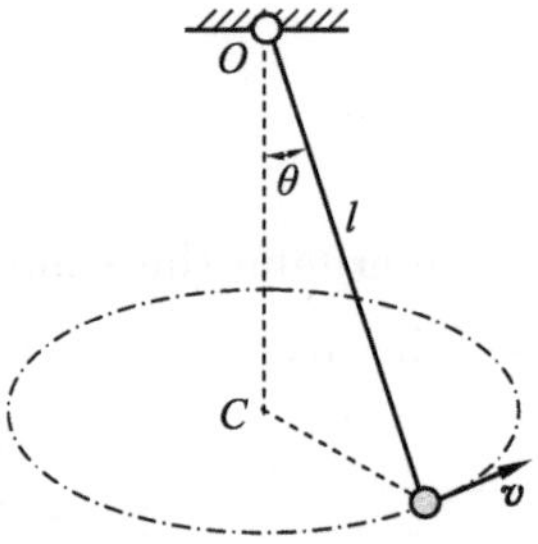

Fig. 13-17

13-6 In Fig. 13-18, the homogenous rod AB has mass of $m=4$ kg, placed on the smooth horizontal surface. If the end B of the rod is subjected to a horizontal force $F=60$ N, determine the acceleration $\boldsymbol{a}$ and the angle θ between the AB rod and surface.

13-7 In Fig. 13-19, the homogeneous rod AB has a mass of m and the length l, and is hung by two ropes. When one rope breaks suddenly, determine the acceleration of the center of mass C and the tension of the other rope.

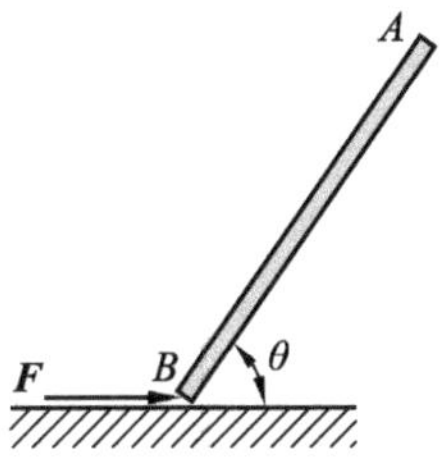

Fig. 13-18

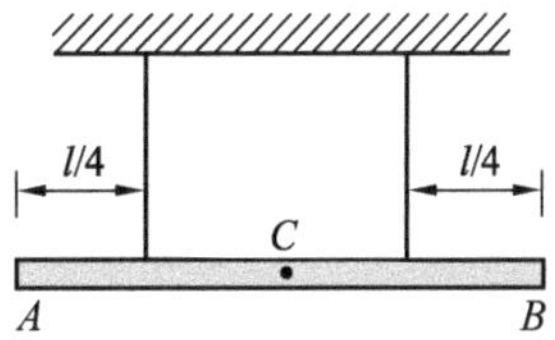

Fig. 13-19

13-8 In Fig. 13-20, the block A of mass $m_1 = 10$ kg is attached to a cord which is cross a negligible pulley and wrapped on the wheel C that has a mass $m_2 = 20$ kg and radius R. If the block A releases from the rest, determine the acceleration of mass center C of the wheel. Neglect the mass of cord. Assume that the wheel rolls without slipping.

13-9 In Fig. 13-21, the pulley B has mass m_2 and radius r. The block A, has mass m_1, is attached to an inextensible cord which is wrapped on the pulley. If the block A falls from the rest, determine the reaction forces in the fixed support C. Neglect the mass of the cable and friction.

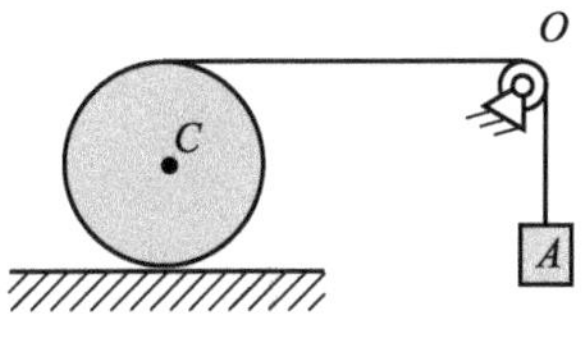

Fig. 13-20

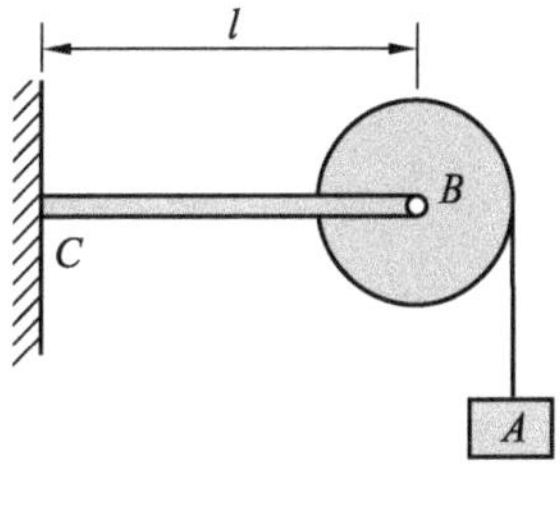

Fig. 13-21

13-10 The homogeneous disk in Fig. 13-22 has mass m and radius r, is hinged at O and supported at point B. If the support B is suddenly removed, determine the acceleration of the mass center C and the constraint reaction at hinge O using the dynamic static method.

13-11 The masses of two blocks in Fig. 13-23 are m_1 and m_2 ($m_1 > m_2$), respectively. They are hanging on two coaxial tub wheels with radius r and R. The moment of inertia of the two wheels with respect to the axis of rotation O is J_O. If they are released from rest, determine the angular acceleration of the wheels and reaction forces of support O.

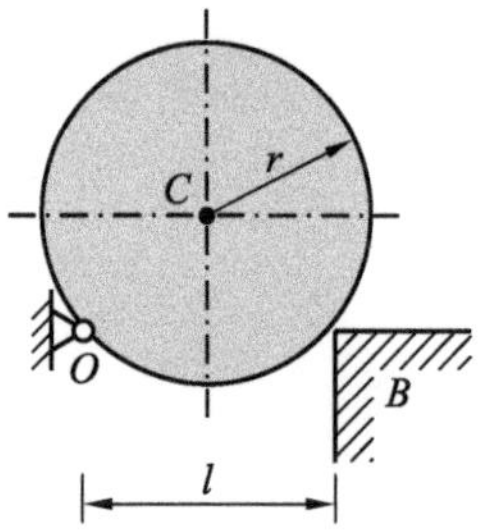

Fig. 13-22

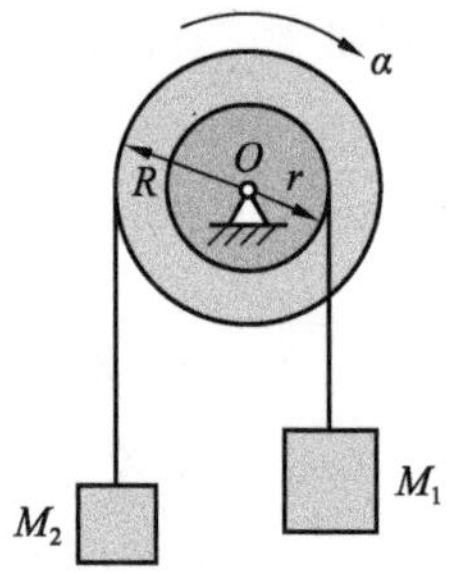

Fig. 13-23

13-12 In Fig. 13-24, the crank OA has mass of m_1 and length of r, rotates courtclockwise about axis O with constant angular velocity ω. The slider rod BC of mass m_2 is at vertical up-down motion. Determine the reaction forces of support O and couple moment $\boldsymbol{M}$ at the instant $\theta = 30°$.

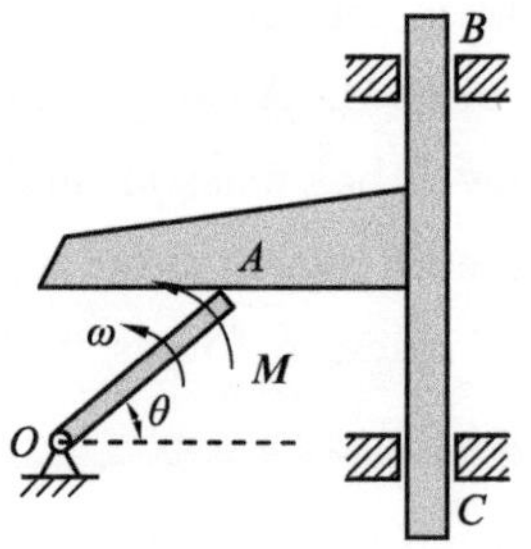

Fig. 13-24

Chapter 14

Principle of Virtual Work

Objectives

In this chapter, the concept of virtual work with virtual displacements is introduced, and the principle of virtual work only the equilibrium in the state of rest of the systems is derived and illustrated. And we will study the applications of the principle of virtual work. It provides an alternative method for solving problems involving the equilibrium of a particle, a rigid body, or a system of connected rigid bodies.

14.1 Virtual Displacements and Virtual Work

14.1.1 Virtual Displacements

A real displacement of particles constituting a system happens over a finite time. But a *virtual* or an *imagined* displacement of a body in static equilibrium is instantaneous and is consistent with the constraints on the system, which indicates a displacement or rotation that is *assumed* and *does not* actually exist, as shown in Fig. 14-1. That is, *virtual displacements* are displacements (or rotations) that are: ① fictitious, i. e. , do not exist in reality; ② infinitesimally small; ③ geometrically (kinematically) admissible, i. e. , consistent with the constraints of the system. In order to distinguish virtual displacements from real displacements $d\boldsymbol{r}$, we denote virtual displacements as $\delta \boldsymbol{r}$, i. e. , with the symbols δr and $\delta\theta$ respectively, taken from the calculus of variations.

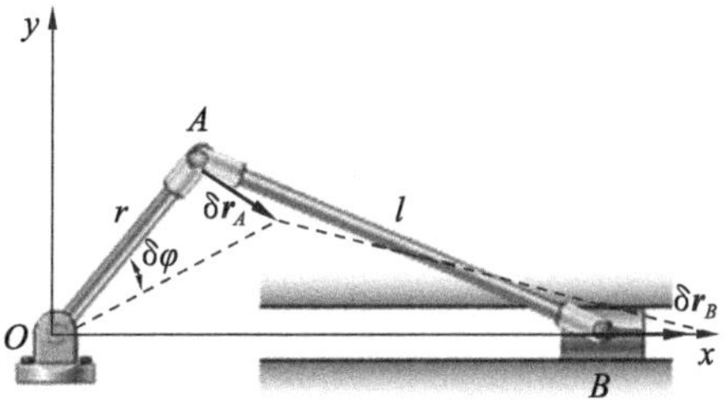

Fig. 14-1

14.1.2 Virtual Work

Consider now an *imaginary* or *virtual movement* of a body in static equilibrium, the virtual *work* done by a force having a virtual displacement δr is

$$\delta W = \boldsymbol{F} \cdot \delta \boldsymbol{r} = F\cos\theta\delta r \tag{14.1}$$

Similarly, when a couple undergoes a virtual rotation $\delta\varphi$ in the plane of the couple forces, the virtual work is

$$\delta W = M\delta\varphi \tag{14.2}$$

We now consider the two-sided lever as shown in Fig. 14-2a and calculate the work done during a virtual displacement. A virtual displacement, i. e., a deflection that is consistent with the constraints of the system, is a rotation with an angle $\delta\varphi$ about the support A, Fig. 14-2b. The total virtual work δW done by the two forces F_1 and F_2 is

$$\delta W = F_1 a\delta\varphi - F_2 b\delta\varphi = (F_1 a - F_2 b)\delta\varphi$$

The minus sign in the second term takes into account that the force F_2 acts in opposite direction of the virtual deflection $b\delta\varphi$. In the equilibrium state, the expression in parentheses disappears due to the equilibrium condition of moments $F_2 b = F_1 a$. Therefore, the virtual work vanishes in the equilibrium position, $\delta W = 0$. It should be noted that only the *external* forces F_1 and F_2 enter into the virtual work, whereas the reaction force in A does not contribute.

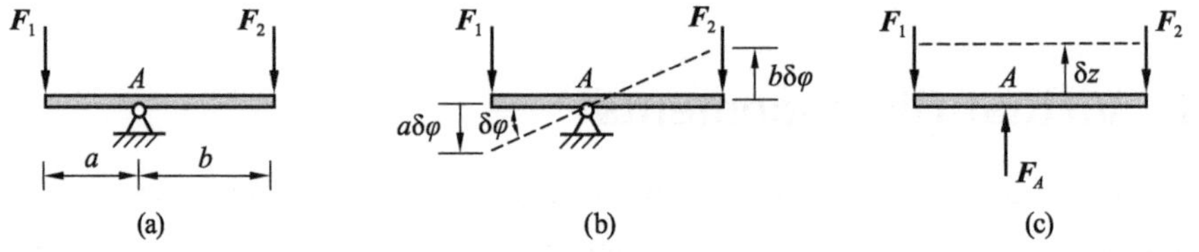

Fig. 14-2

14.1.3 Ideal Constraints

In the vector mechanics, the constraints were classified about more criteria:

① Simple constraints (removing one degree of freedom) or multiple (removing more degrees of freedom).

② Unilateral (eliminating the displacement in one sense of a direction) or bilateral (eliminating the possibilities of displacements in the both senses of a direction).

③ Ideal constraints (punctual and frictionless) or real (with friction). In this book all the constraints will be considered simple, bilateral and ideal constraints.

Consider virtual displacement of the particle P. This displacement being compatible with the constraint represents the simple constraint, therefore it will be perpendicular on the normal reaction force corresponding to the constraint. In this way we have

$$\delta W_{\mathrm{N}} = \boldsymbol{F}_{\mathrm{N}} \cdot \delta \boldsymbol{r} = 0 \tag{14.3}$$

Meaning that *the virtual work of the reaction forces of the ideal constraints is always equal to zero*. This statement and relation represents the first form of the principle of virtual work. During a virtual displacement of the system, the total virtual work of the constraint forces vanishes:

$$\delta W = \sum \delta W_{Ni} = \sum \boldsymbol{F}_{Ni} \cdot \delta \boldsymbol{r}_i = 0 \tag{14.4}$$

The constraints such as smooth fixed surface, smooth hinge, two force member, non-extendable flexible cable and fixed end are all ideal constraints.

14.2 Principle of the Virtual Work

Consider an arbitrary system (subjected to ideal constraints) consisting of particles, *for each particle*, in equilibrium, acted by applied forces $\boldsymbol{F}_i$ as well as the reaction forces $\boldsymbol{F}_{Ni}$ of ideal constraints, with which we can express the equilibrium of the particle.

$$\boldsymbol{F}_i + \boldsymbol{F}_{Ni} = 0 \tag{14.5}$$

We multiply of the particle with the virtual displacement $\delta \boldsymbol{r}_i$, the vector equation of equilibrium is

$$\boldsymbol{F}_i \cdot \delta \boldsymbol{r}_i + \boldsymbol{F}_{Ni} \cdot \delta \boldsymbol{r}_i = 0 \tag{14.6}$$

so that the total work done of the system of particles is

$$\sum \boldsymbol{F}_i \cdot \delta \boldsymbol{r}_i + \sum \boldsymbol{F}_{Ni} \cdot \delta \boldsymbol{r}_i = 0 \tag{14.7}$$

Since the reaction forces of ideal constraints (e. g. normal reaction, tension, rigid body constraints etc.) do not do any work, Eq. (14.4).

The Eq. (14.7) can be rewritten as

$$\sum \boldsymbol{F}_i \cdot \delta \boldsymbol{r}_i = 0 \quad \text{or} \quad \sum \delta W_{Fi} = 0 \tag{14.8}$$

namely the final form of the *principle of virtual work*: *the necessary and sufficient condition as a particle having ideal constraints to be in equilibrium is that the virtual work of the active forces to be equal to zero for any virtual displacement of the body*.

The principle of virtual work is also often referred to as the *principle of virtual displacements*.

It analogously is valid for *rigid bodies*, where the forces can be reduced in a few points, through elementary transformations is obtained the equilibrium conditions under the form,

$$\delta W = \sum \boldsymbol{F}_i \cdot \delta \boldsymbol{r}_i + \sum \boldsymbol{M}_i \cdot \delta \varphi = 0 \tag{14.9}$$

where M_i represents the concentrated couples and $\delta\varphi$ is the virtual rotation of the body. The principle of virtual work can be derived from the equilibrium conditions, conversely, the equilibrium conditions can be derived from the principle of virtual work. Therefore, the entire statics can be based either on the equilibrium conditions or on the principle of virtual work.

Example 14-1

The simply supported beam AB of span 5 m in Fig. 14-3a is carrying a load of $F=2$ kN at a distance 3 m from end A. Determine the reactions at pin end A and B, by using the principle of the virtual work.

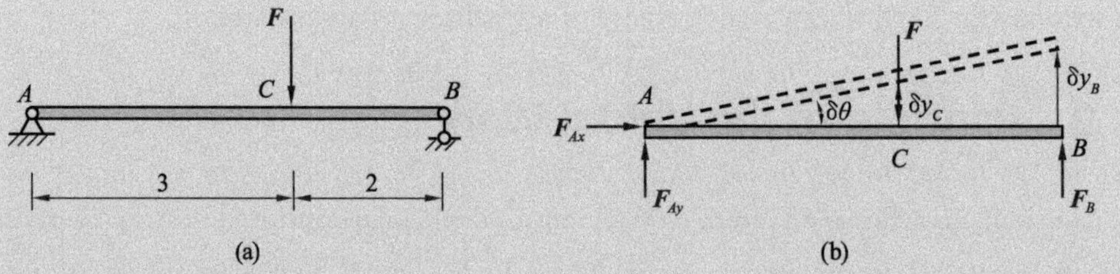

Fig. 14-3

Solution:

The free-body diagram in Fig. 14-3b, the origin of coordinates is established at the fixed pin support A, the beam is given a virtual rotation $\delta\theta$ about point A, only the active forces $\boldsymbol{F}$, and the reactive force $\boldsymbol{F}_B$ do work (The reactive forces $\boldsymbol{F}_{Ax}$ and $\boldsymbol{F}_{Ay}$ are fixed.). The virtual upward displacements of the beam at point B and at point C are

$$\delta y_B = 5 \cdot \delta\theta,\ \delta y_C = 3 \cdot \delta\theta \tag{a}$$

The force $\boldsymbol{F}_B$ would do positive work since the force and its corresponding displacements have the same sense. The force $\boldsymbol{F}$ would do negative work since the force and its corresponding displacements have the opposite sense. Hence, the virtual-work equation for the displacement $\delta\theta$ is

$$\delta W = F_B \delta y_B - F \delta y_C = 0 \tag{b}$$

Substituting Eq. (a) into Eq. (b) in order to relate the virtual displacements to the common virtual displacement $\delta\theta$ yields

$$F_B(5\delta\theta) - F(3\delta\theta) = (5F_B - 3F)\delta\theta = 0$$

Notice that $\delta\theta \neq 0$, excluding $\delta\theta$ and solving, yields

$$F_B = 1.2 \text{ kN} \qquad \textit{Ans.}$$

And

$$F_{Ay} = F - F_B = 2 - 1.2 = 0.8 \text{ kN} \qquad \textit{Ans.}$$

Example 14-2

A screw press loaded with a couple $M=2Fl$ is depicted in Fig. 14-4. The height of the screw-thread is h. Determine the pressure on body. The screw press moves without friction when turned.

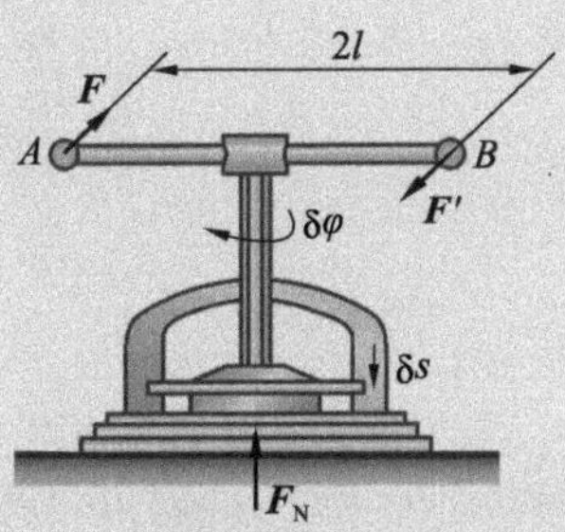

Fig. 14-4

Solution:

When the screw rotates a virtual displacement $\delta\varphi$ and the press plate undergoes a virtual displacement δs, according to the principle of virtual work, the system is in equilibrium when the total work done by force $\boldsymbol{F}_N$ and torque M vanishes:

$$\sum \delta W = -F_N \delta s + M\delta\varphi = 0$$

Here, the directions of M and φ coincide, whereas $\boldsymbol{F}_N$ and δs have opposite directions. The virtual displacements $\delta\varphi$ and δs are not independent. A rotation of $\Delta\varphi=2\pi$ will raise the screw by the height $\Delta s=h$ of the screw-thread. Therefore,

$$\frac{\delta\varphi}{2\pi}=\frac{\delta s}{h}$$

Thus, the principle of virtual work yields

$$\sum \delta W = \left(2Fl - \frac{F_N h}{2\pi}\right)\delta\varphi = 0$$

With $\delta\varphi \neq 0$, excluding $\delta\varphi$ and solving, yields

$$2Fl - \frac{F_N h}{2\pi} = 0$$

$$F_N = \frac{4\pi l}{h}F \qquad \textit{Ans.}$$

Example 14-3

Determine the horizontal force $\boldsymbol{F}_{Bx}$ that the fixed pin at B must exert on rod BD in order to hold the mechanism shown in Fig 14-5a in equilibrium. Neglect the weight of the members.

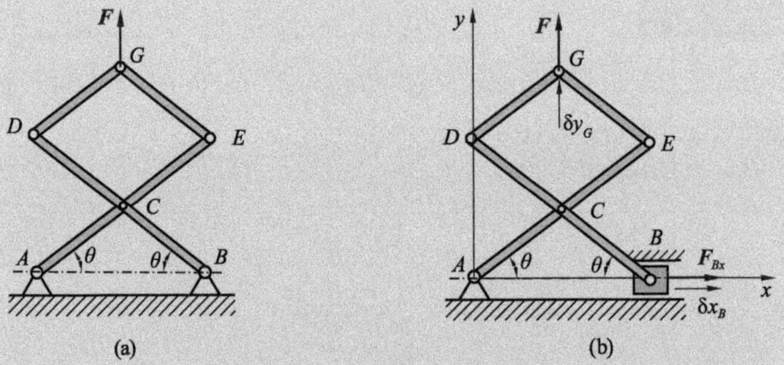

Fig. 14-5

Solution:

The reaction $\boldsymbol{F}_{Bx}$ can be obtained by *releasing* the pin constraint at B in the x direction and allowing the frame to be displaced in this direction, as shown in Fig. 14-5b, the origin of coordinates is established at the fixed pin support A, when θ undergoes a *positive* virtual displacement $\delta\theta$, only $\boldsymbol{F}_{Bx}$, and $\boldsymbol{F}$ do work. The position of $\boldsymbol{F}$ and $\boldsymbol{F}_{Bx}$ can be specified by the position coordinates y_G and x_B. Expressing these position coordinates in terms of θ and taking the derivatives yields:

$$x_B = 2l\cos\theta,\ \delta x_B = -2l\sin\theta\delta\theta \tag{a}$$

$$y_G = 3l\sin\theta,\ \delta y_G = 3l\cos\theta\delta\theta \tag{b}$$

Since the forces and their corresponding virtual displacements would have the same sense. Hence, the virtual-work equation is

$$\delta W_F = F_{Bx}\delta x_B + F\delta y_G = 0 \tag{c}$$

Substituting Eq. (a) and Eq. (b) into Eq. (c) in order to relate the virtual displacements to the common virtual displacement $\delta\theta$ yields

$$[F_{Bx}\cdot(-2l\sin\theta) + F\cdot 3l\cos\theta]\delta\theta = 0$$

Since $\delta\theta \neq 0$, excluding $\delta\varphi$ and solving, yields

$$-F_{Bx}\cdot 2l\sin\theta + F\cdot 3l\cos\theta = 0$$

$$F_{Bx} = \frac{3}{2}F\cot\theta \qquad \textit{Ans.}$$

Note: In the case of a rigid system, one or several supports are removed and replaced by the support reactions. These reactions then will be considered to be external loads and consequently taken into account in the principle of virtual work.

Example 14-4

Two frictionless blocks of mass m each are connected by a massless rigid rod AB. The system is constrained to move in the vertical plane. Determine the required force $\boldsymbol{F}$ in Fig. 14-6 needed to maintain equilibrium.

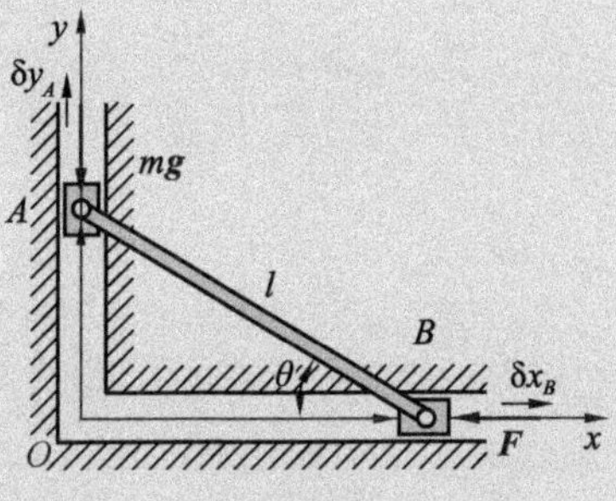

Fig. 14-6

Solution:

The origin of coordinates is established as shown in Fig. 14-6, when θ undergoes a virtual displacement $\delta\theta$, the gravity $\boldsymbol{mg}$ and force $\boldsymbol{F}$ do work. The position of $\boldsymbol{F}$ and $\boldsymbol{mg}$ can be specified by the position coordinates x_B and y_A. Expressing these position coordinates in terms of θ and taking the derivatives yields:

$$x_B = l\cos\theta,\ \delta x_B = -l\sin\theta\delta\theta \tag{a}$$

$$y_A = l\sin\theta,\ \delta y_A = l\cos\theta\delta\theta \tag{b}$$

Since the forces and their corresponding virtual displacements would have the opposite sense. Hence, the virtual-work equation is

$$\delta W = -F\delta x_B - mg\delta y_A = 0 \tag{c}$$

Substituting Eq. (a) and Eq. (b) into Eq. (c) in order to relate the virtual displacements to the common virtual displacement $\delta\theta$ yields

$$F \cdot l\sin\theta - mg \cdot l\cos\theta = 0$$

$$F = mg\cot\theta \qquad \textit{Ans.}$$

Example 14-5

Two beams AC and CD are hinged at point C. These are supported on rollers at the left and right ends (point A and D). A fixed pin support is provided at point B, as shown in Fig. 14-7a. If a force of 700 N acts at point E, determine the reaction force at the support B, using the principle of virtual work.

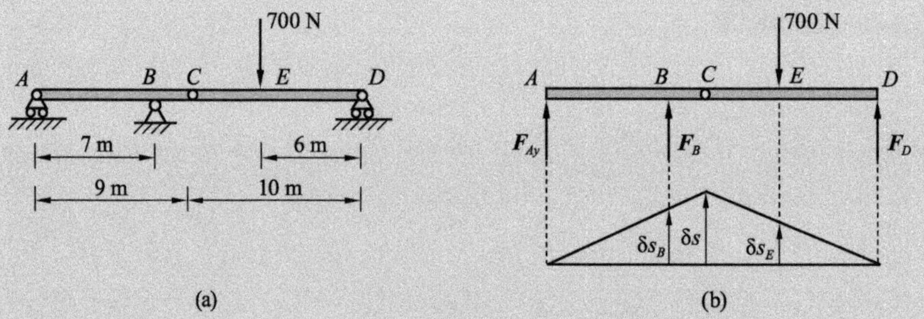

Fig. 14-7

Solution:

The reaction $\boldsymbol{F}_B$ can be obtained by *releasing* the pin constraint at B in the y direction and allowing the point B to be displaced in this direction, as shown in Fig. 14-7b, the support reaction $\boldsymbol{F}_B$ has to be treated as an external load in the principle of virtual work, only $\boldsymbol{F}_B$, and $\boldsymbol{F}$ do work. Let δs be the upward virtual displacement of the beam at C. From two similar triangles, then virtual displacement of point B and E is

$$\delta s_B = \frac{7}{9}\delta s$$

$$\delta s_E = \frac{3}{5}\delta s$$

Since the force of 700 N is opposite to the corresponding virtual displacement δs_E and hence does negative work. Thus,

$$\delta W = F_B \delta s_B - F_E \delta s_E = 0$$

Relating each of the virtual displacements to common virtual displacement δs yields

$$\left(\frac{7}{9}F_B - \frac{3}{5}F_E\right)\delta s = 0$$

Since $\delta s \neq 0$, we have

$$\frac{7}{9}F_B - \frac{3}{5}F_E = 0$$

$$F_B = \frac{27}{35}F_E = 540 \text{ N} \qquad \textit{Ans.}$$

Exercises

14-1 In Fig. 14-8, the beam AC is subjected to a couple $\boldsymbol{M}$ and a point load $\boldsymbol{F}$ at point C. Determine the reactions at the support A using the principle of virtual work.

14-2 In Fig. 14-9, two beams AC and CF are hinged at C and supported at A, D and F. The beams are subjected to two forces $F_1 = 15$ kN and $F_2 = 12$ kN. Using the principle of virtual work, find the reaction at D.

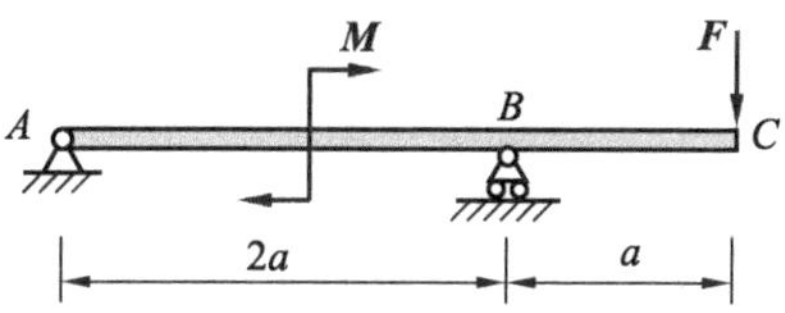

Fig. 14-8

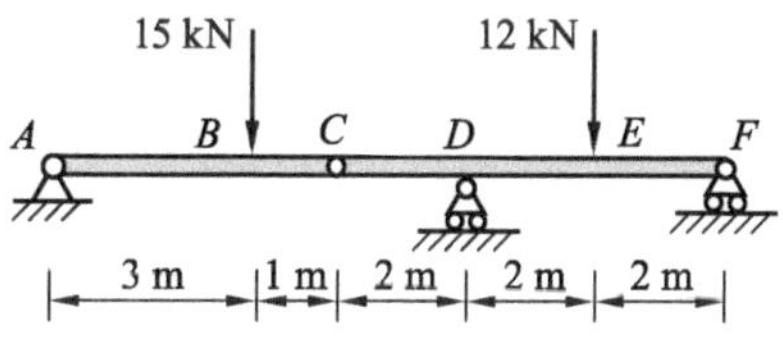

Fig. 14-9

14-3 In Fig. 14-10, determine the required magnitude of force $\boldsymbol{F}$ to maintain equilibrium of the two-member linkage at $\theta = 60°$. Each link has a mass of 20 kg.

14-4 Using the method of virtual work, determine the magnitude of force $\boldsymbol{F}$ required to hold the 50 kg smooth rod in equilibrium at $\theta = 60°$, Fig. 14-11.

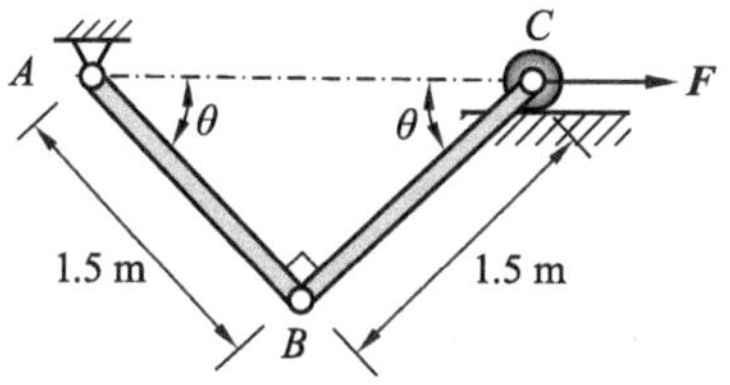

Fig. 14-10

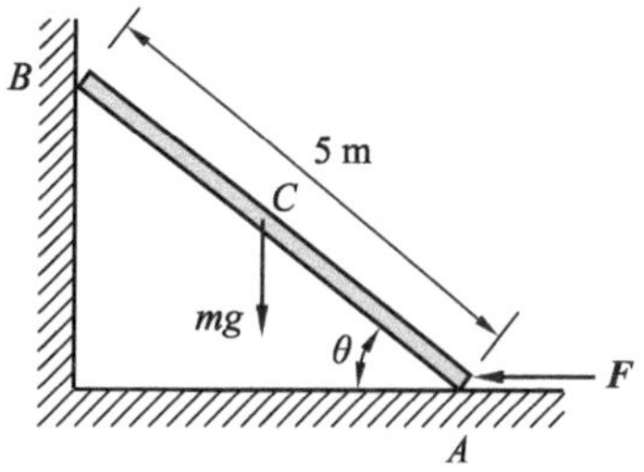

Fig. 14-11

14-5 In Fig. 14-12, the slider crank mechanism consists of the crank AC and the link rod BC. If a force F is acting at slider B, determine the moment M to keep the system in equilibrium at an arbitrary angle θ. Neglect the weight of the members.

14-6 The pin-connected mechanism is constrained by a pin at A and a roller at B, Fig. 14-13. Determine the force F that must be applied to the roller to hold the mechanism in equilibrium when $\theta = 30°$. In Fig. 14-14, the spring is unstretched when $\theta = 45°$. Neglect the weight of the members. All members have same length $l = 0.5$ m, $k = 50$ N/m.

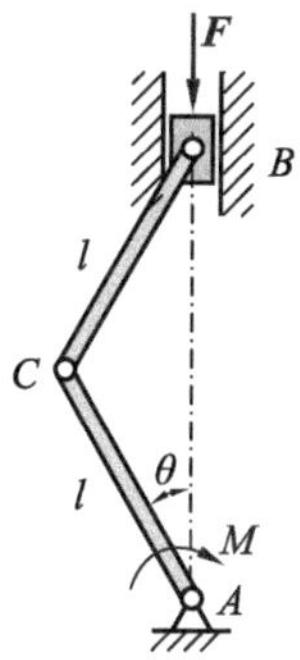

Fig. 14-12

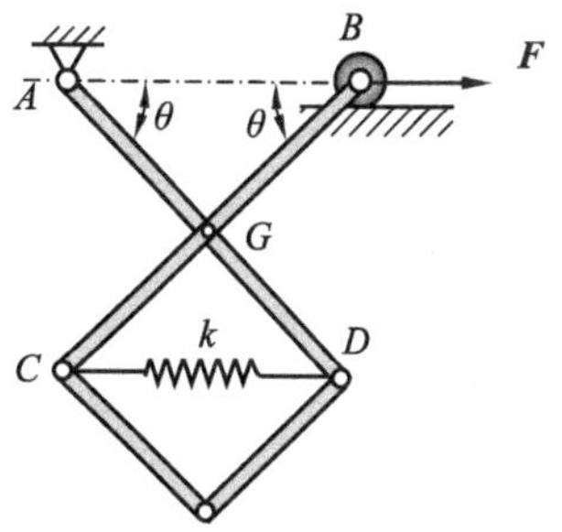

Fig. 14-13

14-7 In Fig. 14-14, the uniform rod AB has a weight of 10 N and a length of 0.6 m, is subjected to a moment of $M = 10$ N · m. Determine the force developed in the spring required to keep the rod AB in equilibrium when $\theta = 30°$. Set $k = 15$ N/m.

14-8 If the spring has a stiffness k and an unstretched length l_0, determine the force F when the mechanism is in the position, Fig. 14-15. Neglect the weight of the members.

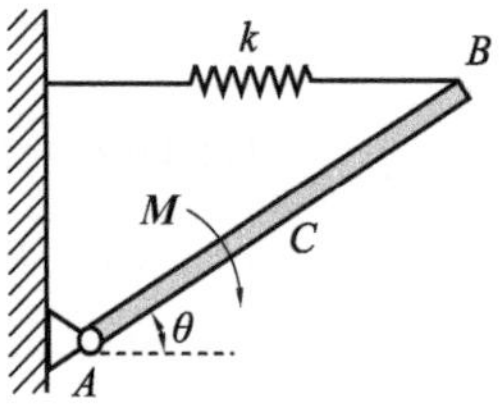

Fig. 14-14

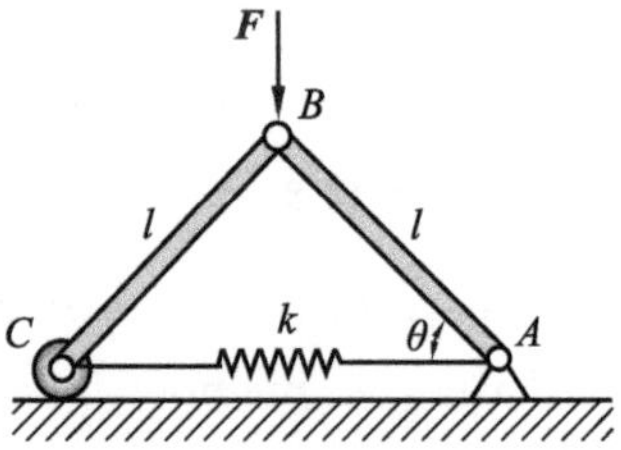

Fig. 14-15

14-9 In Fig. 14-16, a weight of 1 000 N resting over a smooth surface inclined at 30° with the horizontal, is supported by a block of weight G resting on a smooth surface inclined at 45° with the horizontal. By using the principle of virtual work, calculate the value of G.

14-10 If box has a mass of 10 kg, in Fig. 14-17, determine the couple moment $\boldsymbol{M}$ needed to maintain equilibrium when $\theta = 60°$. Neglect the mass of the members. Set $AB = CD = l = 0.45$ m.

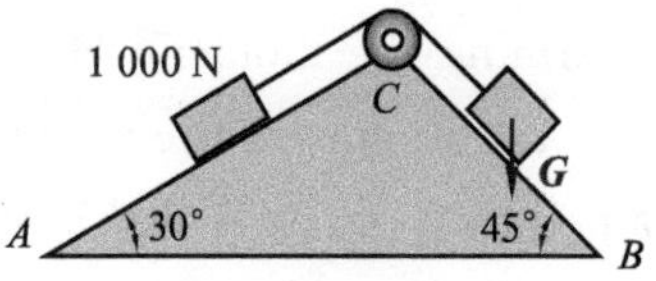

Fig. 14-16

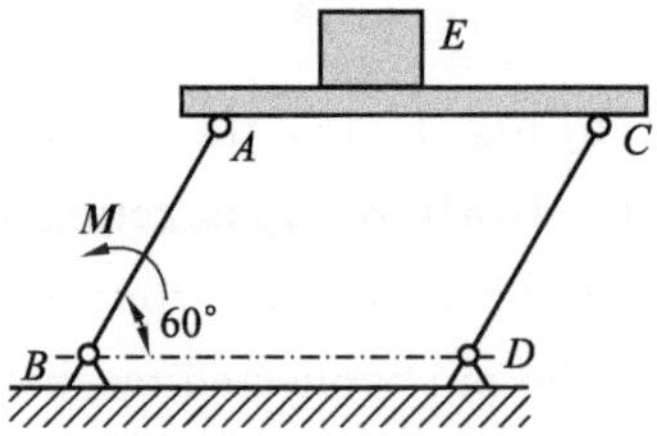

Fig. 14-17

14-11 A block of weight $G = 5$ kN is raised by a system of pulleys, in Fig. 14-18. Using the method of virtual work, determine the force $\boldsymbol{F}$ to hold the weight in equilibrium.

14-12 As shown in Fig. 14-19, the mechanism consists of the bar AB and OC, the collar A can slide over rod OC. If a force $\boldsymbol{F}$ is acting at the end of bar AB, determine the force $\boldsymbol{F}_1$ to keep the system in equilibrium at an arbitrary angle φ. Neglect the weight of the members.

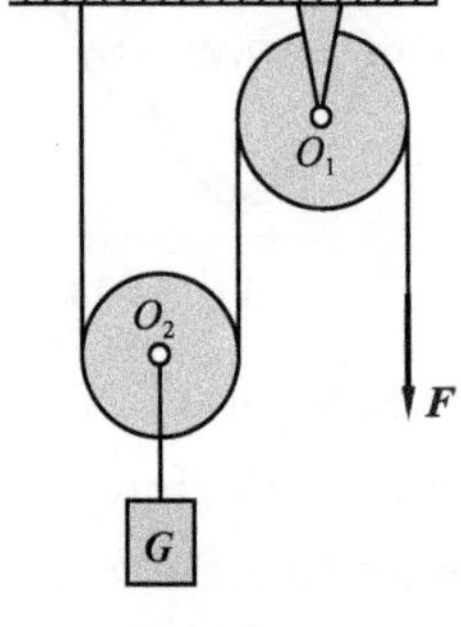

Fig. 14-18

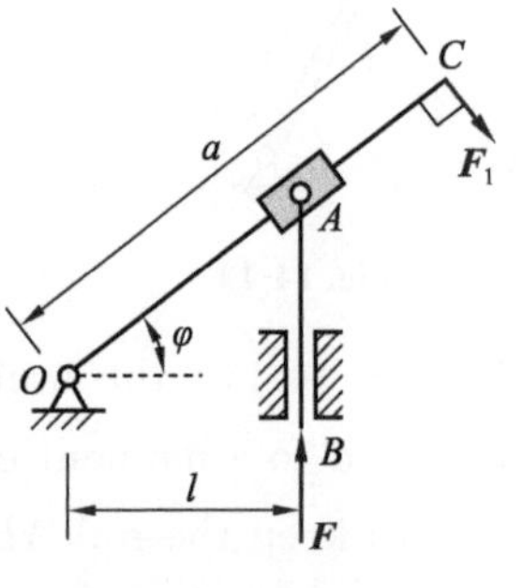

Fig. 14-19

References

[1] Dietmar G, Werner H, Schröder J, et al. Engineering Mechanics 1:Static[M]. 2th ed. New York: Springer, 2009.

[2] Dietmar G, Werner H, Schröder J, et al. Engineering Mechanics 3:Dynamic [M]. 2th ed. New York: Springer, 2009.

[3] Hibbeler R C. Engineering Mechanics:Statics[M]. 12th ed. London: Pearson Prentice Hall,2010.

[4] Hibbeler R C. Engineering Mechanics: Dynamics [M]. 12th ed. Pearson Prentice Hall, 2010.

[5] Awrejcewicz J. Classical Mechanics:Static[M]. Springer,2012.

[6] Awrejcewicz J. Classical Mechanics:Dynamics[M]. Springer,2012.

[7] Khurmi R S, Gupta J K. A Textbook of Engineering Mechanics[M]. 14th ed. New Delhi: S Chand & Company Ltd. , 2005.

[8] Department of theoretical mechanics, Harbin university of technology. Theoretical Mechanics[M]. 8th ed. Beijing: Higher Education Press,2018.

[9] Sun B C. Theoretical Mechanical Basis[M]. Beijing: National Defence Industry Press, 2014.

Answers

Chapter 1

The answers are omitted.

Chapter 2

2-1. $F_R=17.1$ kN, $\alpha=41°$.

2-2. $F_R=185.2$ N, $\alpha=67°$.

2-3. $F_T=21.2$ N, $F_N=7.07$ N.

2-4. $F_{BA}=282.9$ kN, $F_{BC}=200$ kN.

2-5. $F_{BA}=54.6$ kN, $F_{BC}=74.6$ kN.

2-6. $F_A=80$ kN.

2-7. $M_O=-989.8$ N·m.

2-8. $F=W\tan\alpha$.

2-9. $M=100$ N·m.

2-10. $F_A=F_B=\frac{M}{l}$.

2-11. $F_A=F_B=200$ N.

2-12. $F_A=F_C=\frac{M}{2\sqrt{2}}a$.

2-13. $F'_R=150$ kN, $M_O=-900$ N·mm.

2-14. $F_R=5.93$ kN, $\alpha=77.8°$, $M_A=-34.8$ kN·m.

2-15. $F_B=\frac{3}{2}F-\frac{1}{4}qa+\frac{1}{2a}M$, $F_{Ax}=0$, $F_{Ay}=-\frac{F}{2}+\frac{5}{4}qa-\frac{M}{2a}$.

2-16. $F_{Ax}=0$, $F_{Ay}=-2\,000$ N, $F_B=1\,500$ N.

2-17. $F_{Ax}=-(F_1+F_2\sin\theta)$, $F_{Ay}=F_2\cos\theta$, $M_A=-(bF_1+F_2l\cos\theta)$.

2-18. $F_{AB}=3.54$ kN, $F_{Cx}=2.5$ kN, $F_{Cy}=-0.5$ kN.

2-19. $F_{Ax}=2\,400$ N, $F_{Ay}=1\,200$ N, $F_{BC}=848.5$ N.

2-20. $F_s=5.2$ kN, $F_A=17.3$ kN, $F_B=24.9$ kN.

2-21. $F_D=F_E=150$ N, $F_{Ax}=120$ N, $F_{Ay}=90$ N.

2-22. $F_B=F_C=\frac{\sqrt{2}}{4}F$, $F_{Ax}=\frac{1}{4}F$, $F_{Ay}=\frac{3}{4}F$.

2-23. $F_{Ax}=-\frac{5}{2}F$, $F_{Ay}=-\frac{1}{2}F$, $F_{Bx}=\frac{5}{2}F$, $F_{By}=\frac{7}{2}F$, $F_{Cx}=\frac{5}{2}F$, $F_{Cy}=\frac{3}{2}F$.

2-24. $F_{Ax}=0$, $F_{Ay}=-\frac{a}{b}F$, $M_A=\frac{ac}{b}F$, $F_B=\frac{a+b}{b}F$.

2-25. $M_A=\frac{1}{2}qa^2$, $F_{Ax}=\frac{qa}{2}\tan\theta$, $F_{Ay}=\frac{1}{2}qa$, $F_C=\frac{qa}{2\cos\theta}$.

2-26. $M_A=0$, $F_{Ax}=0$, $F_{Ay}=\frac{1}{2}F$, $F_C=F_B=\frac{M}{2a}=\frac{F}{2}$.

2-27. $F_{Ax}=-2.015$ kN, $F_{Ay}=-1$ kN, $F_{Ex}=2.015$ kN, $F_{Ey}=2$ kN.

2-28. $F_{Ax}=1\ 200$ N, $F_{Ay}=150$ N, $F_B=1\ 050$ N, $F_{BC}=-1\ 500$ N.

2-29. $F_{Ax}=-F$, $F_{Ay}=-F$, $F_{Bx}=-F$, $F_{By}=0$, $F'_{Dx}=2F$, $F'_{Dy}=F$.

2-30. $F_{AD}=-318$ N, $F_{AB}=225$ N, $F_{BD}=0$, $F_{BC}=225$ N, $F_{CD}=318$ N.

2-31. $F_{CD}=500$ N, $F_{AD}=400$ N, $F_{BC}=500$ N, $F_{AC}=F_{AB}=0$.

2-32. $F_{CD}=-\frac{\sqrt{3}}{2}F$.

2-33. $F_1=-5.33F$, $F_2=2F$, $F_3=-1.667F$.

2-34. $F_{BC}=800$ N, $F_{GE}=800$ N, $F_{GC}=500$ N.

Chapter 3

3-1. $F_{ED}=10.06$ N, $F_{AE}=F_{BE}=5.51$ kN.

3-2. $F_{AD}=1\ 125$ N, $F_{AC}=844$ N, $F_{AB}=506$ N.

3-3. $F_{AD}=F_{BD}=-26.4$ kN, $F_{CD}=33.5$ kN.

3-4. $M_z=-101.4$ N·m.

3-5. $M_x=1.9$ kN·m, $M_y=0.63$ kN·m, $M_z=0.95$ kN·m.

3-6. $M_O=46.2\boldsymbol{i}-46.2\boldsymbol{j}$ kN·m.

3-7. $F_R=-1\ 400$ N, $x=3$ m, $y=2.5$ m.

3-8. $F_A=102.2$ N, $F_B=159.34$ N, $F_C=138.46$ N.

3-9. $F_{AB}=0.75$ kN, $F_{CD}=3$ kN, $F_{EF}=2.25$ kN.

3-10. $F_T=200$ N, $F_{Bz}=0$, $F_{Bx}=0$, $F_{Ax}=86.6$ N, $F_{Ay}=150$ N, $F_{Az}=100$ N.

3-11. $F_{Ax}=475$ N, $F_{Az}=125$ N, $F_{Bx}=25$ N, $F_{Bz}=1\ 125$ N, $F_{Cy}=450$ N, $F_{Cz}=250$ N.

3-12. $F_1=-F$, $F_2=0$, $F_3=F$, $F_4=0$, $F_5=-F$, $F_6=0$.

3-13. $y_C=\frac{4r}{3\pi}$.

3-14. $y_C=-\frac{a}{7}$.

3-15. $x_C=110$ mm.

3-16. $y_C=17.8$ mm.

3-17. $x_C=-\frac{Ra^2}{2(\pi R^2-a^2)}$, $y_C=0$.

3-18. $y_C=\frac{h(2ah-3\pi r^2)}{6(ah-2\pi r^2)}$.

Chapter 4

4-1. $F=20$ N.

4-2. $F_s=173.2$ N.

4-3. At rest.

4-4. $F=155$ N.

4-5. $e \leqslant \frac{\mu_s d}{2}$.

4-6. $F=500$ N.

4-7. $s=0.456l=1.824$ m.

4-8. $x=120$ mm.

4-9. $F=350$ N.

4-10. $b=110$ mm.

4-11. $F=45$ N.

4-12. $M=F(R\cos\theta-r)$.

Chapter 5

5-1. $a=-2\ \mathrm{m/s^2}$.

5-2. $s=-32$ m.

5-3. $v=35.8$ m/s, $a=17.9\ \mathrm{m/s^2}$.

5-4. $v=36.1$ m/s, $a=36.5\ \mathrm{m/s^2}$.

5-5. $y_A=e\sin\omega t+\sqrt{R^2-e^2\cos^2\omega t}$, $v_A=\omega\cos\omega t+\frac{e^2\omega\sin 2\omega t}{2\sqrt{R^2-e^2\cos^2\omega t}}$.

5-6. $x=(R+R)\cos 2\omega t$, $y=R\sin 2\omega t$;

$v=2R\omega$, $a=4R\omega^2$.

5-7. $\rho=1\,280$ m.

5-8. $v=1.8$ m/s, $a=1.2\ \mathrm{m/s^2}$.

5-9. $\rho=100$ m.

5-10. $v=38.7$ m/s.

5-11. $a_B=1.42\ \mathrm{m/s^2}$.

5-12. $v=7.2$ m/s, $a=1.91\ \mathrm{m/s^2}$.

Chapter 6

6-1. $v_C=0.8$ m/s, $a_C^t=0.4\ \mathrm{m/s^2}$, $a_C^n=3.2\ \mathrm{m/s^2}$.

6-2. $\varphi=\omega t=\frac{1}{30}t$ rad, $x^2+(y+0.8)^2=1.5^2$.

6-3. $v_O=0.707$ m/s, $a_O=3.331\ \mathrm{m/s^2}$.

6-4. $v_A=v_B=v_C=v_D=0.6$ m/s, $a_A=1.2\ \mathrm{m/s^2}$.

6-5. $\omega=10t^2$ rad/s.

6-6. $v_A=22$ m/s, $a_A^t=12\ \mathrm{m/s^2}$, $a_A^n=242\ \mathrm{m/s^2}$.

6-7. $\omega=20t$ rad/s, $\alpha=20\ \mathrm{rad/s^2}$;

$a_B^t=10\ \mathrm{m/s^2}$, $a_B^n=200t^2\ \mathrm{m/s^2}$.

6-8. $\omega=-19.2$ rad/s, $\alpha=-183\ \mathrm{rad/s^2}$.

6-9. $v=21.4$ m/s, $a^t=6.75\ \mathrm{m/s^2}$.

6-10. $v_{AB}=\omega l\cos\theta$, $a_{AB}=-l\omega^2\sin\theta$.

6-11. $v_A=\sqrt{2}\omega R$, $a_A^n=\sqrt{2}\omega^2 R$, $a_A^t=\sqrt{2}\alpha R$;

$v_B=2\omega R$, $a_B^n=2\omega^2 R$, $a_B^t=2\alpha R$.

6-12. $v_A=\omega(r+a)$, $a_A^n=\omega^2(r+a)$, $a_A^t=\alpha(r+a)$;

$v_B=\omega(2r+a)$, $a_B^n=\omega^2(2r+a)$, $a_B^t=\alpha(2r+a)$.

Chapter 7

7-1. $v_a=6.6$ m/s.

7-2. $\omega_1=\omega_0$.

7-3. $\omega_2=2$ rad/s.

7-4. $\omega_2=1.5$ rad/s.

7-5. $v_A=\dfrac{val}{x^2+a^2}$.

7-6. $v_{AB}=\dfrac{\sqrt{2}}{2}\omega e$.

7-7. $v_r=128$ km/h, $\beta=\arctan\dfrac{v_e}{v_a}=38.7°$.

7-8. $v_C=\dfrac{va}{2l}$.

7-9. $v_{BC}=1$ m/s, $a_{BC}=34.6$ m/s^2.

7-10. $v_{AB}=v\tan\theta$, $a_{AB}=a\tan\theta$.

7-11. $\omega_{OA}=1$ rad/s, $\alpha_{OA}=0.268$ rad/s^2.

7-12. $v_{BC}=0.173$ m/s, $a_{BC}=0.05$ m/s^2.

7-13. $v_{CD}=0.1$ m/s, $a_{CD}=0.346$ m/s^2.

7-14. $\omega_{EC}=0.866$ rad/s, $\alpha_{EC}=0.134$ rad/s^2.

7-15. $\omega_{DE}=3$ rad/s, $\alpha_{DE}=-5$ rad/s^2.

7-16. $v_M=2$ m/s, $a_M=4\sqrt{5}$ m/s^2.

7-17. $v_M=17.3$ cm/s, $a_M=35$ cm/s^2.

7-18. $v_{AB}=\omega l\tan\theta$, $a_{AB}=-\omega^2 l\left(1+\dfrac{1}{\rho_A\cos^3\theta}-\dfrac{2}{\cos^2\theta}\right)$.

Chapter 8

8-1. $v_B=2\sqrt{2}$ mm/s, $\omega_{AB}=2$ rad/s.

8-2. $v_C=1.38$ m/s.

8-3. $\omega_{BC}=10$ rad/s.

8-4. $\omega_{BC}=15$ rad/s, $\omega_D=52$ rad/s.

8-5. $v_B=6.24$ m/s, $\omega_{AB}=12$ rad/s.

8-6. $v_B=2.513$ m/s.

8-7. $v_C=0.0707$ m/s.

8-8. $\omega_{BC}=5.3$ rad/s, $\omega_{AB}=5.3$ rad/s.

8-9. $\omega_B=7.25$ rad/s.

8-10. $v_B=0.693$ m/s, $v_C=0.8$ m/s, $\omega_{AB}=1.732$ rad/s, $\omega_{BC}=1.333$ rad/s.

8-11. $v_E=4.76$ m/s, $\theta=40.9°$, $\omega_{AB}=6$ rad/s.

8-12. $\omega_2=15$ rad/s.

8-13. $\omega=\frac{3v}{2r}$.

8-14. $v_A=\sqrt{2}\omega a$, $v_B=\omega a$, $\omega_{AB}=\frac{\omega}{2}$, $\omega_{BC}=\omega$.

8-15. $\omega_{AB}=\omega=3$ rad/s, $\omega_{O_1B}=5.2$ rad/s, $\alpha_{O_1B}=19.76$ rad/s^2.

8-16. $\omega_{O1}=\frac{\sqrt{2}\omega}{2}$, $\alpha_{O1}=0.5\omega^2$.

8-17. $\omega_{AB}=2$ rad/s, $v_B=2.83$ m/s, $a_B=-5.656$ m/s^2.

8-18. $v_B=\sqrt{3}\omega_0 r$, $a_B=-\frac{1}{3}\omega_0^2 r$.

8-19. $v_O=\omega r$, $a_O=\alpha r$, $v_P=0$, $a_P=\omega^2 r$.

8-20. $a_B=9.25$ m/s^2, $\theta=71.1°$.

8-21. $a_A=-l\omega^2$, $\alpha_{AB}=0$.

Chapter 9

9-1. $T=15.30$ kN.

9-2. $F=m\alpha r+mg\mu_k\cos\theta+mg\sin\theta$.

9-3. $v=15.6$ m/s.

9-4. $F_T=176$ N.

9-5. $t=\sqrt{\frac{(m_1+m_2)h}{(m_1-m_2)g}}$.

9-6. $F_{AB}=\frac{\sqrt{3}}{2}mg$.

9-7. $\omega\leqslant\sqrt{\frac{g\mu_s}{r}}$.

9-8. $F_T=114$ N.

9-9. $v=49.5$ m/s.

9-10. $F_{BC}=mg\sin\theta$.

9-11. $a_t=-4.905$ m/s^2, $\rho=188$ m.

Chapter 10

10-1. $p_a=\frac{1}{6}m\omega l$, $p_b=m\omega R$, $p_c=mv_C$.

10-2. $p=3.46$ kg · m/s.

10-3. $v_2=17.9$ m/s.

10-4. $v_2 = 20$ m/s, $F_T = 2\,000$ N.

10-5. $a = \dfrac{(m_1 + m_2) g f - m_2 b}{m_1 + m_2}$.

10-6. $F_{Ox} = m_3 \dfrac{R}{r} a\cos\theta + m_3 g\sin\theta\cos\theta$;

$F_{Oy} = (m_1 + m_2 + m_3) g - m_3 g\cos^2\theta + m_3 \dfrac{R}{r} a\sin\theta - m_2 a$.

10-7. $v_{c2} = -10$ m/s.

10-8. $v_2 = 0.5$ m/s, $F_{avg} = -1.88$ kN.

10-9. $v_{p2} = 46.75$ m/s.

10-10. $v_2 = 0.833$ m/s, $F_s = 138.6$ N.

10-11. $F_{x\max} = \dfrac{1}{2}(m_1 + 2m_2 + 2m_3) l\omega^2$.

10-12. $4x^2 + y^2 = l^2$.

10-13. $\Delta x = 0.577$ m.

10-14. $\Delta x = -0.266$ m.

10-15. $v_c = 3.478$ m/s.

10-16. $\Delta x = -0.138$ m.

10-17. $x_A = 170$ mm, $x_B = 90$ mm.

Chapter 11

11-1. $L_O = \dfrac{3}{2} m\omega R^2$, $L_O = \dfrac{1}{2} mR^2\omega$, $L_O = \dfrac{1}{3} m\omega l^2$.

11-2. $L_O = (J_O + m_A R^2 + m_B r^2)\omega$.

11-3. $v = 2.08$ m/s.

11-4. $v_2 = 15$ km/s.

11-5. $\omega_2 = 48$ rad/s.

11-6. $\omega_2 = 28.6$ rad/s.

11-7. $J_z = \dfrac{2}{5} mR^2$.

11-8. $J_x = 101 \times 10^6$ mm^4.

11-9. $\alpha_1 = \dfrac{MR^2}{J_1 R^2 + J_2 r^2}$, $\alpha_2 = \dfrac{MRr}{J_1 R^2 + J_2 r^2}$.

11-10. $\alpha = 20.77$ rad/s^2, $F_{Ox} = -96$ N, $F_{Oy} = 32.3$ N.

11-11. $\alpha = 8.26$ rad/s^2, $F_A = 167$ N, $F_{Oy} = 324$ N.

11-12. $F = 269.3$ N.

11-13. $F_{Ox} = 0$, $F_{Oy} = 37.5$ kN.

11-14. $L_O = 20$ kg · m^2/s.

11-15. $a_C = \dfrac{2M}{3mr}$, $F_s = \dfrac{2M}{3r}$.

11-16. $\omega = 1.05$ rad/s.

11-17. $a_{Cx}=3.482\ \mathrm{m/s^2}$.

11-18. $a_A=a_B=\frac{4}{7}g\sin\theta$, $F_T=-\frac{1}{7}mg\sin\theta$.

11-19. $a_C=\frac{4}{5}g$.

11-20. $a_A=\frac{m_1g(R+r)^2}{m_1(R+r)^2+m_2(R^2+\rho^2)}$.

11-21. $F_N=\frac{mg}{1+3\sin^2\theta}$.

Chapter 12

12-1. $W=505$ J.

12-2. $W=109.7$ J.

12-3. $T=\frac{1}{18}ml^2\omega^2$, $T=\frac{3}{4}mR^2\omega^2$, $T=\frac{3}{4}mv^2$.

12-4. $T=\frac{3}{2}m_1v_O^2+mv_O^2$.

12-5. $\delta=0.066\ 7$ m.

12-6. $v=8.1$ m/s.

12-7. $s=157.31$ m.

12-8. $x=2.57$ m.

12-9. $\omega=9.59$ rad/s.

12-10. $\omega=0.477$ rad/s.

12-11. $\omega=6.11$ rad/s.

12-12. $v_2=1.5$ m/s.

12-13. $v=2\sqrt{\frac{(m_2+m_3-2m_1)gh}{8m_1+2m_2+7m_3}}$.

12-14. $v_0=\sqrt{\frac{2Ms-2m_1grs\sin\theta}{(m_1+m_2)r}}$, $a=\frac{M-m_1gr\sin\theta}{(m_1+m_2)r}$.

12-15. $v=2\sqrt{\frac{m_3gh}{3m_1+m_2+2m_3}}$, $a=\frac{2m_3g}{3m_1+m_2+2m_3}$.

12-16. $\omega=\frac{2}{r}\sqrt{\frac{M-m_2gr(\sin\theta+\mu_k\cos\theta)}{m_1+2m_2}\varphi}$, $\alpha=\frac{2[M-m_2gr(\sin\theta+\mu_k\cos\theta)]}{(m_1+2m_2)r^2}$.

12-17. $P=37.4$ W.

12-18. $P=172$ kW.

12-19. $F_T=\frac{M(m_1+2m_2)}{2(m_1+m_2)R}$.

12-20. $F_E=\frac{m_1\sin\theta-m_2}{m_1+m_2}m_1g\cos\theta$.

12-21. $\omega=\sqrt{\frac{3g(m_1+2m_2)\sin\theta}{(m_1+3m_2)l}}$, $\alpha=\frac{g(3m_1+6m_2)\cos\theta}{2l(m_1+3m_2)}$.

Chapter 13

13-1. $F_{IO}^{t}=\frac{1}{6}m\alpha l$, $F_{IO}^{n}=\frac{1}{6}m\omega^{2}l$, $M_{IO}=\frac{1}{9}ml^{2}\alpha$;

$F_{IO}^{t}=mR\alpha$, $F_{IO}^{n}=mR\omega^{2}$, $M_{IO}=\frac{3}{2}mR^{2}\alpha$.

13-2. $F_{I}^{t}=m\alpha d$, $F_{I}^{n}=m\omega^{2}d$.

13-3. $F_{IC}=mR\alpha$, $M_{IC}=\frac{1}{2}mR^{2}\alpha$.

13-4. $F_{NA}=\frac{m(gc-ah)}{b+c}$, $F_{NB}=\frac{m(gb+ah)}{b+c}$.

13-5. $F_{T}=1.96$ N, $v=2.1$ m/s.

13-6. $a=15$ m/s^{2}, $\theta=33.2°$.

13-7. $a_{C}=\frac{3g}{7}$, $F_{T}=\frac{4}{7}mg$.

13-8. $a_{C}=2.8$ m/s^{2}.

13-9. $F_{Cx}=0$, $F_{Cy}=\frac{3m_{1}m_{2}+m_{2}^{2}}{m_{2}+2m_{1}}g$, $M_{C}=\frac{3m_{1}m_{2}+m_{2}^{2}}{m_{2}+2m_{1}}ag$.

13-10. $a_{C}=\frac{gl}{3r}$, $F_{Ox}=\frac{mgl}{6r^{2}}\sqrt{4r^{2}-l^{2}}$, $F_{Oy}=mg\left(1-\frac{l^{2}}{6r^{2}}\right)$.

13-11. $\alpha=\frac{m_{2}gr-m_{1}gR}{J_{O}+m_{1}R^{2}+m_{2}r^{2}}$, $F_{Ox}=0$, $F_{Oy}=\frac{(m_{1}R-m_{2}r)^{2}g}{J_{O}+m_{1}R^{2}+m_{2}r^{2}}$.

13-12. $F_{Ox}=-\frac{\sqrt{3}}{4}m_{1}r\omega^{2}$, $F_{Oy}=(m_{1}+m_{2})g-\frac{1}{4}r\omega^{2}(m_{1}+2m_{2})$,

$M=\frac{\sqrt{3}}{4}[(m_{1}+2m_{2})g-m_{2}r\omega^{2}]r$.

Chapter 14

14-1. $F_{A}=-\frac{Fa+M}{2a}$.

14-2. $F_{D}=22.9$ kN.

14-3. $F=56.6$ N.

14-4. $F=142$ N.

14-5. $M=2Fl\sin\theta$.

14-6. $F=7.95$ N.

14-7. $F_{B}=42$ N.

14-8. $F=2k\tan\theta(2l\cos\theta-l_{0})$.

14-9. $G=707$ N.

14-10. $M=22.1$ N·m.

14-11. $F=2.5$ kN.

14-12. $F_{1}=\frac{Fl}{a\cos^{2}\varphi}$.